Edited by
Pieter Imhof and
Jan Cornelis van der Waal

**Catalytic Process Development
for Renewable Materials**

Related Titles

Xie, H., Gathergood, N.

The Role of Green Chemistry in Biomass Processing and Conversion

2013
ISBN: 978-0-470-64410-2

Harmsen, J., Powell, J. B. (eds.)

Sustainable Development in the Process Industries

Cases and Impact

2010
ISBN: 978-0-470-18779-1

Tao, J., Kazlauskas, R. J. (eds.)

Biocatalysis for Green Chemistry and Chemical Process Development

2011
ISBN: 978-0-470-43778-0

Walker, D.

The Management of Chemical Process Development in the Pharmaceutical Industry

2010
ISBN: 978-0-470-65251-0

Jameel, F., Hershenson, S. (eds.)

Formulation and Process Development Strategies for Manufacturing Biopharmaceuticals

2010
ISBN: 978-0-470-11812-2

Edited by Pieter Imhof and Jan Cornelis van der Waal

Catalytic Process Development for Renewable Materials

WILEY-VCH

The Editors

Dr. Pieter Imhof
Avantium
VP Strategic Account Management
Zekeringstraat 29
1014 BV Amsterdam
The Netherlands

Dr. Jan Cornelis van der Waal
Avantium
Zekeringstraat 29
1014 BV Amsterdam
The Netherlands

All books published by **Wiley-VCH** are carefully produced. Nevertheless, authors, editors, and publisher do not warrant the information contained in these books, including this book, to be free of errors. Readers are advised to keep in mind that statements, data, illustrations, procedural details or other items may inadvertently be inaccurate.

Library of Congress Card No.: applied for

British Library Cataloguing-in-Publication Data
A catalogue record for this book is available from the British Library.

Bibliographic information published by the Deutsche Nationalbibliothek
The Deutsche Nationalbibliothek lists this publication in the Deutsche Nationalbibliografie; detailed bibliographic data are available on the Internet at http://dnb.d-nb.de.

© 2013 Wiley-VCH Verlag GmbH & Co. KGaA, Boschstr. 12, 69469 Weinheim, Germany

All rights reserved (including those of translation into other languages). No part of this book may be reproduced in any form – by photoprinting, microfilm, or any other means – nor transmitted or translated into a machine language without written permission from the publishers. Registered names, trademarks, etc. used in this book, even when not specifically marked as such, are not to be considered unprotected by law.

Print ISBN: 978-3-527-33169-7
ePDF ISBN: 978-3-527-65666-0
ePub ISBN: 978-3-527-65665-3
mobi ISBN: 978-3-527-65664-6
oBook ISBN: 978-3-527-65663-9

Cover Design   Simone Benjamin, McLeese Lake, Canada
Typesetting   Toppan Best-set Premedia Limited, Hong Kong
Printing and Binding   Markono Print Media Pte Ltd, Singapore
Printed in the Federal

Contents

The Next Feedstock Transition *XIII*
Preface *XV*
List of Contributors *XVII*

1 **The Industrial Playing Field for the Conversion of Biomass to Renewable Fuels and Chemicals** *1*
Leo E. Manzer, Jan Cornelis van der Waal, and Pieter Imhof
1.1 Introduction *1*
1.2 The Renewables Arena *2*
1.3 Renewable Fuels *9*
1.4 Renewable Chemicals *18*
1.5 Conclusions *22*
References *22*

2 **Selecting Targets** *25*
Gene Petersen, Joseph Bozell, and James White
2.1 Introduction *25*
2.2 Target Selection Can Focus on Specific Structures or General Technologies *28*
2.3 Previous Selection Efforts *29*
2.4 Corroboration of the Value of Screening Studies *37*
2.5 The Importance of Outcomes and Comparisons of Outcomes *38*
2.6 Evaluation Processes Can be Comprised of a Variety of Criteria *40*
2.6.1 Feedstock and Intermediate Availability *40*
2.6.2 Existing Biorefining Infrastructure Dictates Chemical or Biochemical Processes to be Evaluated *41*
2.6.3 Market Drivers *42*
2.6.4 R&D Drivers *44*
2.6.5 Other Screening Opportunities *44*
2.6.6 Other Portfolio Opportunities – Biomass Produced Oils *45*
2.7 Catalysis Aspects *46*

2.8	Conclusions 48
	References 48

3	**The Development of Catalytic Processes from Terpenes to Chemicals** 51
	Derek McPhee
3.1	Introduction 51
3.2	Strain Engineering for the Production of Terpenes 52
3.3	Terpene Building Blocks of Commercial Interest 55
3.4	Sesquiterpenes as Chemical Building Blocks: β-Farnesene 56
3.5	Polymers 58
3.5.1	Differential Scanning Calorimetry 63
3.5.2	Gel Permeation Chromatography 63
3.5.3	Thermal Gravimetric Analysis 64
3.5.4	Tensile Strength 65
3.6	Lubricants 66
3.7	Conclusions 75
	References 76

4	**Furan-Based Building Blocks from Carbohydrates** 81
	Robert-Jan van Putten, Ana Sousa Dias, and Ed de Jong
4.1	Importance of Furans as Building Blocks 81
4.2	Sources of Carbohydrates 82
4.2.1	Storage Carbohydrates 82
4.2.1.1	Sucrose 82
4.2.1.2	Starch 84
4.2.1.3	Inulin 84
4.2.2	Structural Carbohydrates 84
4.2.2.1	Cellulose 87
4.2.2.2	Hemicelluloses 87
4.2.3	Aquatic Carbohydrates 90
4.2.3.1	Macroalgae 91
4.2.3.2	Brown Macroalgae 91
4.2.3.3	Microalgae 91
4.2.3.4	Green Algae 92
4.2.4	Conclusions on Carbohydrate Feedstocks 92
4.3	Carbohydrate Dehydration 92
4.3.1	Introduction 92
4.3.2	Commercial Furfural Production and Applications 93
4.3.3	Furfural Formation from Pentose Feedstock 95
4.3.4	Production Systems of Furfural 99
4.3.5	Heterogeneous Catalysts 101
4.3.6	5-Hydroxymethylfurfural Formation from Hexose Feedstock 105

4.4	Conclusions and Further Perspectives *110*	
	References *111*	
5	**A Workflow for Process Design – Using Parallel Reactor Equipment Beyond Screening** *119*	
	Erik-Jan Ras	
5.1	Introduction *119*	
5.2	The Evolution of Parallel Reactor Equipment *120*	
5.3	The Evolution of Research Methodology – Conceptual Process Design *121*	
5.4	Essential Workflow Elements *126*	
5.4.1	Catalyst Testing Equipment *126*	
5.4.2	Kinetics and Pseudo-Kinetics *129*	
5.4.3	Statistical Design of Experiments *130*	
5.4.4	Data Analysis *136*	
5.4.5	Example of PCA Applied to Catalysis *137*	
5.4.6	Example of PLS Applied to Diesel Properties *141*	
5.5	Other Examples of Parallel Reactor Equipment Applied Beyond Screening – Long-Term Catalyst Performance *143*	
5.6	Concluding Remarks *147*	
	References *147*	
6	**Braskem's Ethanol to Polyethylene Process Development** *149*	
	Paulo Luiz de Andrade Coutinho, Augusto Teruo Morita, Luis F. Cassinelli, Antonio Morschbacker, and Roberto Werneck Do Carmo	
6.1	Introduction *149*	
6.1.1	Overview of Braskem Activities and History *149*	
6.1.2	Why Renewable Polymers and Why Green Polyethylene? *149*	
6.2	Ethanol and Brazil *150*	
6.3	Commercial Plants for Ethanol Dehydration *152*	
6.3.1	Salgema 100 kty Plant *152*	
6.3.2	Triunfo 200 kty Plant *153*	
6.3.3	MEG Plants *154*	
6.3.4	Announced Renewable Polymer Projects *155*	
6.4	Legislation and Certification *155*	
6.4.1	Ethanol Suppliers Code of Conduct *155*	
6.5	Process Description *156*	
6.5.1	Reaction *156*	
6.5.1.1	Catalysts *157*	
6.5.1.2	Side Reactions *158*	
6.5.1.3	Fixed Bed, Isothermal Reaction *159*	
6.5.1.4	Fixed Bed, Adiabatic Reaction *159*	
6.5.1.5	Fluidized Bed Reaction *160*	
6.5.2	Removal of Impurities *161*	
6.5.2.1	Unreacted Ethanol and Oxygenates *161*	

6.5.2.2	CO_2 and Acids *161*
6.5.3	Ethylene Purification *161*
6.6	Polymerization *162*
6.7	Conclusion *162*
	Acknowledgments *162*
	References *163*

7	**Fats and Oils as Raw Material for the Chemical Industry** *167*
	Aalbert (Bart) Zwijnenburg
7.1	Introduction – Setting the Scene, Definitions *167*
7.2	Why Fats and Oils Need Catalytic Transformation *168*
7.2.1	Carboxylic Acids *168*
7.2.2	Alcohols *168*
7.2.3	Amines and Amides *168*
7.2.4	Esters *168*
7.3	Catalytic Process Development – Conceptual *171*
7.3.1	Biology or Chemical Routes? *171*
7.3.2	How to Select between Slurry and Fixed-Bed Operations? *172*
7.3.3	How to Choose between Nickel and Palladium? *174*
7.4	Fatty Alcohols: Then and Now, a Case Study *175*
7.4.1	Catalyst Selection *176*
7.4.2	Slurry versus Fixed-Bed Processes *177*
7.5	Conclusion and Outlook: Development Challenges for the Future *178*
	References *179*

8	**Production of Aromatic Chemicals from Biobased Feedstock** *183*
	David Dodds and Bob Humphreys
8.1	Introduction *183*
8.2	Chemical Routes to Aromatic Chemicals from Biomass *184*
8.2.1	Process Chemistry *186*
8.2.1.1	Pyrolysis *186*
8.2.1.2	Hydrogenation and Hydrogenolysis *186*
8.2.1.3	Catalytic Reforming *188*
8.2.1.4	Zeolite Treatment *188*
8.2.2	Technology Examples *189*
8.2.2.1	Conversion of Biomass-Derived Sugars to Aromatics including BTX *189*
8.2.2.2	Pyrolysis of Solid Biomass to Aromatic Chemicals *190*
8.2.2.3	Upgrading Bio-Oils to Aromatics *192*
8.2.2.4	Aromatic Chemicals from Other Renewable Raw Materials *192*
8.2.3	Summary *193*
8.3	Biological Routes to Specific Aromatic Chemicals *194*
8.3.1	PTA via PX *194*
8.3.1.1	PX via Isobutanol and Isobutylene *194*

8.3.1.2	Valine Pathway to Isobutylene *194*
8.3.1.3	Direct Biological Isobutylene Production *196*
8.3.1.4	Biological Oxidation of PX to PTA *198*
8.3.2	Aromatics via HMF Production *198*
8.3.2.1	Preparation of PTA via HMF *200*
8.3.2.2	Yield Summary of HMF Routes to PTA *202*
8.3.3	Limonene to PTA *203*
8.3.4	The Common Aromatic Pathway *203*
8.3.4.1	Background *203*
8.3.4.2	Other Aromatic Compounds from the Common Aromatic Pathway *210*
8.3.5	Other Routes to Aromatic Compounds *215*
8.3.5.1	Tetrahydroxybenzene and Pyrogallol *215*
8.3.5.2	Phloroglucinol *217*
8.3.5.3	Chalcones, Stilbenes, Vanillin and Lignans *217*
8.4	Lignin – The Last Frontier *220*
8.5	Considerations for Scale-Up and Commercialization *222*
8.6	Conclusion *224*
	References *224*

9 Organosolv Biorefining: Creating Higher Value from Biomass *239*
E. Kendall Pye and Michael Rushton

9.1	Introduction *239*
9.2	Concepts and Principles of Biorefinery Technologies *241*
9.2.1	Types of Biorefineries for Biomass Processing *241*
9.2.1.1	Biorefineries Employing Thermochemical Treatment of Biomass *242*
9.2.1.2	Biorefineries Using Physical and Chemical Pretreatment with Biochemical Processing *243*
9.3	Catalytic Processes Employed in Biorefining *245*
9.3.1	Catalysis in Biorefineries Employing Gasification and Pyrolysis *245*
9.3.2	Catalysts in Anaerobic Digestion Biorefineries *246*
9.3.2.1	Catalysts in Non-Thermochemical Biorefineries *246*
9.4	An Organosolv Biorefinery Process for High-Value Products *247*
9.4.1	Guiding Principles of the Lignol Organosolv Biorefinery *250*
9.4.2	Applications and Markets for Organosolv Biorefinery Products *251*
9.4.2.1	Native Lignin – Its Properties and Composition *251*
9.4.2.2	Lignin from Other Processes *251*
9.4.2.3	HP-L™ Lignin – Organosolv Lignin from the Lignol Biorefinery *252*
9.4.2.4	Lignin Derivatives *252*
9.4.3	HP-L Lignin Properties *253*
9.4.4	Current Applications and Market Opportunities for HP-L Lignin *253*
9.4.4.1	New Product Opportunities for Lignin Derivatives *254*
9.4.4.2	Market Drivers for Commercial Use of HP-L Lignin and Other Bio-Products *256*

9.4.4.3	Application Strategies for the Lignol Biorefinery Process	257
9.4.4.4	Development of the Lignol Biorefinery Process	258
9.5	Conclusions	260
	References	261

10 Biomass-to-Liquids by the Fischer–Tropsch Process 265
Erling Rytter, Esther Ochoa-Fernández, and Adil Fahmi

10.1	Basics of Fischer–Tropsch Chemistry and BTL	265
10.1.1	The FT History and Drivers	265
10.1.2	Reactions	266
10.1.3	Mechanisms and Kinetics	268
10.1.4	Products	269
10.1.5	Fischer–Tropsch Metals	270
10.1.6	The Biomass-to-Liquid FT Concept	271
10.2	Cobalt Fischer–Tropsch Catalysis	272
10.2.1	Catalyst Preparation and Activation	272
10.2.2	Catalyst Activity	274
10.2.3	Selectivity	275
10.2.4	Activity Loss	276
10.2.5	Commercial Formulations	277
10.3	Fischer–Tropsch Reactors	279
10.3.1	Reactor Selection	279
10.3.2	Tubular Fixed-Bed	279
10.3.3	Slurry Bubble Column	280
10.4	Biomass Pretreatment and Gasification	282
10.4.1	Pretreatment of Biomass	282
10.4.2	Biomass Gasification	284
10.4.3	Entrained-Flow Gasifier	286
10.4.4	Fluidized-Bed Gasifier	287
10.4.5	Plasma Gasifier	288
10.4.6	Gasification Pilot and Demonstration Projects	288
10.4.6.1	Entrained-Flow Gasifiers	290
10.4.6.2	Fluidized-Bed Gasifiers	291
10.4.6.3	Plasma Gasifiers	291
10.4.6.4	Steam Reforming	292
10.4.7	Syngas Composition	292
10.5	Biomass-to-Liquids Process Concepts	293
10.5.1	Example of Process Flow-Sheet	293
10.5.2	Gas Conditioning and Clean-Up	294
10.5.3	BTL Mass and Energy Balance	295
10.5.4	CO_2 Management	298
10.5.5	Upgrading and Products	299
10.5.6	Production Cost	300
10.6	BTL Pilot and Demonstration Plants	301
10.7	XTL Energy and Carbon Efficiencies	303

10.8	BTL Summary and Outlook *304*	
	References *305*	

11	**Catalytic Transformation of Extractives** *309*	
	Päivi Mäki-Arvela, Irina L. Simakova, Tapio Salmi, and Dmitry Yu. Murzin	
11.1	Introduction *309*	
11.2	Fine and Special Chemicals from Crude Tall Oil Compounds *313*	
11.2.1	Sitosterol Hydrogenation and Its Application in Food as a Cholesterol-Suppressing Agent *313*	
11.3	Fine and Special Chemicals from Turpentine Compounds *317*	
11.3.1	Isomerization of Monoterpenes and Their Derivatives *317*	
11.3.2	Oxidation of Monoterpenes *327*	
11.3.3	Hydrogenation of Monoterpenes *329*	
11.3.4	Epoxidation of Monoterpenes *330*	
11.3.5	Hydration of Monoterpenes *332*	
11.3.6	Esterification and Etherification of Monoterpenes *333*	
11.3.7	Aldol Condensation of Monoterpene Derivatives *334*	
11.4	Conclusions *335*	
11.5	Acknowledgment *336*	
	References *336*	

12	**Environmental Assessment of Novel Catalytic Processes Based on Renewable Raw Materials – Case Study for Furanics** *341*	
	Martin K. Patel, Aloysius J.J.E. Eerhart, and Deger Saygin	
12.1	Introduction *341*	
12.2	Energy Savings by Catalytic Processes *343*	
12.3	LCA Methodology *346*	
12.4	Case Study: Energy Analysis and GHG Balance of Polyethylene Furandicarboxylate (PEF) as a Potential Replacement for Polyethylene Terephthalate (PET) *348*	
12.5	Discussion and Conclusions *352*	
	References *352*	

13	**Carbon Dioxide: A Valuable Source of Carbon for Chemicals, Fuels and Materials** *355*	
	Michele Aresta and Angela Dibenedetto	
13.1	Introduction *355*	
13.2	The Conditions for Industrial Use of CO_2 *356*	
13.2.1	Environmental Issue *356*	
13.2.2	Energy Issues *357*	
13.2.3	Economic Issues *359*	
13.3	Carbon Dioxide Conversion *359*	
13.3.1	Carbonates *359*	
13.3.1.1	Organic Molecular Compounds *360*	
13.3.1.2	Synthesis of Acyclic Carbonates via Carboxylation of Alcohols *361*	

13.3.1.3	Synthesis of Carbonates via Transesterification or Alcoholysis of Urea *364*	
13.3.1.4	Synthesis of Cyclic Carbonates and Polymers *365*	
13.3.2	Carbamates and Polyurethanes *366*	
13.3.2.1	Synthesis of Molecular Carbamates *366*	
13.3.2.2	Indirect Synthesis of Carbamates *369*	
13.4	Energy Products from CO_2 *371*	
13.5	Production of Inorganic Carbonates *373*	
13.6	Enhanced Fixation of CO_2 into Aquatic Biomass *374*	
13.7	Conclusion and Future Outlook *378*	
	References *379*	

Index *387*

The Next Feedstock Transition

The last century has witnessed the dramatic growth of the energy and chemicals industry, fueled by a steep rise in world population, exploding demand for products, breakthroughs in catalysis and polymer sciences, and finally the switch from coal to relatively cheap oil and gas as a feedstock. In the coming decades, further growth and change are to be expected, induced by a continuing rise in world population, the sustainability imperative, substantial progress in bio- and nanosciences, and a gradual switch from fossil-based resources to renewable resources.

To provide a general backdrop to the material presented in this book I should like to offer a few insights gathered during my career in one of the leading energy and chemicals companies, Royal Dutch Shell. Let us start with reminding ourselves that the golden age of chemocatalysis was brought to the fore by the availability of abundant and relatively cheap oil and gas – created by biomass decomposition over millions of years. These resources are characterized by, relative to coal, low to medium range molecular weights, high H/C ratios, and chemical stability upon storage. Over time, the industry has developed an impressive series of thermal and chemocatalytic processes to convert these resources into fuels, lube oils, and chemical building blocks, such as syngas, olefins and aromatics. In a next set of processes, the chemical building blocks are then functionalized to yield aldehydes, acids, esters, amides, aliphatic oligomers and polymers for use in manufacturing of the chemical products that enabled our modern ways of living. Molecular transformations include oxidation, amination, (de-)hydrogenation, desulfurization, denitrogenation, (hydro-)cracking, hydroformylation, polymerization, and so on. The chemical stability of the fossil resources – brought along by the predominance of C–H and C–C bonds in the feedstocks – forces the industry to apply high temperatures and pressures in (especially) the primary conversion processes. The resulting capital intensity has driven the industry to building ever larger plants enjoying very high space time yields and efficient energy management.

It is my firm conviction that the very extensive knowledge and experience base gained through the conversion and upgrading of fossil resources will be a key contributor to the technology revolution needed to move towards an economy that is based on biomass. Firstly, a number of existing conversion technologies will – with some minor adaptations – be employed for the production of alkanes and oxygenates from 'green' syngas, for the upgrading of biomass-derived oils and for the production of chemical intermediates from bio-alcohols and bio-acids. Secondly, a new generation of catalytic processes are or will be developed to specifically convert biomass components, such as sugars, cellulose and lignin into chemical intermediates currently employed by the industry. Thirdly, bio- and chemocatalysts will play a role in the generation of a set of 'green' intermediates and end products that make better use of the high oxygen contents and low aromaticity of biomass.

Frequently, the conversion of a specific source of biomass to one or more products sold to end consumers will include a series of thermal, biocatalytic and thermocatalytic transformations. The logistics of biomass accumulation, storage, and transport are very different from those of fossil resources, and this will have consequences for the choice of technology and size of plants. One might imagine that biocatalytic processes will be employed in smaller scale distributed plants that can handle the seasonality and variability of biomass, and that chemocatalytic processes will be employed in large scale central plants making full use of economies of scale. A concerted approach and extensive technology development will be essential to optimize the value chain from biomass to products to use.

In the coming years, the push for renewable feedstocks will change the landscape of the process industry. It will be fascinating to watch how bio- and chemocatalysis will complement each other in creating green value chains that displace existing value chains based on fossil resources. The authors of the contributions of this book demonstrate that many scientists and technologists will not only watch this story unfold but will actively shape its outcome by leveraging the inherent strengths of chemocatalysis in imaginative and productive ways.

CTO of Royal Dutch Shell (2006–2009) *Jan van der Eijk*

Preface

The strong incentives to switch from traditional raw materials to renewable resources, as well as the requirements to come to selective and efficient conversion processes, have prompted us to compose this book. It gives an overview of several processes, either in developmental or commercial stage, to highlight the characteristics of chemical catalytic processes based on renewables, and the steps that need to be taken to come to an economically and ecologically viable, and attractive process. We have chosen to start it from an intermediate chemicals perspective, that is, those renewable products that are formed by biological or thermal conversion processes, like fermentation, pyrolysis, or selective combustion.

The renewables area is a fast developing one, in which each step of process development workflow needs to be reconsidered and is open for choice. We have striven to include all aspects of the catalytic process development workflow, that is, from idea, research and process development, up to commercialization, as well as the application of advanced methodologies and technologies. The workflow includes idea generation and concepts, design of experimentation and experimental approach, catalyst discovery and screening, catalyst optimization, process design, reaction kinetics, process optimization, reaction modeling, up-scaling, and life cycle analysis. We have chosen to include several industrial contributions, to demonstrate the current state of technologies, but also to be faced with the implications of introducing a new generation of materials and environment.

After an introduction of the industrial playing field in Chapter 1 and the selection of target molecules in Chapter 2, the subsequent chapters are devoted to various groups of target renewable molecules and processes, like terpenes (Chapter 3), furans (Chapter 4), ethanol (Chapter 6), fats and oils (Chapter 7), aromatics (Chapter 8), biorefining (Chapter 9), syngas (Chapter 10), extractives (Chapter 11), and carbon dioxide (Chapter 13). In Chapter 5 methodologies for the development of catalytic processes are described, whereas in Chapter 12 the application of life cycle analysis to renewables is covered.

This book is directed toward decision makers, chemists, and engineers in strategic, research, development, and engineering departments in renewables, petrochemicals, fine chemicals, refining, and the biotechnology industry, as well as academic researchers and students in chemical engineering, chemistry and biology. Moreover, it is useful for policy makers, involved in the bio-based economy and biology, and industrial organizations linked to the renewables industry.

We trust that this book, dealing with the various aspects of the process development workflow, will enable the reader to make optimal choices, and select the right technologies and partners to do so, by understanding the pros and cons of the different scenarios and ways to improve the process, amongst others by parallelization of experimentation and different phases of the workflow.

We would like to sincerely acknowledge our colleagues and friends, who have contributed with passion and expertise to this book. Moreover, Dr. Martin Lok deserves special attention and recognition for his valuable contributions during the review process. And finally, our thanks go to Lesley Belfit, Stefanie Volk and Claudia Nussbeck from Wiley-VCH for their assistance in preparing this book.

March 2013

Pieter Imhof
Jan Cornelis van der Waal

List of Contributors

Paulo Luiz de Andrade Coutinho
Braskem
Av das Nacoes Unidas, 8501
25th floor
05425-070 Sao Paulo, SP
Brazil

Michele Aresta
University of Bari
Interuniversity Consortium on
Chemical Reactivity and
Catalysis-CIRCC
Department of Chemistry
Via Celso Ulpiani 27
70126 Bari
Italy

Joseph Bozell
University of Tennessee
Department of Forestry, Wildlife and
Fisheries
Knoxville, TN 37996
USA

Luis F. Cassinelli
Braskem
Av das Nacoes Unidas, 8501
24th floor
05425-070 Sao Paulo, SP
Brazil

Angela Dibenedetto
University of Bari
Interuniversity Consortium on
Chemical Reactivity and
Catalysis-CIRCC
Department of Chemistry
Via Celso Ulpiani 27
70126 Bari
Italy

David Dodds
Dodds & Associates LLC
4686 Druids Glen
Manlius, NY 13104
USA

Aloysius J.J.E. Eerhart
Copericus Institute
Department of Science, Technology
and Society (STS)
Room 305, Budapestlaan 6
3584 CD Utrecht
The Netherlands

Adil Fahmi
Statoil
Arkitekt Ebbels veg 10
7005 Trondheim
Norway

Bob Humphreys
PO Box 709
109 Lido Blvd
Point Lookout, NY 11569
USA

Pieter Imhof
Avantium Technologies
Zekeringstraat 14
1014 BV Amsterdam
The Netherlands

Ed de Jong
Avantium
Zekeringstraat 29
1014 BV Amsterdam
The Netherlands

Päivi Mäki-Arvela
Åbo Akademi University
Biskopsgatan 8
20500 Turku
Finland

Leo E. Manzer
LLC
Catalytic Insights
Wilmington, DE 19803
USA

Derek McPhee
Amyris, Inc.
5885 Hollis St. Suite 100
Emeryville, CA 94608
USA

Antonio Morschbacker
Braskem
Av das Nacoes Unidas, 8501
24th floor
05425-070 Sao Paulo, SP
Brazil

Dimitry Yu. Murzin
Åbo Akademi University
Biskopsgatan 8
20500 Turku
Finland

Esther Ochoa-Fernández
Statoil
Arkitekt Ebbels veg 10
7005 Trondheim
Norway

Martin K. Patel
Copericus Institute
Department of Science, Technology
and Society (STS)
Room 305, Budapestlaan 6
3584 CD Utrecht
The Netherlands

Gene Petersen
Biomass Technologies Office
US Department of Energy
Golden, CO 80401
USA

Robert-Jan van Putten
Avantium
Zekeringstraat 29
1014 BV Amsterdam
The Netherlands

E. Kendall Pye
Lignol Innovations Ltd
Unit 101 – 4705 Wayburne Drive
Burnaby, V5G 3L1 BC
Canada

Erik-Jan Ras
Avantium Chemicals
Zekeringstraat 29
1014 BV Amsterdam
The Netherlands

Michael Rushton
Lignol Innovations Ltd
Unit 101 – 4705 Wayburne Drive
Burnaby, V5G 3L1 BC
Canada

Erling Rytter
Statoil
Arkitekt Ebbels veg 10
7005 Trondheim
Norway

Tapio Salmi
Åbo Akademi University
Biskopsgatan 8
20500 Turku
Finland

Deger Saygin
Copericus Institute
Department of Science, Technology
and Society (STS)
Room 305, Budapestlaan 6
3584 CD Utrecht
The Netherlands

Irina L. Simakova
Boreskov Institute of Catalysis
pr. Lavrentieva 5, 630090
Russia

Ana Sousa Dias
Avantium
Zekeringstraat 29
1014 BV Amsterdam
The Netherlands

Augusto Teruo Morita
Braskem
Av das Nacoes Unidas, 8501
05425-070 Sao Paulo, SP
Brazil

Jan Cornelis van der Waal
Avantium Technologies
Zekeringstraat 14
1014 BV Amsterdam
The Netherlands

Roberto Werneck Do Carmo
Braskem
Av das Nacoes Unidas, 8501
24th floor
05425-070 Sao Paulo, SP
Brazil

James White
LLC
3RiversCatalysis
Richland, WA 99352
USA

Aalbert (Bart) Zwijnenburg
Johnson Matthey
Wardstrasse 17
46446 Emmerich am Rhein
Germany

1
The Industrial Playing Field for the Conversion of Biomass to Renewable Fuels and Chemicals

Leo E. Manzer, Jan Cornelis van der Waal, and Pieter Imhof

1.1
Introduction

The world is changing its chemistry backbone. In recent years, it has become clear that the feedstocks of choice for the chemical industry can no longer be just petrochemical-based and that alternatives from biomass will need to become readily available in the near future. This has been driven not only by rising fuel prices and the location of much of the petroleum reserves but also the developing insights as to the effects of additional CO_2 on the changing climate of our planet. Many countries are now pursuing the development of new technology for the conversion to fuels and chemicals of renewable feedstocks from biomass that can be grown locally and harvested, such as wood, agricultural crops, algae, municipal waste and animal residue.

In this book we will focus in great detail, from an industrial perspective, on chemical and process technology routes, and the many challenges for these new platform chemicals, their production routes and their potential compared to biological and traditional oil-based technologies. In this chapter an overview of the most important industrial research efforts and strategies is given. The change to a biomass-based economy faces several challenges as the chemistries involved and new platform chemicals that will become available can differ radically from those used for petroleum feedstocks. In addition, it is clear to all involved that a considerable research effort still lies ahead of us.

In this, it is exemplary to consider the petrochemical industry. It has taken several decades to develop technology for the conversion of crude oil and gas into transportation fuels and chemicals and this research effort is still ongoing. In contrast to the exploration of crude oil and the production of fuels and chemicals, which typically involve cracking of carbon–carbon bonds, heterogeneous acid catalysis and hydrotreatment, the renewable feedstocks require a new collection and distribution system and the conversion of wet, highly functional carbohydrates into products that will ideally be compatible with existing refinery operations. As a result, new catalysis and biology need to be developed to convert these oxygenated feedstocks to useful products.

Catalytic Process Development for Renewable Materials, First Edition. Edited by Pieter Imhof and Jan Cornelis van der Waal.
© 2013 Wiley-VCH Verlag GmbH & Co. KGaA. Published 2013 by Wiley-VCH Verlag GmbH & Co. KGaA.

1.2
The Renewables Arena

Many companies are active in the emerging renewables arena. In this chapter, several of the large multinational and several of the smaller start-up companies will be discussed in more detail. A nice way to identify the key-players based in the field is by looking at the yearly Biofuels Digest polls among its readers and the list of the top 50 companies that its readers deem most active in the renewable fuels and chemicals area [1]. The top 50 companies from the 2010–2011 survey are shown in Table 1.1. It is also interesting to note that the focus of these companies differs both in feedstocks used and in products made, exemplifying that the use of biomass is a general concept and is in good agreement with environmental effort to create biodiverse, sustainable agricultural landscapes.

An alternative way is to look at the most promising bio-renewable platform chemicals derived from the various processes to see which companies have either production capacity already or have reported development of new processes. Table 1.2 shows a list of products with strong growth potential in the chemical industries and some promising ones that are well advanced in the pipeline [2].

The companies shown in Tables 1.1 and 1.2 also reveal the broad and diverse interest in biorenewable fuels and chemicals. The effort is carried by both established

Table 1.1 The top 50 bioenergy companies, compiled by Biofuels Digest 2010–2011.

1	Amyris		26	Algenol
2	Solazyme		27	ZeaChem
3	POET		28	PetroAlgae
4	LS9		29	Neste
5	Gevo		30	Synthetic Genomics
6	DuPont Danisco		31	LanzaTech
7	Novozymes		32	Iogen
8	Coskata		33	OriginOil
9	Codexis		34	Range Fuels
10	Sapphire Energy		35	ExxonMobil
11	Virent		36	Cargill
12	Mascoma		37	SG Biofuels
13	Ceres		38	Butamax
14	Cobalt Technologies		39	Terrabon
15	Honeywell's UOP		40	Cosan
16	Enerkem		41	Verenium
17	BP Biofuels		42	Waste Management
18	Genencor		43	IneosBio
19	Petrobras		44	Dynamic Fuels
20	Abengoa Energy		45	Fulcrum Bioenergy
21	Qteros		46	KL Energy
22	Joule Unlimited		47	KiOR
23	Shell		48	Chevron
24	Bluefire Renewables		49	Monsanto
25	Rentech		50	Inbicon

Table 1.2 Production of platform chemicals [2].

Cn	Products with strong growth potential		Biobased chemicals in the pipeline	
	Chemical	Company	Chemical	Company
1	Methanol	BioMCN, Chemrec	Formic acid	Maine BioProducts
2	Ethylene	Braskem, DOW, Songyuan Ji'an Biochemical	Ethyl acetate	Zeachem
	Ethanol	Many	Glycolic acid	Metabolix Explorer
	Ethylene glycol	India Glycols Ltd, Greencol Taiwan	Acetic acid	Wacker
3	Lactic acid	Purac, NatureWorks, Galactic	Acrylic acid	Cargill, Perstorp, OPXBio, DOW
	Glycerol	Many	Propylene	Braskem
	Epichlorohydrin	Solvay, DOW	3-Hydroxypropionic acid	Cargill
	1,3-Propanediol	DuPont/Tate&Lyle	n-Propanol	Braskem
	Ethyl lactate	Vertec BioSolvents		
	Propylene glycol	ADM	Isopropanol	Genomatica
4	n-Butanol	Cathay Industrial Biotech	1,4-Butanediol	Genomatica
	iso-Butanol	Butamax, Gevo	Methyl methacrylate	Lucite
	Succinic acid	DSM, BioAmber, Myriant		
5	Furfural	Many	Itaconic acid	Itaconix
	Xylitol	Lenzing	Isoprene	Goodyear/Gencor, Amyris
			Levulinic acid	Maine BioProducts, Avantium, Segetis
6	Sorbitol	Roquette, ADM	Adipic acid	Verdezyne, Rennovia, BioAmber, Genomatica
	Isosorbide	Roquette	FDCA	Avantium
	Lysine	Draths	Glucaric acid	Rivertop, Genencor
	Caprolactam	DSM		
N	PHA	Telles	*para*-Xylene	Gevo, Draths, Annellotech, Virent
	Fatty acid derivatives	Croda	Farnesene	Amyris

large multinationals and young inventive start-ups, but the diversity is also clear in other aspects. For example, renewable fuels are obtained from different feedstocks, such as algae, oils, and carbohydrates to challenging ones such as lignin, lignocellulosics and municipal solid wastes. Clearly, with such a broad range of feedstocks many different technologies need to be developed and applied. Currently fermentation of carbohydrates to ethanol is by far the biggest process, but other fermentations, like Gevo's butanols and Amyris's Farnesenes, are now commercialized. However, chemical catalysis will quickly pick up from there and convert the fermented products further. The announcement of The Coca Cola Company [3] of collaboration with Gevo, Virent and Avantium to develop 100% biobased beverage bottles is based on three different catalytic technologies.

One thing the tables above do not reflect well is the considerable impact that existing, well-established technologies have in the renewable area. The fermentation of carbohydrates to alcoholic beverages (6.1 l of pure alcohol per person aged over 15 year worldwide or 32 Mtonne per year total [4]), and the use of vegetable oils that can be saponified to soaps, emulsifiers and other chemicals (around 14% of world production of vegetable oils or an equivalent of 16.8 Mtonne per year [5]).

When each entry in Tables 1.1 and 1.2 is categorized by feedstock use, product made, type of processing, and geographic location, one can draw certain conclusions as to the state and the direction for the current chemicals from renewables efforts. In this analysis, no differentiation between company size and/or development/maturity state of the process is made and the numbers presented are based purely on the number of companies involved. With respect to the geographic spread of companies involved in the renewable areas (see Figure 1.1), it is clear that the United States of America, with two-thirds of the companies being located there, is leading. At first sight, the contribution of Asia appears to be small. However, many of the larger industrial conglomerates are collaborating very actively with smaller start-up companies all over the world. In addition, this survey did not include the already existing biorenewable efforts in which Asia has a long-standing tradition, especially palm oil and alternative sugar crops, such as rice and tapioca, which offer great potential as feedstocks for biorenewable fuels and chemicals (Figure 1.1).

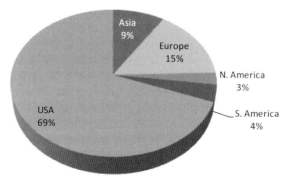

Figure 1.1 Geographic distribution of companies in the renewable area.

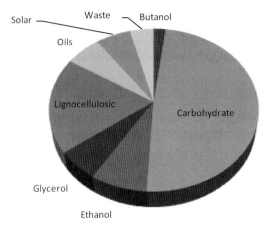

Figure 1.2 Distribution of feedstock use in the renewable area by number of companies. Carbohydrates refer to readily available sugars (glucose, fructose, sucrose, xylose). Oils refer to vegetable oils.

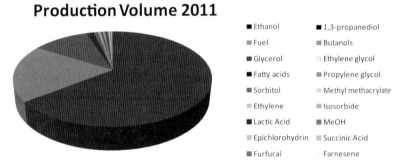

Figure 1.3 Distribution of renewable products made scaled to reported production volume. The production volume in kt a^{-1} for the five largest products is also shown.

With respect to the biomass feedstock used, it is clear from Figure 1.2 that carbohydrates, including lignocellulosics, make up 74% of the starting material; even more if one considers that ethanol and butanol as the starting product in most cases originate from carbohydrates by an earlier fermentation process. The concept of using intermediates derived from biomass is called secondary-derived biobased platform chemicals. Ethanol and butanol, from fermentation of sugars, are examples, but it is not restricted to this. Dohy [2] identified several groups, that is, biogas, syngas (Chapter 10) and H$_2$, pyrolysis oils (Chapter 8), vegetable oils (Chapter 7), lignin (Chapter 9) and the C5 and C6 sugars (Chapters 4 and 8). Fermentation (Chapters 3 and 6) and catalytic conversion of the latter two (Chapter 4) gives rise to an even more diverse group of products, as was identified by Werpy [6] (Figure 1.2). In addition the use of CO$_2$ as feedstock is addressed in Chapter 11.

In Figure 1.3 the distribution of renewable products is shown. From this figure it is clear that the products industry currently still focuses on ethanol and fuel

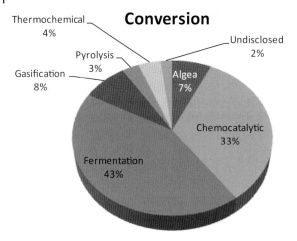

Figure 1.4 Distribution by type of conversion process used by number of companies.

production. There is a strong correlation with the size of the company and the production of these two main products. Smaller, often venture capital backed companies tend to focus on drop-in solutions, hence ethanol, biodiesel and other fuels. This probably reflects the perceived reduced risk profile for these drop-in products. The established low-key technology needed, and the small production volumes. However, one should consider that obtaining a market share can still constitute a high risk since one is competing against almost fully optimized processes employing depreciated capital.

It is also very instructive to consider the type of process used to produce the biorenewable product. From Figure 1.4 it is clear that the most commonly used type of processing is a biological type of conversion, that is fermentative and/or enzymatic (43 processes, 52.1 Mtonne per year) or algae (7 processes, 0.2 Mtonne/year). Not surprisingly, a very strong correlation with ethanol can be observed here. Similarly, chemocatalytic processes are dominated by the FAME (fatty acid methyl ester) biodiesel production. If these two fuel products are excluded from the analysis, the contribution of fermentative processes to the other products is still very high (42%). The remainder of the processes used are bio-renewable chemicals divided between thermal processes and chemocatalytic processes. Especially, the overall low usage of chemocatalytic routes today is surprising as this is in stark contrast with current day petro-based chemicals production where almost all process and products involve catalysis at some point.

It is also refined from our data that the more specialized products in the graph are generally produced by larger multinationals which have the financial backbone and market outlets, or by small companies that have access to a unique technology, often fermentative/enzymatic in nature. Both parties have found each other and many collaborations and joint ventures have since been announced. An overview of these collaborations is given in Table 1.3.

Table 1.3 Overview of efforts, collaborations and joint ventures of some large, multinational oil companies mid-2011.

Company	Field	Partner company	Since	Status
Abengoa energy	Ethanol (carbohydrate)	–		3 plants
BP	Algae	Martek Biosciences	2009	
	Butanol	Butamax	2003	
	Ethanol (cellulosic)	JV with Dupont	2009	
	Diesel (oil seeds)	Verenium (acquired) TERI	2006	
Chevron	Algae	Solazyme	2008	
	Diesel (oil seeds)	Bioselect	2007	1 plant
	Ethanol (carbohydrate)	LS9	2011	
ConocoPhilips	Algae	Colorado Center for Biorefining and Biofuels	2007	
	Biofuel research		2007	2 plants
	Diesel (oils seeds)	Iowa State University Tyson Foods	2007	
DuPont	Butanol	Butamax JV with BP	2003	
Exxon	Algae	Synthetic Genomics, Inc.	2009	
	Biodiesel	–	2008	
	Gasification	–	2008	
Honeywell UOP	Pyrolysis	Envergent JV Ensyn Corp.	2008	
JX Nippon oil	Diesel (oil seeds)	JV Petronas, Toyota		Pilot
Mitsubishi	Methyl methacrylates	Lucite (acquired)	2008	
Mitsui	Ethanol (carbohydrate)	Dow, Hitachi Zosen	2011	
	Ethanol (syngas)	LanzaTech	2011	
	Ethanol (waste)	Sime Darby	2010	
	Succinic acid	BioAmber	2011	
Neste	Diesel (oil seeds)			2 plants
Petrobras	Ethanol (carbohydrate)	Biocombustível		7 plants
	Diesel (oil seeds)	Biocombustível	2010	1 plant
	Diesel (oil seeds)	BSBIOS	2009	
	Algae	Galp Energia		
	Biofuel (cellulosic)	JV with Eni BiChem		
Sinopex	Diesel (oil seeds)	JV CNOOC, Novozymes		
Shell	Valeric esters	–	Ended 2011	
	Algae	Cellana	Ended 2009	1 plant
	Gasification	Choren	Ended 2009	
	Ethanol (cellulosic)	Iogen	2010	
	Ethanol (carbohydrate)	Raizen	2009	
	Ethanol (waste)	JV with Cosan	2006	
	Cellulosic to fuel	Codexis Virent, Cargill	2010	
Total	Ethanol (carbohydrate)	Coskata	2010	

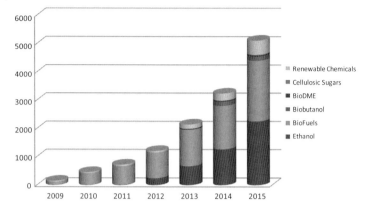

Figure 1.5 Prognoses of growth in biorenewable products in total kt a^{-1} by new technologies [1].

Today's status of the biorenewable fuels and chemicals world is one of direct drop-in replacements and thus strongly tied to fermentation of sugars to ethanol and transesterification of vegetable oils to biodiesel. What is also clear from our survey is that the use of secondary derived biobased platform chemicals is still very limited, suggesting that considerably more effort will be needed in this area for the chemical industry to be able to offer a full biomass-to-product portfolio. In Figure 1.5 the projected cumulative production volume of new biorenewable chemicals and fuels processes, based on announced production volumes employing novel technologies, is shown. The total production volume of 5.1 MTonne per year, however, is still small compared to the current total production volume for ethanol and FAME in 2010, which exceeded 80 Mtonne per year.

Though the advance of new biobased products seems small, it should be realized that these are only the front-runners of a whole new generation of materials. Many of the technologies are not yet mature enough for commercial (pilot-) plant production. At least 25 research efforts in particular bio-renewable chemicals are reported for the 2011–2015 period. In Figure 1.6 these 25 bio-renewable chemicals are shown, with the number of independent research efforts for each identified. Clearly, ethanol and fuels are still the biggest contributors, but it is without doubt that an impressive number of widely diverse chemicals will become available in the near future (Figure 1.6).

The world today is seeing more and more processes that are using bio-renewables. Certain production routes are well-established with large production volumes, such as the traditional fermentation of sugars to ethanol or the transesterification of fatty acids to FAME. However, it is clear that on the horizon a diverse spectrum of molecules will become available. Unlike the current petrochemical-based chemicals industries, which use almost exclusively chemocatalytic and thermal processes, the role of fermentation and enzymatic conversions will be much more marked in the bio-refinery of the future. Still, we have a long way to go. Today, processes are focused on drop-in replacements or on the conversion of biomass

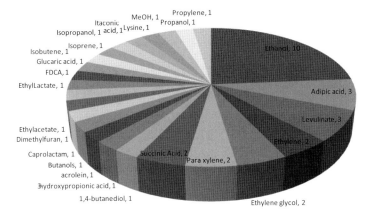

Figure 1.6 Announced research efforts until 2015 in new biobased renewable chemicals. The number in each pie indicates the number of companies researching and/or producing the renewable chemicals.

in secondary derived biobased platform chemicals. Future research will most likely start to focus on further conversions of these platform chemicals, and will make use of technology closer to the catalytic processes currently in place.

In the next two sections recent developments in the renewable fuels and in the renewable chemicals areas will be discussed. Although most players clearly state what their aim is, it is also clear that an overlap exists between the two areas. Where needed the effort of such players is divided over the two sections.

1.3
Renewable Fuels

The eventual transition from petroleum-based transportation fuels to renewable fuels provides a huge boost for new catalysis and process research. The existing renewable gasoline market is entirely dominated by ethanol, with butanol or its isomers as an emerging alternative, both of which are obtained by fermentation of carbohydrates. The existing renewable diesel market is dominated almost entirely by the fatty acid ester derived products. As was already noted in the introduction, most of the producers of these biofuels appear to be small companies, but one should realize that most large multinationals have active collaborations or joint ventures with these companies. In the previous section an overview of the efforts of multinationals in the biofuels area was given (Table 1.3). From this list it is clear that almost all large oil companies embrace the biofuel concept. However, other than production in first generation biofuels, that is ethanol and FAME diesels, most research efforts are directed through collaboration with new, smaller enterprises.

It has also been recognized that the use of food feedstocks is not desirable in the long term. New fuels and technology based on the use of non-food feedstocks, lignocellulosic and waste streams in particular, are under development and will hopefully be commercialized in the near future. However, it is important to recognize that the technical challenges for the conversion of carbohydrates are significant, even with pure carbohydrates such as glucose.

Fractionation of lignocellulosics is well known and widely applied in the production of pulp and paper. The oldest processes known to efficiently obtain fermentable carbohydrates from lignocellulosic materials are the Giordani–Leone process with concentrated sulfuric acid [7], and the Rheinau–Bergius process (1931) with concentrated hydrochloric acid [8], both currently researched and applied by companies like BlueFire Ethanol, Green Sugar and HCl Cleantech; and the Organosolve (1971) [9], first applied in the Alcell process and currently researched by Lignol. However, techno-economic aspects have showed in the mid-1970s that these processes are not cost effective against producing carbohydrate streams from easily hydrolyzable starches such as from corn. New processes for lignocellulosic fractionation will have to be developed that can handle cheap feedstocks such as straw and other waste streams, but even then the nature of these feedstocks will remain challenging, as illustrated in Figure 1.7, and robust conversion processes need to be developed in parallel [10]. It is readily apparent that the sugar-containing

Simulated liquor from dilute-acid pretreated com stover

Actual liquor from dilute-acid pretreated com stover

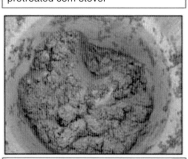

−25% solids dilute-acid pretreated corm stover slurry (recently mixed drum)

−25% solids dilute-acid pretreated corm slurry (drum unmixed – 1 yr)

Figure 1.7 Photos of solutions derived from lignocellulosic biomass and pure carbohydrate, kindly supplied by Dr. E. Wolfrum of NREL, www.nrel.gov.

feedstocks will also contain biomass residues, tars, oligomers, inorganic metals from nutrient solutions, and sulfur, nitrogen and protein from fermentation-derived feedstocks. This mix of components will pose significant challenges in achieving high conversion, selectivity and long lifetimes of new catalytic systems or fermentative processes. Using unpurified hydrosylates/carbohydrate solutions derived from lignocellulosic feedstocks with minimal processing as feedstock will require even more catalyst and process development before implementation as the feedstock in a bio-refinery.

The most commonly used alcohol for blending with gasoline is ethanol. The production of ethanol from readily available carbohydrates is well known, especially in Brazil from sugar cane and in the USA from corn. Several other companies focus on the production of ethanol from lignocelluloses or from waste streams rich in carbohydrates, as discussed above. However, ethanol has many limitations, such as low energy density, hydrophilicity, corrosivity, miscibility/mixture stability, and cannot be transported in existing pipelines. Bio-butanol is an attractive alternative to ethanol with fewer implications. The benefits of butanol over ethanol as a fuel are shown in Table 1.4. Butanol offers better safety, improved fuel economy, can be blended with gasoline in high concentrations, and used without vehicle modifications. Another important advantage is that butanol can be transported in existing pipelines.

An overview of companies that are active in this field is shown in Table 1.5. Large companies like BP and DuPont are moving rapidly to introduce butanol.

Table 1.4 Representative properties of butanol versus ethanol and gasoline [11].

Property comparison	EtOH	BuOH	Gasoline
Energy content (BTU/gal)	78M	110M	115M
Reid V.P. @ 100°F (psi)	2.0	0.33	4.5
Motor octane	92	94	96
Air-to-fuel ratio	9	11	12–15

Table 1.5 Some companies actively involved in bio-butanol.

Company	Biofuel
Butamax (BP/DuPont)	iso-butanol
	n-butanol
	2-butanol
Gevo	iso-butanol
Metabolic Explorer	n-butanol
Cobalt Biofuels	n-butanol
Green Biologics Ltd.	n-butanol
Tetravitae Bioscience	n-butanol (ABE)
Butalco	n-butanol

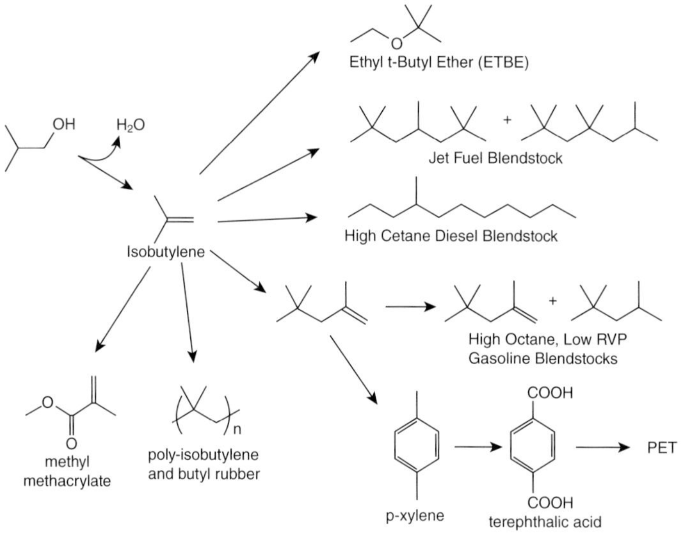

Figure 1.8 Some derivatives available from bio-isobutanol, adapted from Bernacki [12].

Smaller start-up companies, like Gevo and Cobalt Biofuels, have received significant venture capital funding and are making great progress with demonstrations in large pilot plants. Gevo Inc. was founded in 2005 and issued an IPO (initial public offering) in February 2011. At the start of the company, Gevo licensed intellectual property developed by Jim Liao at UCLA and Frances Arnold at Caltech. Gevo's fermentation processes can be retrofitted into existing ethanol plants with a limited amount of capital. The first of these retrofits was at St. Joseph, Missouri (2009) with an organism capable of making butanol. Of the three butanol isomers that their technology can produce, isobutanol is particularly attractive since it can be readily dehydrated to isobutylene, and offers an opportunity as a drop-in substitute as a renewable feedstock in existing processes, as shown in Figure 1.8. These cover a wide range of renewable chemicals and renewable fuels, such as jet fuel, high-octane gasoline, solvents, renewable terephthalic acid for PET bottles and butyl rubbers. In 2010 they announced a joint venture with Lanxess for the production of butenes, and in 2011 a collaboration with The Coca Cola Company on PET for beverage bottles [3].

Founded with technology developed at UC Berkeley and with initial venture capital funding another successful start-up is Amyris, Inc., which issued an IPO in September 2010. Their product platform is based on carbohydrate fermentation to the isoprene trimer, β-farnesene (Figure 1.9), which is insoluble in water and separates easily from the fermentation liquid [13]. After hydrogenation an attractive diesel fuel is obtained. However, the unsaturation in the β-farnesene also provides possibilities for a variety of monomers, polymers and specialty chemicals, in particular surfactant for use in soaps and shampoos, a cream used in lotions, a number of lubricants, and also the fully hydrogenated farnesane for solvent applications.

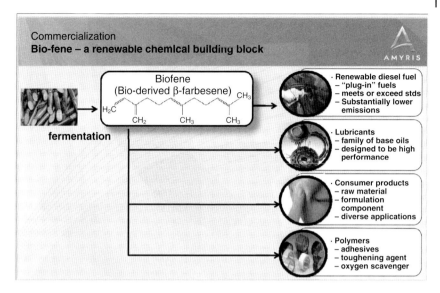

Figure 1.9 Family of products derived from Amyris Biofene, kindly supplied by Neil Reninger, CTO of Amyris, www.amyris.com.

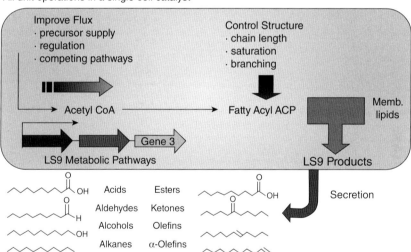

Figure 1.10 Potential biobased products produced by LS9, Inc., kindly supplied by S. del Cardayre, VP R&R from LS9, www.ls9.com.

Another bio-company, which converts carbohydrate to hydrocarbon products that readily separate from the fermentation broth, is LS9 Inc. founded in 2005. They have developed biology for the fermentation of carbohydrates to linear hydrocarbons, as shown in Figure 1.10. Organisms that naturally produces lipids have been genetically altered such that they are able make a variety of linear hydrocarbons.

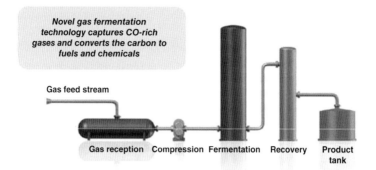

- **Innovative:** Microbe uses gases as its sole source of energy
- **Proprietary:** LanzaTech has filed 58 patents, including two proprietary microbe patents
- **Not just about ethanol**
- **Integrative:** Direct production of fuels and chemicals (2,3 Butanediol, Isoprene, Propanol, Butanol, MEK); multi step production of chemicals and chemical intermediates (olefins)
- **Thermo Chemical Opportunities:** 2,3 Butanediol produced through the LanzaTech Process can be used to make true "Drop in" hydrocarbin fuels (gasoline, diesel, jet fuel).

Figure 1.11 The LanzaTech process for converting CO-rich streams to ethanol and 2,3-butanediol, kindly supplied by S. Simpsoon, CSO and co-founder of LanzaTech, www.lanzatech.co.nz.

Further modifications to the organism can yield an even wider array of products, including alcohols, olefins, ketones and aldehydes.

Though ethanol and other alcohols can readily be obtained by fermentation, it has been restricted to the use of sugar feedstocks, be it sugar cane, corn or lignocellulose. An alternative is being commercialized by LanzaTech Inc., a start-up company founded in 2005 in New Zealand. LanzaTech's commercial plants will effectively convert a variety of non-food, low-value gas feedstocks into bioethanol and other platform chemicals. They have developed a proprietary technology that allows fermentation of CO-rich streams into ethanol and other products like 2,3-butanediol. They have been operating a pilot plant at a steel mill in New Zealand and have just announced construction of a 100 000 gallon per year demo plant in collaboration with Bao Steel in China. Following a successful scale-up of their demo facility, commercial plants are planned for China, South Korea and India. Although, initially using CO-rich gases, a by-product from steel manufacturing, it is easy to envisage that their process can be coupled to bio-mass gasification and thus use non-carbohydrate sources as well (Figure 1.11).

The future of biofuels will, however, be beyond ethanol, butanol and FAME-based biodiesel. Next generation fuels will be hydrocarbon fuels from non-food, lignocellulosic feedstocks. Several options are under investigation, but key in all of these is the removal of the abundant oxygen present in the biomass molecules, especially carbohydrates.

Biomass gasification potentially provides an excellent method of converting lignocellulosic feedstock into synthesis gas. The production of syngas is technically feasible but presents a number of challenges to remove tars and inorganics, which are severe poisons to downstream Fischer–Tropsch or higher alcohol synthesis catalysts [14]. An excellent overview of potential processes to convert lignocellulosic feedstocks to advanced biofuels can be found in a report from a DOE workshop in 2007 [15]. New gasifiers are under development around the world since the variable composition of biomass is significantly different from natural gas or coal, the traditional sources of syngas. An early pioneer in the field was Range Fuels, which built a demo plant in Georgia to gasify wood chips to syngas followed by an alcohol synthesis primarily to methanol and ethanol. Unfortunately, the plant did not work as expected and was closed [16]. Currently, Ineos New Plant Bioenergy, LLC (a division of Ineos Chemical Company) is constructing a commercial demonstration facility in Indian River County, Florida to ferment synthesis gas derived from lignocellulosic feedstocks into ethanol. Hence, a large-scale operation to validate the potential for this thermochemical and biological route should be demonstrated [17].

One concept under consideration by many companies is field densification, or partial dehydration of lignocellulosic materials. A preferred process, called flash pyrolysis, involves heating the biomass very quickly ($\sim 100\,°C\,s^{-1}$) to a temperature of about 400 °C [18]. During this heating step, volatile gases are produced, along with a solid char (from the lignin) and a wet oil called pyrolysis oil. In general, the char and the volatile gases that are co-produced are burned to provide the heat needed for the process. The oil tends to be very acidic and corrosive, highly colored, immiscible with hydrocarbon fuels, and contains about 20–30% residual water. It is also relatively unstable and its composition changes with time. Stabilizing and upgrading of the pyrolysis oil prior to shipment is one of the many challenges to commercial use. The pyrolysis treatment will significantly deoxygenate the pyrolysis oil. UOP and Ensyn have formed a joint venture (JV) called Envergent Inc. with the purpose of developing technology to hydrogenate the pyrolysis oil obtained by the RPT (rapid thermal processing) process of Ensyn to a useful hydrocarbon fuel. In their current process woody biomass is brought in contact with the hot sand at about 510 °C for 2 s to produce pyrolysis oil (Figure 1.12). Dynamotive is another fast pyrolysis company that has several commercial plants making pyrolysis oil for energy use and is developing a two-stage hydrogenation process that first stabilizes the oil by reducing water and oxygen content and subsequently the first product to give a gasoline/diesel fuel.

An alternative approach in pyrolysis is pursued by KiOR Inc. [19], by using a catalyst during the pyrolysis step. The effect is very significant since the catalyst allows the pyrolysis of the biomass at a lower temperature and at the same time reduces the oxygen content of the pyrolysis oil. Their biomass catalytic cracking process (BCC) produces a less-acidic oil that separates from the aqueous phase and can be readily processed to produce fuels (Figure 1.13).

Other companies are actively investigating the option of chemo-catalytic routes to convert carbohydrate-rich streams to fuel intermediates. In general, the strategy

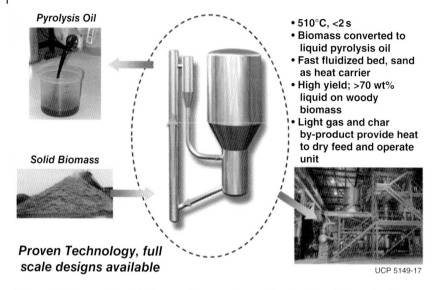

Figure 1.12 Honeywell/UOP's Invergent's process for production of pyrolysis oil, kindly supplied by J. Holmgren, formerly Honeywell UOP, www.uop.com.

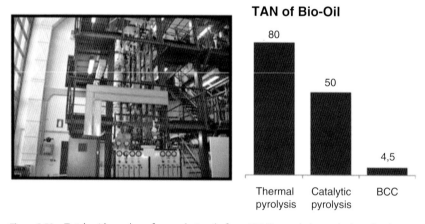

Figure 1.13 Total acid numbers for pyrolysis oils from KiOR's catalytic pyrolysis technology (BCC), kindly supplied by P. O'Conner, founder of KiOR, www.kior.com.

chosen here is to leave the carbon backbone of the carbohydrate molecule intact, and remove the oxygen atoms, either by dehydration, decarbonylation/decarboxylation or by hydrogenation. The dehydration route is the most preferred as it is not accompanied by loss of carbon (decarbonylation/decarboxylation) or usage of hydrogen (hydrogenation).

Dehydration of carbohydrates leads to furfural from C5 carbohydrates and hydroxymethyl furfural (HMF) from C6 carbohydrates. The latter rapidly reacts further to give levulinates. Both furfural, HMF and levulinates can serve as the starting point for a new class of biofuels. Currently no commercial process is known for HMF, but levulinates are produced by Biofine Renewables, LLC, directly from biomass (see more details in Section 1.4). Several groups in academia are working on related processes with HMF, such as the dimethylfuran or nanones from the Dumesic group [20] and the methyl furans (RWTH-TMFB). Industrially, Avantium and Shell, both from the Netherlands, have claimed biofuels via furfural, HMF and Levulinate pathways. A group at the University of Maine has conducted work in the production of higher alcohols liquid biofuel via acidogenic digestion and chemical upgrading of industrial biomass streams. The resulting fuels have a very low acid number [21].

Shell International has claimed the use of tetrahydrofurfuryl ethers derived from hydrogenation and etherification of furfural [22]. More recently, their focus has shifted to valeric acid based fuels (see Figure 1.14) still using levulinate as the starting point [23]. The levulinates are typically obtained from C6 carbohydrates, via HMF as the intermediate, a process currently being developed by BioFine, or from C5 via furfural by the process developed by the Quaker Oats Company in 1922. Shell has claimed routes from furfural by hydrogenation to furfuryl alcohol followed by a rearrangement to levulinate. The possibility to use both C5 and C6

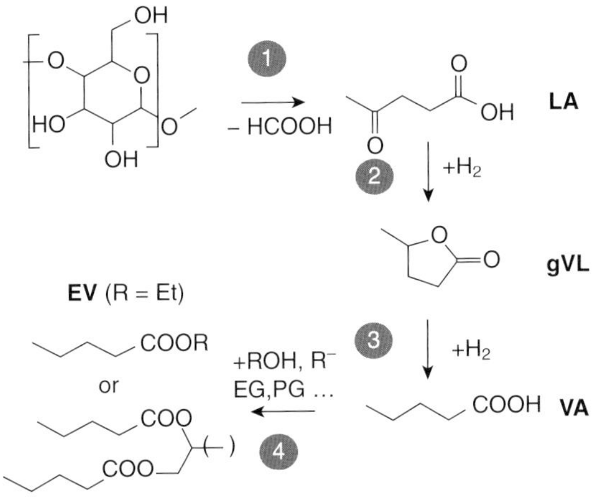

	1. Hydrolysis	2. Hydrogenation	3. Hydrogenation	4. Esterification
Catalyst	H_2SO_4	Pt/TiO_2	Pt/ZSM-5	IER
Selectivity	50–60%	>95%	>90%	>95%
Productivity	>0.1 h^{-1}	>10 h^{-1}	>1 h^{-1}	>0.02 h^{-1}
Concentration	<5%	>90%	>50%	>50%

Figure 1.14 GVL-based fuels [23]. LA = levulinic acid, gVL = gamma-valerolactone, VA = valeric acid and EV = ethyl valerate.

carbohydrates fits very nicely with the potential to use non-food lignocellulosic feedstocks for these fuels.

In the process described by Avantium Chemicals, carbohydrates are dehydrated in alcoholic solutions to give the stable ethers of hydroxymethyl furfural and furfural. Further processing yields ethers of furfuryl alcohol and bishydroxymethylfuran and the corresponding tetrahydro equivalents. Depending on the type of alcohol used the ethers and bis-ethers can be applied as gasoline or as diesel [24].

It remains clear that a large source of biomass will be needed to make a significant contribution to society's need for transportation fuels. A 2005 report from the USDA, updated in 2012 [25], suggested that the US had about 1.366 billion tons of agricultural and forest residue available but that this amount of biomass would satisfy only about 30% of the US transportation fuel needs. An issue that holds merit for most biomass sources is that it has a high water content. Economical transportation over long distances is usually considered uneconomical and it is estimated that conversion units will need to be located within about 50 to 80 miles or less, depending on the type of biomass. Much of the biomass available is located at significant distances from any potential biorefinery and end-users, thus requiring small-scale conversion to condensed, water-free intermediates before it may be economically transported.

1.4
Renewable Chemicals

Several companies focus on chemicals derived from biomass, even more than for fuels, for which other clean energy sources and uses can be envisaged, such as electric cars via solar panels. The production of renewable chemicals inherently requires carbon atoms from the biomass. Several companies are working on biological, catalytic or thermochemical routes to chemicals as they generally have a higher value than fuels. In some cases the products are direct drop-in replacements, while other companies obtain new chemicals and products. An important and commonly cited report was prepared by PNNL and NREL [6], which will be discussed in detail in Chapter 3. They proposed 12 molecules (see Table 1.6) that

Table 1.6 The top carbohydrate-derived building blocks outlined by DOE [6].

Top 12 carbohydrate-derived building blocks	
Succinic, fumaric and maleic	Itaconic acid
2,5-Furan dicarboxylic acid	Levulinic acid
3-Hydroxyl propionic acid	3-Hydroxybutyrolactone
Aspartic acid	Glycerol
Glucaric acid	Sorbitol
Glutamic acid	Xylitol/arabinitol

would be attractive building blocks for future renewable chemicals. This report has stimulated a very significant amount of research within the academic and industrial communities.

Several of these building blocks can be made via fermentative or enzymatic routes and are thus outside the scope of this book. Nevertheless, as was already concluded in earlier sections, the role of fermentation in bio-mass conversion will certainly be an important technology in future bio-refineries. Genencor and Rivertop describe enzymatic and thermochemical processes for the carbohydrate acids, gluconic and glucaric acids [26, 27]. Likewise, for the amino acids aspartic and glutamic acids, which are produced largely in China by a variety of companies, the Fufeng group being the largest.

The chemical synthesis and production of some of these building blocks are already well known. In the mid-1950s Roquette in France developed processes for sorbitol and xylitol. These processes are based on the hydrolysis of starches to glucose and subsequent selective hydrogenation to sorbitol and can be modified to give other polyols, that is, maltitol, mannitol, xylitol and arabinitol. The sorbitol is then further converted into the anhydride isosorbide, which is considered a potential diol building block for renewable polyesters [28].

One molecule that is receiving considerable attention is succinic acid. A number of small and large companies (BioAmber, Myriant, DSM, BASF, and others) have announced plans to commercialize biobased succinic acid by fermentation and its derivatives by chemical conversion. The world's first commercial demo-plant of succinic acid was started in France in January 2010 by BioAmber Inc. with nominal capacity of $220\,t\,a^{-1}$, with new full-scale plants being planned in Canada and Thailand. The petrochemical route for producing succinic acid is by hydrogenation of maleic acid, obtained via butane oxidation. A biobased source of succinic acid and its derivatives would be drop-in replacements for the petroleum-derived compound and could even be used to supply bio-maleic acid and maleic acid anhydride by selective oxidation. The markets for products derived from biobased succinic acid (polymers, plastics, solvents, adhesives, and coatings) (Figure 1.15) are significant. So it is not surprising that it has attracted the attention of so many companies.

First described in 1875 by Freiherrn, Grote and Tollens [30], levulinic acid has been of interest for many years as a bifunctional keto-acid but it has never been made on a commercial scale [31]. Biofine Renewables, LLC has developed a thermochemical process that produces levulinic acid, furfural, formic acid and char from a wide variety of carbohydrate-containing biomass feedstocks. The process uses two reactors: in the first the biomass is broken down into small components, such as carbohydrates and hydroxymethylfurfural, which are then further converted in the second reactor to levulinic acid. The conditions employed are extreme, with high temperatures and pressures in the presence of acid catalyst, yielding about 50% levulinic acid. With the ketone and acid group functionality, levulinic acid can be converted to a wide variety of compounds. Currently three demo-plants are operational, the largest in Caserta, Italy with a $50\,t\,d^{-1}$ lignocellulosic feed intake. A study by Hayes [32] suggests that the production of ethyl levulinate can

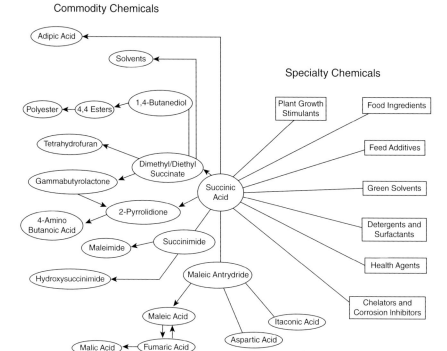

Figure 1.15 Succinic acid as a biorenewable platform chemical [29].

be competitive at a 400 kt a^{-1} scale, if all energy required is generated on-site from waste streams.

DuPont has disclosed a diverse portfolio of products that can be obtained such as *N*-alkylpyrrolidones, monomers for the preparation of thermally stable polymers, ionic liquids and nylon intermediates (Figure 1.16) [31]. While many of these materials are exciting new renewable-based chemicals, none have reached commercial production yet, due to the absence of a high volume levulinic acid production process at sufficiently low cost.

Segetis Inc., a Minnesota-based venture capital backed company, also uses levulinic acid as the basis of their product portfolio [33]. They developed a novel class of ketals formed by the reaction of diols and polyols with the ketone group of levulinic acid. In particular, by using glycerol and levulinic acid, both part of the DOE "Top 12" list (Table 1.6), new ketal products are produced which have excellent functionality and can potentially replace existing petroleum-based solvents, surfactants and plasticizers (Figure 1.17).

Furandicarboxylic acid (FDCA) is another of the DOE Top 12 molecules, and has potential to replace terephthalic acid [34]. It has been shown that this could be prepared by oxidation of hydroxymethylfurfural, HMF. Production of HMF has long been researched, but even with two pilot plants having been commissioned no commercial process has been established that can produce HMF at low cost

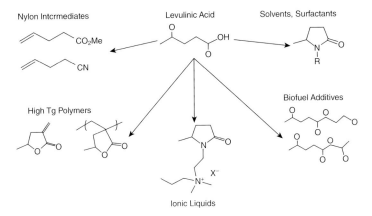

Figure 1.16 Selected derivatives of levulinic acid, adapted from Ritter [31].

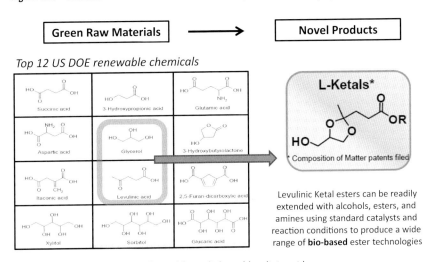

Figure 1.17 Segetis ketals are derived from diols and levulinic acid.

[35]. Avantium is pursuing the use of HMF-ethers as renewable platform chemicals (Figure 1.18) [24] and has recently announced the start-up of a pilot-plant in the Netherlands. An advantage of forming the HMF-ether is that it is much more stable and easier to work with than HMF. Having shown that the HMF-ethers can easily be oxidized to FDCA, polymers of FDCA have now been made. Especially, the FDCA polyesters (PEF) are potential replacements for terephthalic acid esters. The new PEF polymer had a reported glass transition temperature, $Tg = 86\,°C$ versus PET with $Tg = 81\,°C$, whereas the gas barrier properties for CO_2 and O_2 are at least 2 and 9 times better, respectively, compared to PET, allowing it to be used in demanding applications such as bottles and food packaging [24]. The main price drivers for biobased FDCA are the feedstock price and economy of scale. However, at a $>300\,kT\,a^{-1}$ scale the price of FDCA will be <€1000 per ton, and therefore competitive with pTA produced on the same scale.

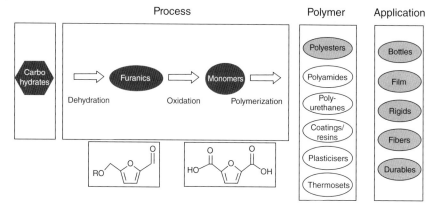

Figure 1.18 The Avantium process for producing furanic polyesters [36].

1.5
Conclusions

The intent of this chapter was to provide examples of the creativity and innovation being developed in the field of bio-renewable fuels and chemicals. Today new start-up companies mostly lead the effort. These companies are playing a key role in the development of renewable fuels and chemicals. Not all will be successful, larger oil and chemical companies will acquire some, and some will grow into new companies. With luck, perseverance, and significant funding, a significant portion of the world's needs for fuels and chemicals will someday be based on renewable feedstock.

In the rest of this book, it will be shown that chemical catalytic processes, in combination with fermentative processes, will play a pivotal role in creating cost-effective and sustainable production methods.

References

1 www.biofuelsdigest.com (accessed July 2011).
2 Dohy, M., de Jong, E., Jørgensen, H., Mandl, M., Philips, C., Pouet, J.C., Skiadas, I., Van Ree, R., Walsh, P., Wellisch, M., and Willke, T. (2009) Adding Value to the Sustainable Utilisation of Biomass, RRB5, Fifth International Conference on Renewable Resources and Biorefineries, Ghent.
3 Neuman, W. (2011) Race to Greener Bottles Could Be Long, Printed on December 16, 2011, on page B1 of the New York edition; Press release 15 december 2011, http://www.thecoca-colacompany.com/dynamic/press_center/2011/12/plantbottle-partnerships.html (accessed July 2011).
4 World Health Organisation (2011) Global status report on alcohol and health.
5 Rupilius, W., and Ahmad, S. (2005) The Changing World of Oleochemicals, Palm Oil Developments 44, http://www.chgs.com.my/chinese/download/Palm%20Oil%20Developments/june%202006/The%20changing%20world%20of%20Oleochemicals.pdf (accessed January 2012).

6 Werpy, T., and Petersen, G. (2004) Top Value Added Chemicals From Biomass. Pacific Northwest National Laboratory (PNNL) and the National Renewable Energy Laboratory (NREL), August 2004.

7 Giordani, M. (1939) *Chim. Ind.*, **21**, 266.

8 Bergius, F. (1937) *Ind. Eng. Chem.*, **29**, 247.

9 Kleinert, T.N. (1971) Organosolv Pulping and Recovery Process. US Pat. Appl. US 3.585.104.

10 Blanch, H.W., Simmons, B.A., and Klein-Marcuschamer, D. (2011) *Biotechnol. J.*, **6**, 1086.

11 Ramey, D.E. (2012) Butanol: The Other Alternative Fuel, http://nabc.cals.cornell.edu/pubs/nabc_19/NABC19_5Plenary2_Ramey.pdf (accessed 3 July 2012).

12 Bernacki, B. (2011) "Biobutanolpotential for biobasedmonomers and polymers" at BioPlastics, Las Vegas.

13 Keasling, J., Martin, V., Pitera, D., Kim, S.-W., Withers, S.T., Yoshikuni, Y., Newman, J., and Khlebnikov, A.V. (2011) Host cells for production of isoprenoid compounds. US Patent Appl. US 201122958, A1 (2001).

14 Phillips, S., Aden, A., Jechura, J., and Dayton, D. (2007) NREL/TP-510-41168, April, 2007.

15 Huber, G.W. (2008) Breaking the Chemical and Engineering Barriers to Lignocellulosic Biofuels: Next Generation Hydrocarbon Biorefineries, National Science Foundation. Chemical, Bioengineering, Environmental, and Transport Systems Division. Washington DC, NSF.

16 (2010) Wall Street Journal, February 10, 2010.

17 Downloaded from news section www.inpbioenergy.net item "09/08/2010 Bioethanol effort gets boost in Florida" on 3 Jul 2012 and downloaded from http://www.ineos.com/new_item.php?id_press=257 on 3 July 2012.

18 Funazukuri, T., Hudgins, R.R., and Silveston, P.L. (1984) Flash Pyrolysis of Cellulose in a Micro Fluidized Bed, *ACS Fuel Meeting Fall, Philadelphia*, **29** (6).

19 O'Connor, P., Stamires, D., and Daamen, S. (2006) Process for the conversion of biomass to liquid fuels and specialty chemicals. US Patent Appl. US 2009090046 (A1), KiOR was formed and funded in 2007.

20 Román-Leshkov, Y., Barrett, C.J., Liu, Z.Y., and Dumesic, J.A. (2007) Production of dimethylfuran for liquid fuels from biomass-derived carbohydrates. *Nature*, **447** (7147), 98.

21 Van Walsum, G.P., and Wheeler, M.C. (2009) Production of Higher Alcohols Liquid Biofuel via Acidogenic Digestion and Chemical Upgrading of Industrial Biomass Streams, http://bcsmain.com/mlists/files/biomass/obpreview2009/biochem/documents/2.3.3.1.pdf (accessed 3 July 2012).

22 Haan, R.J., and Lange, J.-P. (2007) Gasoline composition and process for the preparation of alkylfurfuryl ether. US Patent Appl. US 2011035991 (A1).

23 Lange, J.-P., Price, R., Ayoub, P.M., Louis, J., Petrus, L., Clarke, L., and Gosselink, H. (2010) Valeric biofuels: a platform of cellulosic transportation fuels. *Angew. Chem. Int. Ed.*, **49**, 4479.

24 Gruter, G.-J., and de Jong, E. (2009) Furanics: novel biofuel options from carbohydrates. *Biofuels Technol.*, **1**, 11–17; Gruter, G.J.M., and Dautzenberg, F. (2007) Method for the synthesis of organic acid esters of 5-hydroxymethylfurfural and their use, WO2007104515 (A2).

25 United States Department of Agriculture (2005) Biomass as a Feedstock for a Bioenergy and Bioproducts Industry: The Technical Feasibility of a Billion-Ton Annual Supply, April 2005. Updated in 2011 "U.S. Billion-ton Update, Biomass Supply for a Bioenergy and Bioproducts Industry" http://www1.eere.energy.gov/biomass/pdfs/billion_ton_update.pdf (accessed July 2011).

26 Lantero, O.J. (1995) Process for the preparation of gluconic acid and gluconic acid produced thereby. Canadian Patent Appl. CA 2240294 A1.

27 Kiely, D., Carter, A., and Shrout, D. (1995) Improved oxidation process, WO 9638402 (A1).

28 Flèche, G., Huchette, M. (1986) Isosorbide preparation, properties and chemistry sorbitol, *Starch*, **38** (1) 26; production started in 1954 by Roquette-Frères, in Lestrem, France. http://www.roquette.com/history (accessed July 2011).

29 Zeikus, J.G., Jain, M.K., and Elankovan, P. (1999) Biotechnology of succinic acid production and markets for derived industrial products. *Appl. Microbiol. Biotechnol.*, **51**, 545.

30 Freiherrn, A., Grote, V., and Tollens, B. (1875) Untersuchungen über Kohlenhydrate. I. Ueber die bei Einwirkung von Schwefelsäure auf Zucker entstehende Säure (Levulinsäure). *Justus Liebigs Ann. Chem.*, **175**, 181–204.

31 Ritter, S. (2006) Biorefinery gets ready to deliver the goods. *Chem. Eng. News*, **84**, 47.

32 Hayes, D.J., Ross, J., Hayes, M.H.B., and Fitzpatrick, S. (2006) The biofine process – production of levulinic acid, furfural, and formic acid from lignocellulosic feedstocks, in *Biorefineries: Industrial Processes and Products*, vol. 1 (eds B. Kamm, P.R. Gruber, and M. Kamm), Wiley-VCH Verlag GmbH, Weinheim, pp. 139–164.

33 Wicks, D., Williams, C., and Selifonov, S. (2007) Ketal compounds from polyols and oxocarboxylates, WO 2009032905 (A1); Elifonov, S., Rothstein, S.D., Wicks, D., Mullen, B.D., Mullen, T., Ppratt Jason, D., Williams, C.T., and Wu Chunyong, K. (2007) Polyketal compounds, synthesis, and applications, WO 2009049041 (A2).

34 Gandi, A., and Belgacem, M.N. (1997) *Prog. Polym. Sci.*, **22**, 1203–1379.

35 van Putten, R.-J., van der Waal, J.C., de Jong, E., Rasrendra, C.B., Heeres, H.J., de Vries, H.J. (2012) Hydroxymethylfurfural, a versatile platform chemical made from renewable resources. *Chem. Rev.*, in press, http://dx.doi.org/10.1021/cr300182k.

36 For more information see www.avantium.com and www.yxy.com (accessed July 2012).

2
Selecting Targets
Gene Petersen, Joseph Bozell, and James White

2.1
Introduction

The development of processes for catalytic conversion of renewable feedstocks into chemicals and materials is significantly aided by the rational selection of target products. This selection is important for two obvious reasons. First, the chemistry or biochemistry surrounding any process development is better enabled when the type of product is identified, such as an oxidized molecule, hydrocarbon or polymeric material (or its precursor). Second, the development of chemical and biological tools needed to enable these processes is more germane if the specific chemical nature(s) of the desired end-products are identified. Other factors that contribute to the need to select targets include market drivers or interest in desirable or useful properties in the renewable chemicals or materials produced. For example, the demand for biobased products generated from regenerable biomass resources appears to be growing, occasioned by both market potential and environmental concerns. In addition, some of these chemicals and materials exhibit unique properties of commercial interest currently not readily available from conventional petrochemical sources.

A historical example of the development of chemical processes to produce biobased products was found in the Chemurgy movement in the twentieth century. While the driver for the chemurgical effort was the wider use of agricultural feedstocks, the push to use plant-based carbohydrates, both in monomeric and polymeric form, and plant-based oils required some process development. Targets included replacement of petroleum-based parts in automobiles; as a result, the Ford Motor Company in the mid-twentieth century claimed that there was a bushel of soybeans used in each car manufactured [1]. The onset of World War II in the Pacific Basin limited or even cut off supplies of natural products vital to the military, in particular, rubber. This led to research and development of alternative sources of these materials, such as the isoprenoid components of the desert plant guayule and corn starch for ethanol [1]. These needs also motivated research into better catalytic processes, such as those related to vulcanizing rubber [2]. In the latter part of the twentieth century, there arose interest in biobased

Catalytic Process Development for Renewable Materials, First Edition. Edited by Pieter Imhof and Jan Cornelis van der Waal.
© 2013 Wiley-VCH Verlag GmbH & Co. KGaA. Published 2013 by Wiley-VCH Verlag GmbH & Co. KGaA.

polymers that could substitute for petroleum-based fibers and, hence, targets for polymer precursors were sought that could be produced or derived from biomass itself. Two examples are lactic acid that is generated microbially and then chemically converted into polylactic acid (PLA) [3] and polyhydroxybutyrate derived from storage carbohydrates in microbial systems [4]. These polymeric materials have been proposed as replacements for existing plastics in areas such as packaging [5, 6], and substitutes for petrochemically derived chemical polyesters [7]. More recently, the market is being driven to find alternate sources of strategic raw materials as a hedge against potential shortages in conventional hydrocarbon feedstocks, and an interest in developing sustainable, domestic sources of chemical building blocks.

Whether driven by a desire to develop new markets or to find new uses for regenerable resources, like plant carbohydrates or oils, wood or agricultural residues, the development of biobased products faces the challenge of an overabundance of targets, regardless of whether catalysis will be employed as a process technology or not. The list of biobased products can include structures already made by the chemical industry (and therefore *de facto* commercial opportunities), as well as new structures that arise from the ever-expanding research efforts in carbohydrate and lignin chemistry. Accordingly, biorefinery or biobased product development will greatly benefit from robust screening processes that can maximize the value of expended research and development (R&D) resources and identify those structures that possess the best chance for actual introduction to, and commercial success within the chemicals market. Historically, several approaches to this problem have been taken. An individual molecule or class of molecules can be targeted, and the conversion technology that will achieve a reasonably cost-effective production process for that molecule is then developed. Alternatively, targets that use high carbon efficiency production routes, lessen unwanted byproducts and emissions, and exhibit low life-cycle impacts in comparison to typical chemical processes can also be pursued. Another fundamental approach is to build on the study of chemical and biological catalysis of natural products such as biomass-derived sugars into products because that approach has yielded many general conversion capabilities that are often useful for process development [8]. The key message of this chapter is that the selection of targets is not found in one specific approach or one specific analytical screening process, but is a result of establishing screening criteria that incorporate a range of factors that may differ among potential developers of biorefinery or product development processes. These factors could include assessment of the fit of a product within the business plans of the chemical manufacturers or biorefinery owners and operators, the final product cost and purity, and utility as either a platform or terminal output, the value proposition of incorporating new technology into an integrated operation, or the existence of a feedstock or patent technology unique to a potential user of biomass. In all cases, a key element of these factors is how the criteria for screening is established. To illustrate this with examples from process development in chemical and biological processing, one can compare this to the screening of chemical catalysts or the selection of bacterial species with unique phenotypes. In both of these cases, whether one is using wet chemistry, petri dish approaches, or high throughput screening equipment, the determining factor in finding successful candidates for further development is selection criteria, or selective pressure in the case of microbial species. When applied, the selection protocols are what provide the desired yield of products or candidates.

The selection of candidate biobased products must be closely integrated with the classic value chain for product development shown in the following graphic.

An end-user understands the marketing and sales aspects of their business. A government entity understands what kinds of outcomes they are seeking (viz. biofuels to replace petroleum-derived fuels). Hence, selection criteria need to reflect the value chain of the screening entity. Additional drivers, such as simple discovery research, environmental impact (green chemistry), and strategic national needs all serve as inputs to an evaluation process. The multiple approaches to screening potential candidates will reflect subsets of drivers most pertinent to the end user seeking guidance on the best use of their available resources.

The best screening processes will take into account three basic principles:

1) *Develop a screening process with the end in mind.* Screening for product development requires a different set of criteria than screening for portfolio development (i.e., classes of molecules that could be further evaluated for product development). Similarly, programmatic or strategic outcomes, such as those sought by governmental agencies, create yet another set of criteria. Addressing the desired "value-chain" creates the best screening scenario and resulting outcomes.

2) *Defining the intermediate or source of intermediates will more meaningfully frame the selection criteria.* Woody biomass is different than municipal solid waste plastics and the development of screening criteria would require different considerations, such as the nature of intermediates obtained from fractionating the wood into its components or cracking the plastic back into monomers or polymers. Other renewable resouces could include agricultural residues or even animal manures.

3) *Establishing the desired impact of the selection process is a key determinant in achieving a useful outcome.* For example, reducing process energy intensity or reducing carbon footprints through an alternative processing route (carbon economy) will help create a downselect of greater value. Asking that the downselect guide production of coproducts in a biorefinery producing biofuels is another example of this principle.

In some instances, a fourth principle may be included as part of a selection process, especially for evaluations strongly linked to commercial development.

4) *Integrating process requirements into the selection criteria may be needed to create the most valuable list of candidates.* For example, keeping the various aspects of separation and purification of the product in mind may be a determinant in final selections. Also, issues of unwanted or even toxic byproducts involved in the potential production of the desired candidate molecule can be a deciding determinant in selection.

This chapter will examine concepts around decision-making processes that have developed for choosing promising biobased targets for more detailed investment of R&D efforts for the production of renewable materials, largely chemicals. The focus of the chapter will be on (i) learning from the lessons of previous approaches to screening for biobased chemicals, and (ii) providing some additional guidelines for establishing screening processes for renewable materials. The majority of the discussion covers the first three principles in selecting targets described above. The fourth principle is discussed only minimally, as it is generally most appropriate when downselecting for actual product development. The chapter further focuses on molecules with carbon numbers generally between C2 and C6, and demonstrates how they can be applied to primary outputs of the biorefinery, such as sugars, bio-oils or lignin. The chapter will not discuss C1 chemistry, or the large body of work already existing in Fischer–Tropsch and hydroformylation chemistry other than to note that potential targets could be derived from either CO_2 [9] or synthesis gas [10, 11]. There also exists a very large body of research on plant oil conversions that this chapter will not discuss except in a general way.

2.2
Target Selection Can Focus on Specific Structures or General Technologies

A key determinant in developing a search or selection scenario for process development is establishing the underlying principles upon which a selection is based. A company interested in commercializing a specific, single chemical structure (a *product development* scenario) will incorporate different factors than one interested in understanding conversion or transformational technology applicable to a potentially broad family of products (a *technology development* scenario). In addition, availability of, or access to feedstock or basic intermediates plays a significant role in the selection approach. Although the catalog of renewable resources most usually considered is broad, and includes fractionation products of biomass, such as starch, plant oils, cellulose, hemicellulose, lignin and extractives from lignocellulosics, pyrolysis oils or synthesis gas from biomass gasification processes, not all sources are equally available to all users.

In product development scenarios, biobased targets can be structurally identical to existing hydrocarbon-based chemicals. Production of ethylene from ethanol would be an example of a biobased intermediate producing a structurally identical hydrocarbon [12]. Closely related are processes that may not produce molecules that are structurally identical, but are instead functionally identical. Polylactic acid-based material has shown that it can replace styrene in certain clear packaging situations, as cited above. This approach has significant benefit in having an existing market but also has some challenges, as described in a later section.

Technology development scenarios have selection criteria that typically focus on functional groups within molecules or compounds. Hence, there is more latitude in selecting candidate targets for conducting R&D efforts. For example, the most abundant sugar available from cellulose is glucose, but in a technology development scenario, the screening process would evaluate the potential of whether other sugars such as xylose could be employed, depending on the desired outcome.

2.3
Previous Selection Efforts

Illustration of these principles is given by comparing four available published reports on selection of potential candidates for further development as biobased products.

In the early 1990s, the US Department of Energy's (DOE) Alternative Feedstocks Program (AFP) supported development of chemical products from biomass resources. A need to select appropriate targets was required and a report was issued in 1993 that was a compendium of possible candidates for biological and thermochemical processing [8]. This early attempt at using selection criteria to choose products employed energy displacement, lower environmental impact, and competitive economics as its drivers, with a goal of answering a central question: *given a wide range of possibilities, how can an informed decision be made to determine which materials to investigate more thoroughly?* A unique feature of this methodology was that it was one of the first to employ a systematic, quantitative screening methodology that considered yields of processes, improvements in the technologies, and various feedstocks that had not previously been developed for using renewables as chemical feedstocks. By placing existing data into a consistent methodology, a systematic evaluation of chemicals and materials from renewable resources became possible. The evaluation largely took a product development approach and, as such, was based on comparing the raw material cost contribution for producing a chemical from biomass to that for producing the same chemical from hydrocarbons. In those cases where a direct comparison was not possible (for example, in situations where a petrochemical-based process was not known), a comparison was made between the existing process and processes that could improve production of that material. The study started with 70 promising targets and subjected them to a three-step screening procedure, based on the raw materials cost contribution:

- A raw materials cost assessment of the biobased route was performed.
- A raw materials cost assessment of the corresponding route from petrochemicals was performed.

- The cost of the biobased route was adjusted upward by 30% and was compared in the form of a ratio to the petrochemical route to reflect the risk inherent in new product development. A biobased process was considered promising when the difference in raw material costs for a biomass-based process was around 30% less than costs for a current petrochemical-based process.

By comparing these costs, a ratio significantly less than one indicated that the existing routes to a chemical had a considerable advantage over a new biomass-based process, and ratios significantly greater than one indicated that the biomass-based process could be an attractive candidate for further investigation. The screening methodology used to select and classify biological processes was similar.

As a result of the evaluation process, several processes or materials emerged as potential R&D opportunities, and were categorized into an interesting combination of products and technologies for the near-, mid- and long term, as shown in Table 2.1.

The emphasis on energy impacts by the Department of Energy drove another program within the DOE, the Biological and Chemical Catalysis Technologies Program, to analyze the potential for improved catalytic processes to reduce energy requirements in production of the top 50 commodity chemicals produced in the United States [13, 14]. These early attempts were useful compilations of information and used screening criteria critical to DOE interests in energy efficiency, environmental impact, and preliminary economic evaluation. Nonetheless, these concepts also realize important applicability to the nascent biorefining industry. The AFP report generated significant interest and was the basis for a follow-on report ten years later.

Table 2.1 Biobased product opportunities identified by the DOE Alternative Feedstocks Program.

Near term	Mid term	Long term
Succinic acid	Butanol	Levoglucosan
Organosolv fractionation of lignocellulosics	Sugar reduction and dehydration products from fractionated biomass	Vinylphenol
Acetylated wood	Anthraquinone	Hydroxyacetaldehyde
Starch plastics	Butadiene and C5 hydrocarbons	Polyhydroxybutyrate and valerate
Fast pyrolysis of wood	Acetic acid from biomass syngas	
Benzene, toluene and xylene from wood	Peracetic acid	
Acrylic acid		

Some of DOE's different biomass programs were merged in the early 2000s, corresponding with a rapidly increasing interest in the use of biomass as a source of feedstock for chemical production. This increased attention led to multiple claims of top prospects, as presented to DOE by the industry. Thus, DOE commissioned the National Renewable Energy Laboratory and Pacific Northwest National Laboratory to do an independent screening of the many prospects for chemicals, starting with an evaluation of products that could be derived from biorefinery carbohydrates, and which showed the greatest potential for making an impact in the chemicals market [15]. This analysis employed a value chain approach along with a systematic Delphi analysis of a body of over 300 prospects gleaned from various external reports and analyses. The selection criteria included (i) eliminating targets where an already existing body of technology existed and hence did not justify extensive, high-risk R&D support; (ii) identifying the potential of a chemical target to be a platform chemical suitable for transformation into a wide range of other products; and (iii) classifying the selections by carbon number as typically classified in the petrochemical industry. The report included limited analysis of synthesis gas as a C1 compound but given the extensive technology and literature already available for Fischer–Tropsch chemistry, the analysis referred to an evaluation of products available from chemical or biological transformations of carbohydrates. While the initial downselection produced a list of about 30 useful candidates, the drive to develop a "top-ten" type list dictated further screening, largely to identify potential new candidates as opposed to those that already had a large body of work in existence, such as lactic acid and ethanol. Figure 2.1 outlines the hierarchy of the selection process leading to the downselect to a first cut of 25 potential candidates.

This evaluation was also unique in that it addressed an ongoing tension in evaluation of target opportunities by combining product evaluation and technology evaluation scenarios. For a biorefining industry whose focus remains almost exclusively on biofuel operations making ethanol (EtOH) or biodiesel as a single product, engineering process analysis is ideally suited for identifying price targets, impacts and technologies that offer the best prospects for research investment, since the molecular structure of the desired output is pre-identified [16]. However, when applied to multi-product biobased chemical scenarios, these analysis techniques became less useful because of fundamental differences between fuel and chemical research. When chemicals are included as part of a biorefinery's portfolio, the number of possible outputs soars. The experience of the chemical industry shows that this complexity is best handled by using a small number of broad-based technologies to produce multiple products. However, process analysis becomes complicated, as each target has its own set of costs, depending on the market and application. Nonetheless, target-based approaches using process analysis methodology employed for biofuels had persisted as a means to winnow a huge number of possible biobased chemical targets to a manageable size [17]. Pre-identifying specific desired molecular structures prior to research was perceived to have several advantages, particularly in an industrial setting. It offered defined opportunities

2 Selecting Targets

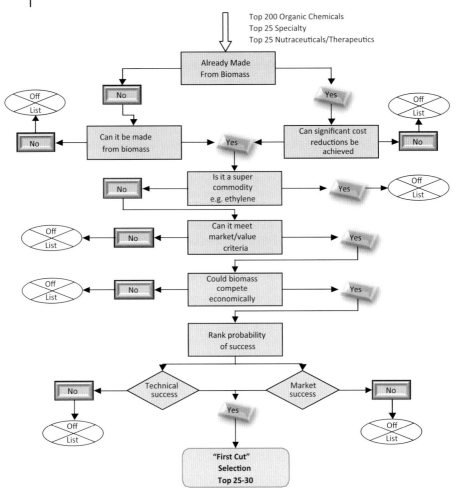

Figure 2.1 Hierarchal down select process from hundreds of candidates to 25–30 potential targets.

for decision makers when prioritizing limited research funds. It also reduced risk, as preliminary process engineering estimates and life cycle analyses (LCA) can address "what if" questions.

Unfortunately, the sheer number of new and existing structural possibilities suggested that the chances of process analysis correctly identifying a commercial winner were small. Process analysis can be limited to structures currently manufactured by the petrochemical industry, but this approach can lead to uneconomical force fits of highly oxidized renewables into processes designed for highly reduced hydrocarbons. In addition, LCA and process analyses employed at the beginning of an evaluation may become moot as new technology is developed, and approaches once thought to be too expensive are rendered viable. In the light of these contrast-

Table 2.2 Top chemical opportunities from carbohydrates as identified in 2004.

Building blocks
1,4-succinic, fumaric and malic acids
2,5-furan dicarboxylic acid
3-hydroxypropionic acid
aspartic acid
glucaric acid
glutamic acid
itaconic acid
levulinic acid
3-hydroxybutyrolactone
glycerol
sorbitol
xylitol/arabinitol

ing approaches to target identification, it was realized that the challenge for incorporating chemical products into the biorefinery was finding the appropriate balance between the clear need for fundamental technology while demonstrating that this technology will lead to identifiable marketplace products.

The 2004 DOE report was one of the first to integrate technology development and product development scenarios by coupling technoeconomic assessment and product identification. This evaluation included a strong emphasis on broad-based technology development for the biorefinery, as well as the use of a rational target selection mechanism. The methodology incorporated key factors for eventual commercial deployment, such as known processes, economics, industrial viability, size of markets, and the ability of a compound to serve as a platform for the production of derivatives. By developing a list of specific structures, this process embraced product identification as a guide for research, and provided a starting point for technoeconomic evaluation. The list of top targets from carbohydrates as identified in the evaluation is shown in Table 2.2.

The members of the list were classified as building blocks, a primary selection criterion, with carbon numbers of 3 to 6. A simple, preliminary analysis of the biological and chemical approaches to these building blocks and their potential conversion to other molecules was presented. When appropriate, similar compounds were rolled up into larger categories, such as 1,4 dicarboxylic acids or sugar alcohols. In addition, star diagrams were included to support the concept of a building block potential.

The attempt to meet the value proposition by applying the first principle in screening cited above is illustrated in the use of star diagrams. Figure 2.2 provides a star diagram for one of the final selections, succinic acid. The analysis went beyond simply identifying a set of related compounds; the routes to the various secondary products were further characterized and categorized. For example, for transformations to 1,4 butanediol, tetrahydrofuran, and gamma butyrolactone

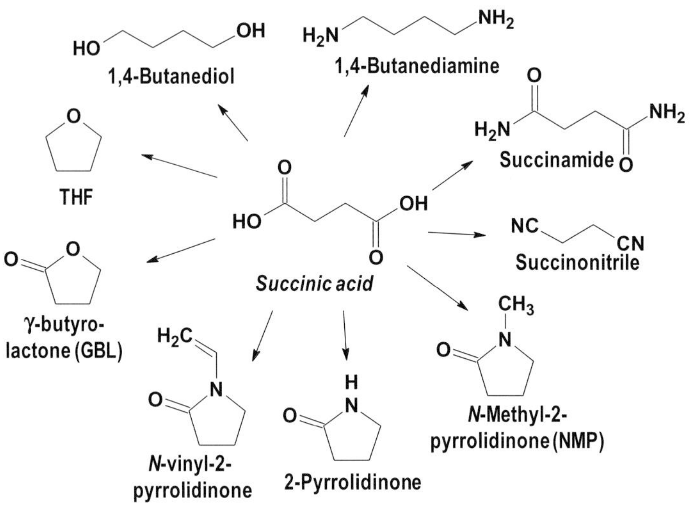

Figure 2.2 Example star diagram for the platform chemical, succinic acid, and some possible derivatives derived therefrom.

(GBL) the report noted that there exist analogous commercial transformations or development of transformations by chemical routes that are expected to require little development. The analysis recognized that the transformation to succinonitrile was practiced commercially at that time. Identifying secondary pathways where more research and development was needed was noted for the production of N-methyl-2-pyrrolidinone (NMP), N-vinyl-2-pyrrolidinone, succinadiamide. and 1,4-diamionobutane. For succinic acid, all transformations were identified as chemical processes. In other top ten candidates biological transformations to secondary products were identified, such as those potential products from microbial conversions of glycerol [15].

In 2007, another analysis similar to the 2004 Top Ten report was generated by the same two DOE laboratories but addressing the issue of lignin [18]. One impetus for this report was the fact that lignin is the only renewable source of aromatic moieties, and aromatics represent a large fraction of commodity chemical products and fuels. Once again, the basic principles of creating a selection process were followed. The value chain was identified and the source of material and its potential use in a biorefinery drove much of the analysis.

The screening protocol that was used attempted to show a value chain approach in the use of lignin and documented many of the potential uses of this unique, but very complicated and amorphous material. One factor that was unique to this analysis was the criterion that the lignin comes from biorefinery sources. Enabling the biorefinery through better use of lignin was a factor in selection. In addition, barriers to use of lignin were identified and evaluated for potential R&D opportunities. Opportunities for the use of lignin or its derivatives in the market were evaluated. Major categories for the use of lignin in a biorefinery setting were outlined

Table 2.3 Results of screening for lignin-based products.

Products potentially derived from lignin conversions

Process heat and power including syngas
Synthesis gas products
Hydrocarbons
Phenols
Oxidized products
Macromolecules

in the report. Each potential product was categorized for its ability to be derived from lignin, including difficulty, potential market volumes, and utility as a building block. That information was presented as a list that was ordered in expected degree of implementation difficulty from simplest to more difficult, and is shown here in Table 2.3.

An attempt to outline forward-looking opportunities was also included in the analysis which opposed the oft-cited criticism of lignin that it is only useful for its fuel value. The need for improved catalysis and novel approaches to processing, converting, separating and purifying lignin was highlighted, as were potential biological and hybrid chemical/biological approaches to facilitate the use of lignin and an assessment of how the network of aromatic rings in lignin might be most effectively used was discussed. This analysis differed from the many, many research articles one can find about lignin because it offered a framework around which to evaluate lignin's potential in the bioindustry.

In 2010 a new report re-evaluated the original 2004 assessment of biobased products from carbohydrates, but with a slightly different perspective [19]. Similar to the Volume II report on lignin, this analysis retained the integration of product development and technology development scenarios, but expanded the initial criteria to include a target's potential impact within a biorefinery operation. The evaluation criteria included evaluation of the level of technology development already in place that would enable facile integration of the conversion operation into a biorefinery. Nine criteria were used in the selection process, and are summarized in the following list. Although product and technology development held the highest significance, other drivers included the ease by which a target can be integrated into the biorefinery. The first five were weighted somewhat more heavily than the last four criteria.

- *The compound or technology has received significant attention in the literature.* A high level of reported research identifies both broad technology areas and structures of importance to the biorefinery.

- *The compound illustrates a broad technology applicable to multiple products.* As in the petrochemical industry, the most valuable technologies are those that can be adapted to the production of several different structures.

- *The technology provides direct substitutes for existing petrochemicals.* Products recognized by the chemical industry provide a valuable interface with existing infrastructure and utility.

- *The technology is applicable to high volume products.* Conversion processes leading to high volume functional equivalents or utility within key industrial segments will have particular impact.

- *A compound exhibits strong potential as a platform.* Compounds that serve as starting materials for the production of derivatives offer important flexibility and breadth to the biorefinery.

- *Scale-up of the product or a technology to pilot, demo, or full scale is underway.* The impact of a biobased product and the technology for its production is greatly enhanced upon scale-up.

- *The biobased compound is an existing commercial product, prepared at intermediate or commodity levels.* Research leading to production improvements or new uses for existing biobased chemicals improves their utility.

- *The compound may serve as a primary building block of the biorefinery.* The petrochemical refinery is built on a small number of initial building blocks: olefins; benzene, toluene, xylene (BTX); methane; carbon, oxygen. Those compounds that are able to serve an analogous role in the biorefinery will be of high importance.

- *Commercial production of the compound from renewable carbon is well established.* The potential utility of a given compound is improved if its manufacturing process is already recognized within the industry.

The criteria are similar to those used in the 2004 report, and are listed roughly in the order of importance as used in this evaluation, although the relative difference between adjacent criteria is small. Cost evaluations, although ultimately an important consideration for commercial utility, were not included in this evaluation. Since the technology base is still developing, cost structures will change as a result of ongoing research activity.

The reader must note that categorizations such as those in the list shown above include some subjectivity because the biorefining industry is in a state of rapid change and expansion. Further, the criteria used in this evaluation may be insufficient for others developing their own list. The fourth principle for selecting targets described above would be a consideration in a downselect for a commercial application. A commercial chemical producer would be assessing issues such as unique market position, proprietary access to a specialized feedstock, experience in the field, specific intellectual property (IP), or existing infrastructure and could end up with a much different group of opportunities. However, the criteria outlined in the list were assessed to offer a reasonable starting point for identifying promising technologies and products for the biorefinery. The outcome of this screening process yielded ten candidates, as shown in Table 2.4.

Table 2.4 Results of screening from value added chemicals from carbohydrates – Top ten revisited.

Potential product targets for biorefinery applications
Ethanol
Furans
Glycerol
Biohydrocarbons
Lactic acid
Succinic acid
Hydroxypropionic acid/aldehyde
Levulinic acid
Sorbitol
Xylitol

2.4 Corroboration of the Value of Screening Studies

Projections made in each of these reports have been corroborated by new developments in chemicals manufacturing. Since 1994, commercial plants producing lactic acid, succinic acid, and potentially other carboxylic acids have been built and larger commodity-scale production facilities are under construction. Myriant, Bio-Amber and Roquette are planning large-scale commercial succinic acid fermentation plants for start up in the next several years in the United States, Canada and Europe [20–22]. Solvay recently announced construction of a commercial epichlorohydrin production facility based on vegetable-oil derived glycerol rather than propylene (Chem. Eng. News, March 5, 2012, p. 15). The recent DOE-sponsored Advanced Research Projects Agency-Energy program (ARPA-E) has an electrofuels component (Chem. Eng. News, Nov 28, 2011, p. 36) and one project is targeting 3-hydroxypropionic acid from carbon dioxide (CO_2) and hydrogen (H_2) as a route first to acrylic acid and eventually to fuels. The use of sugar alcohols to produce polyol chemicals is now part of a commercial facility recently built and operated by Archer Daniels Midland (ADM) in Decatur, IL as part of an overall much larger and well established biorefinery. ADM has also been involved to further the early work of Metabolix to commercialize polyhydroxybutyrate polymers obtained microbially for use in items such as compostable grocery bags and consumer packaging [23]. Sorona® fiber is a new polyester made from the simple diol, 1,3-propanediol which, in turn, is produced biologically from glycerol via sugars. DuPont and Tate and Lyle have developed this technology for commercial applications [24]. The resulting fiber exhibits properties similar to polybutylene terephthalate. The lignin challenge has been engaged by a few companies producing specialty chemicals, such as food flavorings or lignosulfonates for use as surfactants, dispersants and cement additives, but no additional commodity level production systems are in place yet although one start-up company, Lignol, claims to have found potential for lignin as a commercial intermediate [25] but has yet to build a large commercial facility.

Lignol's public discolosures reveal they are developing, in house and with partners, a wide range of lignin applications based on their family of novel HP-L$^{(R)}$ lignin products arising from a unique biorefinery process. The uses closest to commercialization involve incorporating lignin into one or more common chemical resin families – phenolic, isocyanates, epoxy and others – and involve specific applications in wood products, rigid foams, coatings, adhesives and longer term applications involving carbon fibers, thermoplastics, and a few unspecified proprietary areas. These examples illustrate that the screening processes positively identify opportunities for both developing challenging process chemistry through R&D and chemical targets that result in commercialization of either the target compound or one or more of its derivatives. Each of these commercial examples arose from a careful consideration of a value-chain approach, informed by early developmental research and targeted to a market for which there was a potential for growth.

2.5
The Importance of Outcomes and Comparisons of Outcomes

Accurately defining the desired outcome is very important. Bozell and Petersen outlined the difference between developing technology for fuels versus chemical products, and the challenges that production of biobased products face versus fuels production [19]. Research in fuels tends to investigate a wide number of different technologies to produce a single or very small number of pre-identified and commodity-type outputs, whereas biofuels research is convergent. Renewable biofuels production must compete effectively with a giant in-place and highly efficient infrastructure which has been developed by the petrochemical industry over the last 120 years. Thus, if a technology for a biofuel process does not meet predetermined cost targets, it is typically discarded in favor of more economical processes[1]. The outcome is that for biofuels, focus on specific product identifications leads the choice of technology.

However, if chemicals are included as part of the biorefinery's portfolio, the number of possible outputs soars. The experience of the chemical industry shows that this complexity is best handled by using broad-based technologies (selective reductions and oxidations, bond-making/-breaking processes, catalysis, etc.) to produce multiple outputs, that is, chemical production is divergent. These distinctions highlight what was stated earlier, in that more clarity in defining selection criteria and the need to define the end point of the selection process is the key to accomplishing a desired identification of a series of potential bioproducts. Another

---

1) Today, there is a dichotomy between the need for a biobased fuel to meet certain environmental standards versus actual costs. Use of fuels derived from sources such as algae or pyrolysis oils are being evaluated and used for environmental and policy reasons, even though the actual 2012 production costs for applications like jet fuel are much higher than jet fuel derived from crude oil. Thus far these uses are only on a limited basis and the reality is the real costs are largely unknown at present.

point that results from the divergent character of chemicals production is that it is richer in technology driven choices. For example, selective oxidation in the presence of several functional groups of similar reactivity requires more sophisticated chemistry, and may limit the range of materials available from this technology.

The selection criteria in each of the above reviewed reports led to different sets of candidates because the desired outcome was defined differently. In the 1993 report, the basic principles did not limit the assessment to carbohydrates, thus yielding targets not identified in subsequent reports. Later evaluations added selection criteria essentially trying to facilitate more rapid introduction of biobased products into a biorefining operation by highlighting the extent of the existing body of knowledge about each potential target. Further, the Top 10 report for lignin included an evaluation of the potential ease of process implementation in its selection criteria to meet demands for evaluating the programmatic impact of lignin R&D on DOE projects underway at the time. Table 2.5 shows a comparison between three of the reports (excluding the lignin analysis) which highlights similarities but also diverse product choices based on the differences in selection criteria.

Table 2.5 Comparisons between selections of potential biomass derived target candidates for development.

1993 TEA Report	2004 Top Ten	2010 Top Ten	Distinctive differences (if any)
Butanol		Ethanol	Early recognition of value of higher carbon number alcohols. Bioethanol now a potential source of ethylene. Commercially available in large quantities in many biorefineries.
	2,5-Furan dicarboxylic acid	Furans	2010 report broadened to include furan functionality
	Glycerol	Glycerol and derivatives	2010 report included glycerol derivatives
		Biohydrocarbons	Recognition of biorefinery need to produce drop in chemicals and fuels
		Lactic acid	Increased body of conversion technology and in commercial production hence more readily available
Organic acids (succinic, acetic from syngas, acrylic, peracetic)	1,4 Dicarboxylic acids	Succinic acid	2010 narrower selection favored by commercial potential. 1993 selections based on broad potential value in organic acids.
	3-Hydroxypropionic acid	Hydroxypropionic acid/aldehyde	2010 report broadened to include aldehydes
	Levulinic acid	Levulinic acid	
	Sorbitol	Sorbitol	

(Continued)

Table 2.5 (Continued)

1993 TEA Report	2004 Top Ten	2010 Top Ten	Distinctive differences (if any)
	Xylitol/arabitol	Xylitol	Greater availability of xylose feedstock a determinant in 2010 selection
Levoglucosan, hydroxyacetaldehyde	3-hydroxybutryolactone		Early 1993 report recognized value of molecules obtained from processing biomass. Integration in a biorefinery may be less promising but they remain viable building blocks
	Itaconic acid		Integration in a biorefinery less viable but remains a viable building block
	Aspartic acid		Integration in a biorefinery less viable but remains a viable building block
	Glutamic acid		Glutamic is produced commercially but mainly as a commodity food ingredient. Integration in a biorefinery is contingent on finding applications as a viable building block

2.6
Evaluation Processes Can be Comprised of a Variety of Criteria

Each of these reports represents attempts at using various selection criteria to downselect a range of desired outcomes. The importance of maintaining the basic principles cited above is usually necessary for a successful evaluation or screening. In many ways, the processes developed in the four examples cited above were more valuable than the actual candidates chosen in that it forced a correlation of desired outcomes with a rational approach. The criteria employed were selected in order to achieve intended or desired outcomes, but it should be noted that three of the studies were obtained for a government agency. The other was more of an academic analysis of technology readiness tied to the emerging bioindustry. However, any evaluation of targets for eventual commercial production may require the inclusion of a broader range of selection criteria than those employed in these initial reports. The following sections outline additional criteria that could be employed in doing downselects of potential biobased candidates for catalytic chemical processes.

2.6.1
Feedstock and Intermediate Availability

Process evaluation using the basic principles defined earlier may include an assessment of the dependence of the target selection on the nature of the feedstock or intermediates. Well-defined biorefinery process streams (carbohydrates or lignin) available from biomass fractionation or upgrading of these process streams were largely the bases for the previous studies. Each analysis defined its feedstock source and, hence, defined much of what was needed in the ultimate selection criteria. However, other feedstock sources, such as plant oils or pyrolysis oils from thermochemical processing of biomass, can be used, even though their specific structures are much less well defined. An example of how feedstock composition dictates product and process analysis is found in the petrochemical industry's development of oxidative chemistry to convert highly reduced hydrocarbons into products such as organic acids or polyols, that is, these would be reasonable targets to consider in a product evaluation starting from hydrocarbons. In contrast, targeting hydrocarbons from carbohydrates via selective reduction of oxidized biomass fractions has seen limited effort because there is already a plentiful supply of reduced materials from crude oil, or coal processing. Conversely, the use of oxidized molecules, such as carbohydrates, for their inherent properties has long been recognized as a viable pathway to different categories of products, as described in the earlier reports; the pool of potential additional products is not small.

Alternate feedstocks also exist which can incorporate petrochemical-like intermediates derived from biomass, such as synthesis or producer gas. This option has led to screening of options for syngas fermentation. This is a biological catalysis route that recently has matured to the point where construction is progressing on demonstration and commercial plants [26]. Hence, the implementation of principle 2 for use of an unusual feedstock in a biological process was limited to a known, useful end product, ethanol as a biofuel. Engaging in a selection process wherein the selection criteria involved a broader end-use, such as drop-in advanced biofuels or bioproducts (higher alcohols, esters and ethers), could yield another suite of potential bioproduct candidates.

2.6.2
Existing Biorefining Infrastructure Dictates Chemical or Biochemical Processes to be Evaluated

A number of different commercial processes for transforming biomass exist within the chemical industry. This existing infrastructure and corporate experience may be critical to a manufacturer considering how they might expand their biobased product portfolio. There are a large number of starch- and lipid-based biorefineries which produce carbohydrate and lipids of sufficient quantity and purity to allow economical product development. Starch, as a source of glucose, has been of particular interest as the feedstock of choice for many bioproduct developers and various product evaluations have provided a reasonable body of targets from which to choose. Target selection under these conditions is dictated by what the developers perceive as good market opportunities for their current infrastructure. Lignocellulosic biorefineries can draw on the experience of carbohydrate or oil producers and employ similar selection approaches with the restriction that they need to get sugars or other raw materials of sufficient purity to assure good economics.

Ethanol is an example of this type of selection process. The existing corn milling industry provides an abundant and very processable stream of carbohydrates. The existing infrastructure (intermediate availability, availability of effective fermentative organisms, and effective separation technology) allow production of the simple alcohol which, in spite of miscibility issues with gasoline hydrocarbons, provides octane enhancement and also the reduction of certain emissions. Another example of matching technology with available biorefining processes is the production of new intermediates for biofuels production from existing sugar supplies. Two companies, Gevo and Amyris, have developed microbial homo-fermentation routes to isobutanol and farnesene, respectively [27, 28]. They targeted functionally useful intermediates that might be relatively easy to convert to petroleum fuel blend stock (isobutanol for Gevo) or a single C-15 molecule (farnesene for Amyris) that could either be used as a fuel or a chemical building block. Similarly, the company LS9 has developed biological routes to fatty acids and esters that are being utilized as chemical components [29]. These companies chose starch-based sugars as their initial feedstock with an eye towards eventually migrating to lignocellulosic feedstock sources. The end use economic and technology driver dictated the need for fairly clean sugars, whether they come from starch-based feedstocks or lignocellulosic sources. Clean in this context means largely free of contaminants that can inhibit fermentations.

The current interest in advanced biofuels as drop-in replacements within the petrochemical processing value-chain (crude oil, intermediates or finished product) favors hydrocarbon-like molecules, such as fatty acids or plant/algal oils. Hence, biorefineries that produce such intermediates would do well to prepare evaluation criteria that build on their strengths in producing hydrocarbon-like intermediate materials. This opportunity is discussed in more detail later in the chapter.

2.6.3
Market Drivers

Where the feedstock supply is understood and adequate supplies are available a more market-driven selection of targets is possible [30]. Table 2.6 outlines the issues and conditions surrounding the target selection for product development, and serves as a set of axioms for targeting biobased products.

Three classes or types of biobased products exist as depicted in the table, each exhibiting a range of advantages and disadvantages. These external factors affect the selection criteria, particularly in situations placing product development as the top criterion in selection of targets. Two examples of this were mentioned above. ADM has chosen to produce polyols from sugars based on their availability of sugars easily converted into sugar alcohols. Their technology development was driven by the market for polyols. They possessed depreciated capital and a relatively small investment in an operating facility appears to have exhibited to ADM an avenue to a viable commercial operation. The development of Sorona® was driven partially because they found that the polymers made from 1,3-propanediol possessed marketable properties that were functionally equivalent to petrochemical-

Table 2.6 Biobased product axioms.

Type of biobased product	Upside	Downside
Drop in replacement chemicals (PG, acetic acid, acrylic, 1,4-BDO, adipic acid)	• Markets exist (fungible) • Cost structures and growth potential understood • Market risk reduced	• Harsh cost competition (competes on cost alone) • Competes against depreciated capital • Limited market differentiation
Functionally equivalent replacement [Phenolic resins, unsaturated polyesters, 1,3-PDO (Sorona™), PHAs, PLA, HFCS]	• New market opportunities may exist • Competitive petrochemical routes undeveloped • Cost tolerance more likely • Inherent properties of biomass utilized	• Cost competition remains real • Market performance potential unclear • High capital risk • Long commercialization times likely
New functionality or properties (Engineered wood products, isosorbide, 5-HMF, levulinic acid and derivatives, dehydromucic acid)	• New market opportunities may exist • Competitive petrochemical nonexistent • Cost often not an issue (yet) • Inherent properties of biomass utilized	• Capital risks often high • Long commercialization times likely • Market development almost always required & not always clearly defined

PG – polyethylene glycol, BDO – butanediol, PDO – propanediol, PHAs – polyhydroxybutyrates, PLA – polylactic acid; HFCS – High fructose corn sirup, HMF – hydroxymethylfurfural.

derived materials. Another driver for DuPont was to replace chemical processes with biological processes exhibiting a smaller environmental footprint. They and their partners successfully blended microbial glucose conversion to glycerol with microbial glycerol conversion to 1,3 propanediol in a single organism [24, 31]. This effort faced all the challenges listed in the biobased product axioms above but has been successfully implemented, demonstrating how internal company drivers can lead to success. This process essentially involved all the principles 1–4 cited above.

In addition, a recent report by Vennestrøm employed a similar set of guidelines to categorize a potentially valuable list of biobased chemicals [32].

Different companies have diverse market strategies and selection criteria that need to be aligned with business plans if completion of the R&D is expected to occur. If there is not a specific product development driver, but there is a need to develop enabling conversion and transformation technologies, the selection criteria can be more encompassing. However, having some directional indicators for the potential end use is often a good incentive to conduct a thorough technoeconomic and R&D effort.

2.6.4
R&D Drivers

Potential users of biomass may choose to develop enabling technologies instead of trying to identify specific structures for R&D investment. For example, the intent of the DOE Top 10 reports was not to select winners, thus making a technology-driven evaluation more suitable. In these cases, the range of biobased product targets can include higher risk targets that could potentially yield high payoffs in commercial deployment. Also, model compounds can be used as actual potential targets or surrogates for R&D. The 2004 Top Ten report's use of star charts was intended to illustrate the general technical barriers associated with each of the potential building block selections, as explained previously in this chapter. The 2007 lignin report attempted to highlight potential technology needs that could lead to value-added products. The challenges identified in these technology-based approaches are useful in defining R&D directions. For example, 3-hydroxypropionic acid is an appealing target but its production via fermentation is challenging, and initial attempts to pursue this approach have not yielded results that have been commercialized [33, 34]. In other cases, the use of homogeneous catalysis for biobased products is likewise appealing, but there are inherent challenges that must be overcome, such as developing catalysts tailored for highly oxygenated components of biomass, finding systems that operate efficiently in aqueous environments, and operate under milder conditions than processes conventionally used in the petrochemical industry. Hence, there remains a good opportunity for investigative, high-risk R&D with potential high-payoff benefits. Partnerships between universities, government and industry are well suited to undertake such efforts. Thus, selection criteria tailored to R&D drivers probably could be supplied by the range of targets already outlined in the reports published to date. The targets identified in the reports to date already present significant challenges for R&D efforts and may not require additional selection of new targets in order to provide R&D avenues with high impact results.

2.6.5
Other Screening Opportunities

Byproduct uses – A challenge in processes involving renewable feedstocks such as biomass limits the type and amounts of byproducts. In the cases where multiple products are unavoidable, a need develops to use those byproducts to best advantage. Hence, screening for opportunities related to use of these byproducts creates a separate set of criteria. One salient example of this is the production of glycerin from transesterification of plant oils to make fatty acid methyl esters or biodiesel. Hence, looking at potential products from the glycerol component of glycerin invites a set of opportunities for value-added processing and was a factor in the choice of this molecule in three of the reports cited above. As also noted above, some commercial operations using glycerin-based feedstocks have been proposed [19].

Value-added processing – Both glycerol and levulinic acid present a large potential for further use as platform chemicals due to their existing chemical structure and reactive groups. However, commercial applications have been relatively minor, largely due to a combination of factors including lack of cost effective or efficient conversion technologies for the market opportunities that currently exist [35].

Other portfolio opportunities – New portfolio development can often benefit from screening practices even if only to avoid going down pathways with lesser chances of positive outcomes. For example, if algal biomass became a larger part of the production of biobased fuels and products, then product development based on the carbohydrates in algae (mannose, galactose and arabinose) may be important. The Department of Energy published an algal roadmap which noted the potential for producing biobased chemicals from algae along with oils that could be converted into fuels [36]. Some companies are capitalizing on these opportunities to produce specialty chemicals of high value to complement their commodity biofuels production efforts (http://www1.eere.energy.gov/biomass/pdfs/ibr_arra_solazyme.pdf).

Substituting biotechnology for conventional chemistry has generated opportunities for chemicals production. Numerous reports and papers have been published on this potential including a National Academy study [37] and more recently, a report from Wilke and Vorlop [35]. As noted in principle 3, a desire to employ alternative processes to achieve a particular impact such as "green chemistry" can focus the selection criteria.

2.6.6
Other Portfolio Opportunities – Biomass Produced Oils

The demand for advanced biofuels and products beyond simple alcohols has the biomass and renewables communities looking at how biomass produced oils (i.e., pyrolysis oils) can be employed for inclusion in the petrochemical industry. This is not a new area of research but one which is now receiving increased attention as a source of crude oil. The initial level of processing of biomass typically yields a crude biooil with properties similar to petroleum crude but enough dissimilarities to be non-useable as a direct replacement for crude oil due to several factors including mainly high oxygenates contents and storage instability [38, 39]. This particular situation could be a separate case study on how to employ the principles for selection of candidate targets for further development. There are three levels or states of pyrolysis oils: crude pyrolysis oil, upgraded pyrolylsis oils (via hydrotreating, for example, as cited in [39], and finished products such as green diesel or gasoline. The complexity of the mixture of chemicals in pyrolysis oils, for example, creates a daunting challenge. As stated by an experienced process chemist, "there is a world of difference between a valuable mixture of chemicals and a mixture of valuable chemicals"[2]. Hence, the question of where to insert

---
2) Arthur J. Power and Associates, personal communication.

these bio-oils into the typical petrochemical processing chain from crude to finished products invokes all four of the screening principles cited above.

Despite many processing advances over the last 5–10 years, the answer(s) may not be completely clear as yet.

For example, if the crude oil is the feedstock intermediate, it is clear that screening for targets wherein crude bio-oil is directly used may not be worth the effort as it is simply too corrosive or oxygenated to substitute for crude oil at the front end of a refinery, or even in direct applications such as bunker fuel. Fractionation or processes like hydrogenation of the oils to improve material and chemical properties that make the bio-oil more similar to crude oil intermediates creates a better feedstock from which to start a screening process [40]. The end uses are then likely more viable and possible selections can be made. One example of this is the fractionation of pyrolysis oils into water-soluble and non-soluble materials and the extraction of a potential pulp brightening agent from the aqueous fraction [41]. The impact of potentially producing finished products that can directly be used in the transportation market has driven consideration to targets such as green diesel or even gasoline. A design case to evaluate the technical and economic factors in achieving this end use was prepared recently and shows the challenges of such processes [42].

Because there are numerous options to get to these finished products, such as catalytic fast pyrolysis, hydrogenation of pyrolysis oils, or hydrotreating of biomass, there remains a wide gap between research findings and commercial viability. However, once the end use is defined, and a bio-oil intermediate is selected the screening becomes more relevant and choices of targets may well be dependent on ease of process requirements per principle 4. This is an example of where "convergent R&D" is needed to evaluate various process options to produce a limited set of products that could be easily integrated into the existing petrochemical processing infrastructure and hence guide selection criteria.

2.7
Catalysis Aspects

Regardless of the specific outcomes of a screening process it is highly likely that catalysis and catalytic processing will necessarily play a large role in bringing the chemical products to market. In a few cases, it is possible that the required catalysts and the process conditions have already been developed. The expected reality, however, is that efficient and highly economically viable new catalysts will be required and significant efforts may be expended to achieve that end.

However, it is possible even at this stage to use recent history of biomass processing to enumerate some general classes of catalysts and catalytic processing that will be needed once a selection process is completed.

Biomass-derived materials, that is, feedstocks, have several features in common, mostly an abundance of oxygen (OH, carbonyl and carboxylates) and often the presence of highly reactive functionality (aldehyde, olefinic, and carboxylate). The conversion of biomass raw material feeds mostly involves the manipulation,

2.7 Catalysis Aspects

removal and modification of such functionality and also the cleavage of certain carbon–carbon linkages.

With this in mind the following types of highly selective and active catalysis will continue to be a requirement to facilitate economically efficient and attractive conversions:

1) Selective and controllable carbon–oxygen cleavage via hydrogenolysis, for example, removal of the OH on the central carbon of glycerol to yield 1,3-propanediol. A similar logic applies to other polyhydroxy compounds, that is, selective removal of OH on internal carbons. Lactic acid dehydration to acrylic is an example of a difficult dehydration that if made easy might lead to low cost renewable acrylate polymers.

2) Selective and controllable dehydrations, for example, of sorbitol to isosorbide while avoiding formation of mono-cyclic sorbitans, or fructose to HMF without co-production of waste "humins", other polymers and levulinate. Included in this realm is selective conversion of sugars to levulinic acid without significant waste co-product formation. Lactic acid dehydration to acrylic is an example of a difficult dehydration that if made easy might lead to low cost renewable acrylate polymers.

3) Selective and controllable high yield carbon–carbon bond cleavages, for example, hydrogenolysis of the aromatic–aliphatic links in lignin that can yield high yields of single component aromatic products or high yield, high selectivity conversion of sugar alcohols to single product simple polyols such as glycerol, ethylene glycol, or propylene glycol Additional examples may include efficient, selective and high yield, low coke decarboxylations and selected carbon–carbon hydrogenolysis chain breaking of complex vegetable oils.

4) Selective oxidations that effectively discriminate between the terminal and internal hydroxyl groups.

5) Direct conversions of cellulose to sugar alcohols.

Catalytic processes involving enzymes should also be considered. Typically the use of enzymes has been for deconstruction of lignocellulosics to sugars and other components. However, there is potential beyond just hydrolysis of lignocellulosics to employ enzymes as catalytic systems. Examples would be in areas such as achieving more efficient and stable hydrogenases and dehydratases. Nontraditional biocatalysis can include enzymes functioning in organic solvents or supercritical fluids creating a biological homogeneous catalytic system or immobilized enzymes that mimic heterogeneous catalytic systems [43].

There are no doubt others that will be revealed as new compounds of potential value are revealed through the outcomes from individual selection processes. We might suggest that valuable results would accrue from directed research, perhaps via application of high throughput methods, towards the goals shown above. Clearly, selection of appropriate model feeds will become an issue in this sort of effort but this area is beyond the scope of this chapter.

2.8
Conclusions

Criteria-based selection processes are useful tools for expanding the area of renewable feedstocks for fuel and bio-product development and achieve the following:

- Allow for crystallization of R&D targets, hence preserving and allocating resources for the most potentially useful outcomes
- Allow for earlier integration of feedstocks such as lignocellulosics into the value chain for producing fuels and chemicals
- Help define the impacts on a biorefining process such that inclusion or exclusion of a second or third product stream is warranted.

The selection criteria chosen in targeting chemistry or biochemistry for process development must be tailored to a specific set of drivers. The drivers or the framing of the selection criteria will differ from end-user to end-user. In some cases, such as technology development scenarios, the drivers allow a broad range of potential targets. In others, such as product development scenarios, the drivers help narrow the selection to a few desired candidates. Inherent in this process will be a better approach to synthesizing and clarifying the path forward in developing process chemistry.

It must be recognized that there may not be a single and universally applicable set of selection criteria. The optimal criteria will depend substantially on the economic self interests of the evaluator, that is, the companies and industries involved in biorefining and biomass conversions. Governments historically have been poor at predicting, detecting, or even recognizing opportunities or successes but they should be prepared, and indeed proactive, to create and support an environment where innovation is encouraged and rewarded.

The successes of using value-chain selection criteria have been realized in the chemicals market as illustrated in the examples cited above. There is no one approach that will generate the next commercial success simply by using it; however, the various examples mentioned in this chapter provide a glimpse into how window shopping for technology development is better served if there are parameters by which to temper the R&D vision with boundaries that can lead to meaningful outcomes.

References

1 Finlay, M.R. (2004) Old efforts at new uses: a brief history of chemurgy and the American search for biobased materials. *J. Ind. Ecol.*, **7** (3–4), 33–46. Article first published online: 8 Feb 2008.

2 Morton, A.A., Magat, E.E., and Letsinger, R.L. (1947) Polymerization VI. The alfin catalysts. *J. Am. Chem Soc.*, **69** (4), 950–961.

3 Datta, R., and Henry, M. (2006) Lactic acid: recent advances in products, processes and technologies – a review. *J. Chem. Technol. Biotechnol.*, **81**, 1119–1129.

4 Snell, K.D., and Peoples, O.P. (2009) PHA bioplastic: A value-added coproduct

for biomass biorefineries. *BioFPR*, **3**, 456–467.

5 Sinclair, R.G. (1993) US Patent 5,216,050, January, 1993.

6 Sinclair, R.G. (1996) The case for polylactic acid as a commodity packaging plastic. *J. Macromol. Sci. A, Pure Appl. Chem.*, **33** (5), 585–597.

7 Sudesh, K., Abe, H., and Doi, Y. (2000) Synthesis, structure, and properties of polyhydroxalkanoates: biological polyesters. *Prog. Polym. Sci.*, **25** (10), 1503–1555.

8 Bozell, J.J., and Landucci, R. (1993) Alternative Feedstocks Program Technical and Economic Assessment: Thermal/Chemical and Bioprocessing Components. 227 pp., NREL Report No. TP-20646.

9 Petersen, G., *et al.* (2005) Nongovernmental valorization of carbon dioxide. *Sci. Total Environ.*, **338**, 159–182.

10 Phillips, S.D. (2007) Technoeconomic analysis of a lignocellulosic biomass indirect gasification process to make ethanol via mixed alcohols synthesis. *Ind. Eng. Chem. Res.*, **46** (26), 8887–8897. NREL Report No. JA-510-42274 (November 2007).

11 Spath, P., and Dayton, D. (2003) Preliminary Screening – Technical and Economic Assessment of Synthesis Gas to Fuels and Chemicals with Emphasis on the Potential for Biomass-derived Syngas, NREL Technical Report NREL/TP-510-34929, December 2003.

12 Morschbacker, A. (2009) Bio-ethanol based ethylene. *Polym. Rev.*, **49**, 79–84.

13 Tonkovich, A.L.Y. (1994) Impact of Catalysis on the Production of the Top 50 U.S. Commodity Chemicals. Prepared for the U.S. Department of Energy under Contract DE-AC06-76RLO 1830 by Pacific Northwest Laboratory, PNL-9432/UC-401.

14 Lipinsky, E.S., and Ingham, J. (1994) Final Task Report on Brief Characterization of the Top Fifty U.S. Commodity Chemcials. Pacific Northwest Laboratory, Battelle Memorial Institute. September, 1, 1994.

15 Werpy, T., and Petersen, G. (2004) *Top Value Added Chemicals from Biomass. Volume I – Results of Screening for Potential Candidates from Sugars and Synthesis Gas*, U. S. D. o. Energy, DOE/GO-102004-1992.

16 Aden, A., Ruth, M., Ibsen, K., Jechura, J., Neeves, K., Sheehan, J., and Wallace, B. (2002) *NREL report NREL/TP-510-32438*.

17 Little, A.D. (2001) Aggressive Growth in the Use of Bio-derived Energy and Products in the United States by 2010. Prose Summary, Final Report. DOE Contract GS-23F-8003H. October 31, 2001. A.D. Little Reference 71038.

18 Bozell, J.J., Holladay, J.E., White, J.F., and Johnson, D. (2007) Top Value-Added Chemicals from Biomass: Volume II – Results of Screening for Potential Candidates from Biorefinery Lignin. PNNL 16983, October, 2007.

19 Bozell, J.J., and Petersen, G.R. (2010) Technology development for the production of biobased products from biorefinery carbohydrates – the US Department of Energy "Top 10" revisited. *Green Chem.*, **12**, 539.

20 Potera, C. (2005) Making succinate more successful. *Environ. Health Perspect.*, **113** (12), A832–A835.

21 Myriant (2011) DOE One Page Project Description, http://www1.eere.energy.gov/biomass/pdfs/ibr_arra_myriant.pdf (accessed 8 January 2013).

22 BioAmber (2011) Bioamber Siting Biobased Succinic Acid, BDO Plant in Ontario, Canada. http://www.greencarcongress.com/2011/08/bioamber-20110829.html (accessed 8 January 2013). 29 Aug 2011.

23 Tullo, A. (2008) Growing Plastics. *Chem. Eng. News*, **86** (39), 21–25.

24 Kurian, J.V. (2005) A new polymer platform for the future – Sorona™ from corn derived 1,3 propanediol. *J. Polym. Environ.*, **13** (2), 159–167. COE.1007/s10924-005-2947-7.

25 Arato, C., Pye, E.K., and Gjennestad, G. (2005) The lignol approach to biorefining of woody biomass to produce ethanol and chemicals. Twenty-Sixth Symposium on Biotechnology for Fuels and Chemicals ABAB Symposium, 2005, Session 5, 871–882, *Appl. Biochem. Biotechnol*. DOI: 10.1007/978-1-59259-991-2_74.

26 Ineos (2011) DOE One Page Project Description, http://www1.eere.energy.

gov/biomass/pdfs/ibr_arra_ineos.pdf and http://www.ineosbio.com/57-Welcome_to_INEOS_Bio.htm (accessed 8 January 2013).

27 Gevo (2012) Company Web Page, http://www.gevo.com/our-business/our-science-and-technology/ (accessed 8 January 2013).

28 Amyris (2011) DOE One Page Project Description, http://www1.eere.energy.gov/biomass/pdfs/ibr_arra_amyris.pdf (accessed 8 January 2013).

29 LS9 (2012) Using Synthetic Biology to Develop Fuels and Products, Company Web Page, http://www.ls9.com/technology/technology-overview and publications – http://www.ls9.com/newsroom/resources (accessed 8 January 2013).

30 Vijayendran, B. (2010) *J. Bus. Chem.*, **7**, 109–115.

31 DuPont (2012) Fiber Applications, http://www2.dupont.com/Sorona_Consumer/en_US/ (accessed 8 January 2013).

32 Vennestrøm, P.N.R., Osmundsen, C.M., Christensen, C.H., and Taarning, E. (2011) Beyond Petrochemicals: The Renewable ChemicalsIndustry. *Angew. Chem. Int. Ed.*, **50**, 10502–10509.

33 Lio, H., Gokarn, R.R., Gort, S.J., Jessen, H.J., and Selifonova, O.V. (2010) Production of 3-hydroxypropionic acid using beta-alanine/pyruvate aminotransferase. US Patent 7,785,837.

34 Raj, S.M., Rathnasingh, C., Jo, J., and Park, S. (2008) Production of 3 hydroxypropionic acid from glycerol by a novel recombinant *Eschericia coli* BL21 strain. *Process Biochem.*, **43** (12), 1440–1446.

35 Wilke, T., and Vorlop, K.D. (2004) Industrial bioconversion of renewable resources as an alternative to conventional chemistry. *Appl. Microbiol. Biotechnol.*, **66**, 131–142.

36 US DOE (2010) National Algal Biofuels Technology Roadmap. US Department of Energy, Office of Energy Efficiency and Renewable Energy, Biomass Program, http://www1.eere.energy.gov/biomass/pdfs/algal_biofuels_roadmap.pdf (accessed 8 January 2013).

37 Committee on Bioprocess Engineering, National Research Council (1992) *Putting Biotechnology to Work: Bioprocess Engineering*, The National Academies Press, Washington, D.C.

38 Ringer, M., Putsche, V., and Scahill, J. (2006) Large-Scale Pyrolysis Oil Production: A Technology Assessment and Economic Analysis. Technical Report. NREL/TP-510-37779.

39 French, R.J., Hrdlicka, J., and Baldwin, R. (2010) Mild hydrotreating of biomass pyrolysis oils to produce a suitable refinery feedstock. *Environ. Prog. Sustain. Energy*, **29** (2), 142–150.

40 Agblevor, F.A. (2009) Fractional Catalytic Pyrolysis of Biomass. US Patent Application 20090165378.

41 Agblevor, F.A., and Besler-Guran, S. (2001) US Patent 6,193,837. Preparation of brightness stabilization agent for lignin containing pulp from biomass pyrolysis oils.

42 Jones, S.B., Holladay, J.E., Valkenburg, C., Stevens, D.J., Walton, C.W., Kinchin, C., Elliott, D.C., and Czernik, S. (2009) Production of Gasoline and Diesel from Biomass via Fast Pyrolysis, Hydrotreating and Hydrocracking: A Design Case. PNNL-18284, February, 2009.

43 Davison, B.H., Barton, J.W., and Petersen, G.R. (1997) Nomenclature and methodology for classification of nontraditional biocatalysis. *Biotechnol. Prog.*, **13**, 512–518.

3
The Development of Catalytic Processes from Terpenes to Chemicals
Derek McPhee

3.1
Introduction

While many of us working in the field of renewable fuels and chemicals might like to see ourselves as representing the cutting edge of modern science and technology, the truth is that neither the production of fuels nor that of chemicals from renewable feedstocks are new concepts by any means, with the origins of the former going back to ancient times, and the production of other chemicals and products derived from them dating back to at least the late nineteenth century.

People familiar with the field of renewable fuels will certainly recognize feedstocks such as "... tar, rosin, rough turpentine, or spirit, or alcohol, or any kind of oil, fat, or tallow, mineral coal, pitch-pine wood and the knots, birch bark, pumpkin, sun-flower, flax, and other seeds, as well as many other substances", but may not know that the list was written by Samuel Moray (1762–1843), who designed, patented and built several early engines, including the first known American prototype of an internal combustion engine [1].

As far as the creation of renewable chemicals is concerned, the term "chemurgy", referring to the branch of applied chemistry concerned with preparing industrial products from agricultural raw materials, was coined by the Dow Chemical biochemist William J. Hale and used in his 1934 book *"The Farm Chemurgic"* [2]. By then several renewable feedstock-derived products were already being commercialized by early adopters of the technology, such as the Ford Motor Company, whose founder Henry Ford was not only an early champion of ethanol as a fuel, but also a proponent in the 30s of the use of agricultural raw materials, particularly soybeans, as sources of plastic vehicle components, like gearshift knobs and horn buttons. This work culminated in 1941 with the unveiling of Ford's "Soybean Car" at the Dearborn Days, an annual community festival. Although neither the vehicle itself nor any documentation has survived, it is known that it had a lightweight plastic panel body composed of "soybean fiber in a phenolic resin with formaldehyde used in the impregnation" [3].

The circuitous history of how the popularity of renewable resource based chemicals alternatively grew and waned, until after WWII petrochemicals essentially captured the whole chemical feedstock market, is beyond the scope of this chapter, and the reader interested in this history or the current resurgence of biomass-derived chemicals (a Google search of "renewable chemical feedstocks" gives over 5 million hits!) is referred to any of the recent reviews on the topic [4].

The remainder of this article is organized as follows: in Section 3.2 a brief history of Amyris' involvement in this area, and a description of its platform technology for the genetic manipulation of microbial strains to produce terpenes by fermentation are presented. Some references to the use of the technology to produce hydrocarbon-based renewable fuels are also given. Section 3.3 provides a brief overview of the terpenes market in general, its main current sources and some of the cost drivers in the sector. Since Amyris has focused its initial efforts on the production of sesquiterpenes as a chemical feedstock, Section 3.4 reviews current uses of sesquiterpenes as end products or chemical building blocks. Sections 3.5 and 3.6 cover some new sesquiterpene-derived products developed by Amyris, and some of their properties and potential applications. Brief conclusions are finally presented in Section 3.7.

3.2
Strain Engineering for the Production of Terpenes

A comprehensive discussion on the engineering of microbial strains to produce chemicals more generally, and terpenes in particular, as practised by Amyris and other firms, is also beyond the scope of this review, but for context a brief summary of how the technology was developed and has evolved at Amyris is provided in this section.

Amyris was incorporated in 2003, and in early 2005 the company started operations as part of a non-profit three-way partnership with Prof. Jay Keasling's lab at UC Berkeley and The Institute for One World Health, funded by The Bill and Melinda Gates Foundation, with the purpose of developing a genetically engineered pathway for the industrial scale production of artemisinin, an antimalarial drug precursor, via microbial fermentation [5]. Some two years later, and upon successful completion of this project and transfer of the technology to Sanofi-Aventis for scale up and eventual commercialization [6], the company then switched its attention to other potential applications of the platform technology [7], to which it retained the "for profit" rights in all areas outside that of the artemisinin project, such as the production of renewable fuels. Details of this work can be found on the Amyris website [8], in several patents [9] and in the popular media [10].

Since Amyris' approach to choosing what molecules to pursue as targets is somewhat unique, a brief description of the thought process and its validation is given here. Renewable fuels have traditionally been either products that man already knew how to make in abundance via fermentation and had historically already been used as fuels (e.g., alcohol blendstocks like ethanol and butanol), or products available by simple chemical transformation of renewable feedstocks, as is the case of biodiesel, which refers to monoalkyl esters of natural fatty acids, that can be blended with petroleum fuels. Aside from the environmental benefits of using renewable fuels, from a technology point of view both types of renewable fuel have advantages and disadvantages. The alcohols, for example, are good octane enhancers and have now replaced banned lead compounds and ethers, like methyl tert-butyl ether (MTBE), for this purpose. On the other hand, they are hygroscopic, which precludes blending at the refinery for pipeline transport and long term storage, as well as their use in marine engines. Biodiesel also has its advantages and shortcomings. On the positive side it is compatible in blends up to 20% with most current diesel engine technologies and it has high cetane values, while its major shortcoming is its poor cold weather performance due to gelling, which depends on both the blend ratio and the feedstock used.

Rather than develop a technology to produce a renewable fuel and then figure out how to use it in today's engines, Amyris' approach was to focus initially on diesel fuel, since worldwide it represents a larger market, being used essentially for all medium and heavy duty transport and a major portion of the passenger vehicles in most places outside North America. The plan was to start from a clean slate by listing the properties of an "ideal" diesel molecule replacement that could provide all the desirable properties of existing petroleum diesel with the advantages of reduced emissions and environmental footprint. These properties were then matched with those of the terpenes that could be produced using Amyris' platform strain engineering technology. This led to the selection of trans-β-farnesene [(6E)-7,11-dimethyl-3-methylidenedodeca-1,6,10-triene; CAS RN # 18794-84-8, hereafter referred to as simply β-farnesene] as the target molecule, as we anticipated that its fully hydrogenated form 2,6,10-trimethyl-dodecane would meet all the desired criteria. This was subsequently proven by the fact that its blends with petroleum diesel meet or exceed the existing ASTM D975–11 Standard Specifications for Diesel Fuel Oils and equivalent foreign standards (unlike biodiesel, for example, which has its own set of ASTM and EN standards covering different blend ratios). Extensive ensuing engine and vehicle testing has confirmed this and allowed Amyris to obtain EPA registration for its diesel fuel in the US and to actively pursue it elsewhere. Figure 3.1 shows a comparison of some key properties between petroleum diesel, today's biodiesel and Amyris diesel.

Among its many advantages, the following may be listed:

- At least 80% reduction of greenhouse gas emissions relative to petroleum-based diesel (based on California Air Resources Board metrics).

- Zero sulfur content to meet today's stricter emissions requirements.

- Demonstrated reduction of tailpipe hydrocarbon emissions such as NO_x, particulate matter, and carbon monoxide when blended with ultra-low sulfur diesel (ULSD).

Amyris renewable diesel has also been tested by several OEMs (engine and vehicle manufacturers) and has received their enthusiastic support. To date, it is the only fermentation-derived renewable hydrocarbon diesel to receive OEM engine warranties from the Cummins Engine Company and the Mercedes Benz Truck and

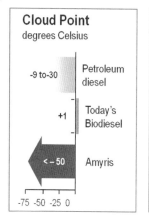 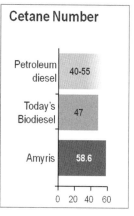 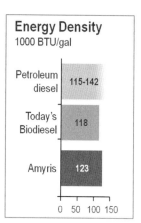

Figure 3.1 Key fuel property comparison.

Bus Company (Brazil). In 2010 a Mercedes-Benz do Brasil Accelo BlueTec EEV Concept Truck running on 100% Amyris No Compromise™ sugarcane-derived diesel won first place honors for design, fuel economy, CO_2 emissions, acceleration, drivability, local emissions and noise emissions at the 10th Michelin Challenge Bibendum, held in Rio de Janeiro, Brazil. The Amyris diesel-fueled truck demonstrated an 88% reduction of particulate matter compared to current emission levels using fossil diesel. In addition, the GHG emission reductions were estimated to be higher than 90%. Similar advantages are seen in what one might consider high performance vehicles[1].

Amyris' core technology that has allowed the development of this product can be illustrated by the diagram shown in Figure 3.2. The mevalonate pathway, also known as the HMG-CoA reductase pathway or isoprenoid pathway, is an important metabolic pathway in all higher eukaryotes and some bacteria [11] for the production of dimethylallyl pyrophosphate (DMAPP) and isopentenyl pyrophosphate (IPP), which serve as the basis for the biosynthesis of the terpenoids. It is also the starting point for the biosynthesis of the steroids.

The diagram shows how, by manipulation of the mevalonate pathway and by selection of appropriate branch nodes in the pathway and synthases, it is possible to produce in a host organism (in this case *Saccharomyces cerevisiae*) commercially viable quantities of a wide range of isoprenoids, ranging in size from C5 (isoprene) through C10 (monoterpenes) to sesquiterpenes (C15) or higher (not shown) through *in vivo* addition of successive incremental C5 building blocks. In Figure 3.2 the abbreviation IS represents isoprene synthase, which converts the equilibrium mixture of DMAPP and IPP into isoprene, ADS is amorphadiene synthase,

---

1) In March 2012 a Total-Amyris sponsored Peugeot 308 HY-4 hybrid-diesel running on a blend of Amyris and petroleum diesel finished 4th (and 1st among non-electric vehicles) in fuel consumption and 5th in the official final standings at the RALLYE MONTE-CARLO DES ÉNERGIES NOUVELLES held over the route of the regular annual Monte Carlo Rally.

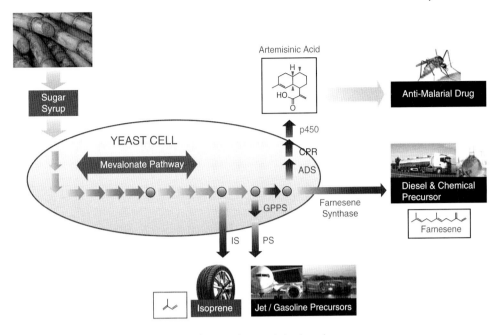

Figure 3.2 The engineered mevalonate pathway and some derived products.

which cyclizes farnesylpyrophosphate to give 1,4-amorphadiene, a bicyclic system with four chiral centers (and precursor to the anti-malarial drug artemisinin), and farnesene synthase is the enzyme that converts farnesyl pyrophosphate into β-farnesene, our diesel precursor and, as described later in this chapter, chemical precursor). The figure also shows how, by incorporation of other enzymes into the pathway, it is possible to further functionalize these isoprenoids (exemplified by the production of the chiral bicyclic sesquiterpene artemisinic acid by oxidation of the 1,4-amorphadiene produced by the action of ADS on the linear intermediate farnesyl pyrophosphate).

3.3
Terpene Building Blocks of Commercial Interest

Aside perhaps from isoprene, of which some 800 000 metric tons (MT) are made and used annually, mainly (ca. 95% of the total) for the synthesis of *cis*-1,4-polyisoprene or synthetic rubber, with the rest going into other co-polymers, no other terpenoids are currently produced or used in any significant amounts in industry. Certainly the often cited US Department of Energy (DOE) report *"Top Value Added Chemicals from Biomass"* [12] does not list any terpenes or terpene derivatives in its final Top-12 list of chemical targets, nor do any appear in the preliminary Top-50 listing (at the time of writing and to the best of the author's knowledge, the final version of this report, which was to include the complete list of 300 chemicals considered by the DOE, has not been published yet).

The other terpenes are used as solvents and ingredients in perfumery, cosmetics, medicine, and a few other applications. The total volume of terpenes used as fragrance ingredients is about 50 000 t a^{-1}, but only a very limited number of them, all monoterpenes (menthol, geraniol, linalool and citronellol, and their respective esters) account for more than 5000 t a^{-1} [13]. While many are still extracted directly from natural sources, almost exclusively from the plant kingdom, some can be produced synthetically from petrochemical-derived building blocks, or semi-synthetically from other readily available terpene-based feedstocks, like turpentine. Because of this relatively small market size and the fact that they are obtained from either natural sources, like gum turpentine, crude sulfate turpentine (CST, a byproduct of the Kraft paper industry), or they share raw materials with much larger industries which drive the raw material costs (e.g., vitamins – vitamin E production for animal feed is much larger than the combined worldwide terpene production and uses mostly synthetic terpene intermediates), terpene supply and prices can be very volatile. As an example, while exports of gum turpentine from China, the world's largest producer, have remain fairly flat since 1995 at somewhere between 200 000–300 000 MT, over this same period the price per MT, after being quite stable at around $500 per MT up to 2004, then doubled in the span of two years, and since then it has undergone a series of spikes and small dips to now stand at around $2000 per MT [14].

Since Amyris' initial focus has been on sequiterpenes (C15 terpenes), before discussing some of the novel terpene derivatives developed by Amyris, a brief review of the current market and uses of these chemicals is in order.

3.4
Sesquiterpenes as Chemical Building Blocks: β-Farnesene

According to the most recent article on terpenoids in the *Kirk-Othmer Encyclopedia of Chemical Technology* [12], although more than 3000 sesquiterpenes are known in Nature, perhaps less than a dozen have any commercial significance at all, mostly as ingredients in the fragrance and food industries (either used as pure compounds or as essential oils composed of mixtures rich in sequiterpenes, for example, sandalwood, patchouli, and vetiver oil) or as building blocks in the synthesis of more complex molecules (linalool, for example, is a key intermediate in some routes to the commercially important vitamins E and K). More details about the important sesquiterpenes and their uses can be found in specialized references such as ([12], p. 52–59) or [15]. Details of their use as building blocks can be found in some reviews on the end products [16].

In parallel to the work on development of renewable fuels, Amyris' chemists embarked on studies of the novel chemicals that could be derived from the fermentation building blocks and their possible applications. In particular we were intrigued by the possibilities of β-farnesene (Figure 3.3).

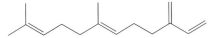

Figure 3.3 Structure of β-farnesene.

This substance is a natural product, best known as the major, and in some cases the only component of alarm pheromones emitted by numerous species of aphids when attacked by predators [17]. Exposure to β-farnesene causes aphids to initiate predator avoidance behaviors and also to produce more winged offspring that can leave the plant [18]. Some plants are also known to emit this substance, purportedly as a feeding deterrent against aphids, which has led to the proposal of using β-farnesene as a crop protection agent, either by application or through production of transgenic plants genetically engineered to produce large amounts of the sesquiterpene [19]. This topic, however, remains controversial, as other authors have reported no evidence of either repellence or reproductive effects [20], or that aphids quickly become accustomed to β-farnesene and return to plants to which it has been applied [21]. To further complicate matters, there is also evidence that β-farnesene-emitting plants may do so to attract natural enemies of aphids as pollinators rather than for defense purposes [22]. In any case, its potential use as a crop protection agent seems limited by its volatility and oxidative instability under field conditions and, consequently, the use of more stable derivatives has been proposed [23].

In the plant kingdom, like many other sesquiterpenes, β-farnesene can be found in a long list of essential oils, in amounts ranging from mere traces to as much as 18% of the total, as in the case of *Alpina galanga* oil [24].

Synthetic β-farnesene can be produced by a variety of routes, although the most common are trimerization of isoprene [25], dehydration of alcohols like nerolidol [26] or farnesol [27], Wittig-type chemistry involving geranyl halides [28], other coupling-type chemistry [29], or homologation of the C10 analog myrcene [30] (the cited references are not meant to be exhaustive, but rather are given as examples).

The most common routes are the first two, although not surprisingly, given the lack of stereocontrol in these processes, the product obtained can best be described as a technical grade feedstock, typically constituted by a mixture of isomers and byproducts. Higher purity materials are available by extensive fractionation of these mixtures, but are logically only available in smallish amounts and are quite expensive. Until fairly recently Kuraray Co. Ltd. (Tokio, Japan) sold a technical grade product (spec. 75% purity – a roughly 1:1 mixture of β-farnesene and α-farnesene) that, according to their product flow chart [31], was prepared in six steps from isoprene, with the final step being a dehydration of nerolidol (the economics of this process were presumably driven by the use of a shared route to isophytol, a key vitamin E intermediate), but production was discontinued sometime in 2007 as a result of the company's rationalization of its product portfolio. We contrast this with Amyris β-farnesene, which is typically obtained as a single isomer directly from the fermentation process in

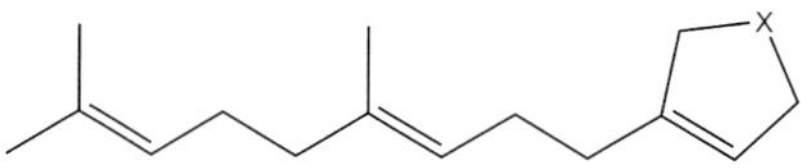

Figure 3.4 β-Farnesene Diels–Alder adducts.

93–95% purity. A simple flash distillation to remove traces of related sesquiterpene hydrocarbons and alcohols that are formed in the biological process increases this to 97–98%.

As far as the chemistry of β-farnesene is concerned, given its scarcity as a raw material up to now, it should come as no surprise that few mentions of its use can be found. A relatively old Givaudan patent [32] claims its use in perfumery, but recent monographs on perfumery raw materials describe this use as "limited" [33]. Given the presence of a conjugated diene in the β-farnesene molecule (Figure 3.3), it is not surprising that most mentions of the compound involve cycloaddition-type chemistry. A 1982 patent [34] describes reactions of β-farnesene with various dienophiles to produce products of the general structure shown in Figure 3.4 (where X is a generic dienophile residue), and claims their use as crop protection agents.

In this same theme a newer patent [35] describes the Diels–Alder reactions of a series of terpenes, including β-farnesene, with hydroxymethacrylates to give adducts that, after hydrogenation, are claimed to find use as high boiling point solvents in a broad range of industrial applications, including vehicle components for ink compositions, electrical insulation oils, dye-solubilizers for pressure-sensitive copying materials, diluents for urethane resins, plasticizers for ethylene–vinyl acetate copolymers and thermoplastic elastomers, heat media and solvents for coating materials, pastes, or brake fluids. A 2007 patent [36] includes β-farnesene in a list of terpenes that were reacted with unsaturated dicarboxylic acids to produce polyurethane reactants. Chiral Diels–Alder reactions [37] and some ene-type chemistry [38] are also reported.

The non-Diels–Alder chemistry of β-farnesene is much more limited. A 1996 patent claims its use in the production of α-tocopherol derivatives [39] and an assortment of functionalization reactions including hydrohalogenation [40], dihydroxylation [41] and reactions with iminobutanoates [42] are reported. The photochemistry of β-farnesene with and without sensitizers has also been published [43]. Finally, its use to prepare useful detergent ingredients via standard hydroformylation has recently been disclosed in a patent application by Proctor and Gamble [44].

3.5
Polymers

On reviewing the existing literature on β-farnesene, Amyris chemists noted the curious absence of literature mentioning its homopolymers. Indeed, a SciFinder

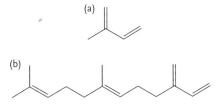

Figure 3.5 Structures of isoprene (a) and β-farnesene (b).

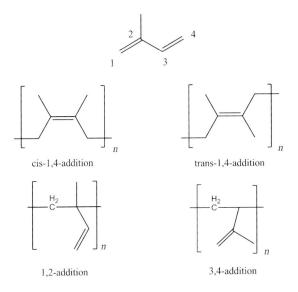

Figure 3.6 Different possible poly(isoprene) isomers.

search on the term "polyfarnesene" gave only one hit! [45] (although in hindsight this should not be so surprising, given the quality of the β-farnesene available heretofore – this has been confirmed through our own work and in our subsequent technical interactions with many of Amyris' commercial partners, who have informed us that most standard polymerization reactions do not work with the technical grade material).

At first glance, and given their obvious structural similarities (Figure 3.5), β-farnesene might be expected to show chemistry similar to that of isoprene, perhaps with some differences due to the presence of the additional double bonds in the sesquiterpene.

Because of its importance in the manufacture of synthetic rubber – "high *cis*-1,4-poly(isoprene)" – and other homo and heteropolymers, the polymerization of isoprene has been extensively studied for over 150 years [46], and it is well known to afford four different products, depending on the carbons linked together (Figure 3.6).

Additional complexity arises from polymer tacticity – the possible different relative orientations of the polymer backbone substituents – giving rise to polymers described as isotactic (all substituents on the same side), syndiotactic (alternating substituents) or atactic (randomly distributed substituents).

Nowadays, the stereoregularity and relative orientations of isoprene homopolymers can be controlled by the selection of appropriate catalyst and reaction conditions. Much of the research on catalysis of the polymerization of isoprene has been driven by the commercial interest, dating back to WWII, in duplicating the cis-1,4-polyisoprene structure of natural rubber [47]. All these early efforts were unsuccessful, and it was not until 1954 when B.F. Goodrich scientists successfully prepared synthetic cis-1,4-polyisoprene using the recently discovered Ziegler transition metal halide (trialkylaluminum/titanium tetrachloride) coordination-type catalysts. Shortly after, Firestone Tire and Rubber disclosed the preparation if cis-1,4-polyisoprene with lithium metal based catalysts. The first commercial scale plants using both types of catalyst came online in the early to mid-1960s ([47] and references therein).

Other newer catalyst systems for producing high cis-1,4-polyisoprene include aluminum hydride (alane)/transition metal halide and lanthanide rare earth/alkyl aluminums with a variety of ligands (these are believed to involve bimetallic Ln-Al species with the halide and ligand atoms bridging between the two metals). Interestingly, alkylaluminum/titanium alkoxides or alkylaluminum/iron acetylacetonate/amine species give high 3,4-polyisoprene [47]. Of less commercial interest, due to the fact they invariably give mixtures of cis- and trans-1,4, 3,4- and 1,2-polyisoprene, are the so-called Alfin catalysts (alkenyl sodium/sodium alkoxide/halide salt), discovered in the late 1940s [47].

Sodium metal-catalyzed polymerization of isoprene was reported as early as 1911. The process can be carried out heterogeneously in hydrocarbon solutions or neat, or homogeneously in polar solvents, but was of limited interest because it invariably produced roughly equivalent mixtures of trans-1,4-polyisoprene and 3,4-polyisoprene, with smaller amounts (up to 25%) of the 1,2-microstructure, but negligible amounts of the more commercially interesting cis-1,4-polyisoprene. As mentioned above, it was only in the 1950s that it was found that finely divided lithium metal in hydrocarbon solvents produces high cis-1,4-polyisoprene, whereas in polar solvents essentially none of this microcrystalline structure is formed. These reactions all feature a more or less long and frequently irreproducible induction period, and in the late 1950s lithium dispersions were largely abandoned in favor of the much easier to handle organolithiums, which also do not display this induction period [47].

Anionic polymerization of isoprene is of great value because it is the only method that allows the preparation of polyisoprenes of predictable molecular weight and microcrystalline structure, which is largely controlled by the stoichiometry, nature of the solvent used and the presence or absence of additives, typically amines or ethers. In addition, anionic polymers do not show a tendency to chain terminate, so the reaction continues as long as there is monomer present.

It also allows the preparation of true block copolymers by addition of different monomers to the growing chains. Butyl lithiums are frequently used in these reactions because they are soluble in both polar and non-polar solvents, thus allowing targeted preparation of different microstructures with a common catalyst by just changing solvents. Catalyst reactivity increases on going from primary alkyl lithiums to secondary and tertiary ones [47].

Cationic polymerizations of isoprene using strong Lewis acids, like BF_3, can produce, in addition to the desirable linear structures, rings and partially saturated chains due to disproportionations. These reactions can be directed towards *trans*-1,4-, 3,4- and 1,2-polyisoprene mixtures (albeit with little or no *cis*-1,4-structure) by carrying out the reactions at low conversions and at low temperatures (−78 °C to room temperature) in hydrocarbon solvents [47].

Finally, it is possible to perform free radical polymerizations of isoprene too, using two-phase emulsion systems and standard free radical initiators (e.g., hydroperoxides). These give low contents of *cis*-1,4-polyisoprene [47].

By analogy, one might expect polymerization of β-farnesene to also afford similarly distinctive microstructures (Figure 3.7), so with abundant high purity β-farnesene in hand, in our initial proof of concept experiments we sought to determine if it was indeed possible to target specific microstructures and chain lengths (molecular weights) in a controlled fashion using standard isoprene polymerization catalysts and conditions, to characterize the products, and finally to examine their properties to establish if they might have any commercial interest.

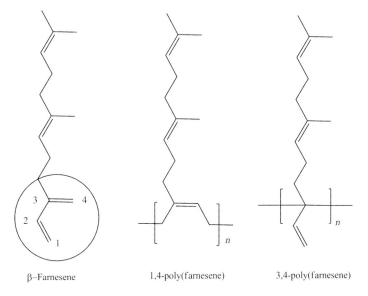

Figure 3.7 Possible poly(farnesene) microstructures.

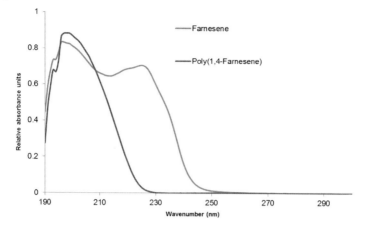

Figure 3.8 UV spectrum of 1,4-poly(farnesene) (1,4-PF).

Given the vast number of catalysts and reactions conditions reported in the literature, as proof of concept we decided to focus only on anionic polymerization using the easy to handle alkyl lithiums [48]. Thus, using standard anionic polymerization conditions (n-BuLi to prepare low molecular weight polymer of $M_n \approx 100\,000$ and sec-BuLi for high molecular weight polymer), and a hydrocarbon solvent (cyclohexane), we were gratified to observe that high concentrations of β-farnesene (12%) could be effectively polymerized to give 1,4-polyfarnesenes [49]. A notable feature of the reactions was the very high conversions, as indicated by the UV spectrum before and after polymerization (Figure 3.8), showing the complete disappearance of the characteristic farnesene band, which suggests a very efficient monomer incorporation and the essentially exclusive formation of polymer product.

Excellent regioselectivity was also indicated by examination of the corresponding ^{1}H-NMR spectrum (Figure 3.9). In this spectrum, the two peaks at 4.85 and 4.81 ppm are associated with the 3,4-PF microstructure while the peaks at 5.17, 5.16, 5.14, and 5.13 are associated with both the 1,4- and 3,4-microstructures, but the area under the curve contributed by the 3,4-structure is equivalent to the area under the 4.85 and 4.81 peaks.

In a 50:50 mixture of the 1,4- and 3,4-microstructures the contributions of both forms to the 5.17, 5.16, 5.14, and 5.13 ppm peaks would be equal. By subtracting out the contribution of the 3,4-microstructure and ratioing the area under the 4.85 and 4.81 peaks with the residual area under the 5.17, 5.16, 5.14, and 5.13 peaks one obtains the relative ratio of the 1,4- and 3,4-forms. The result of this analysis indicates that circa 12% of the monomers in the polymer chain have the 3,4-polyfarnesene microstructure, therefore the 1,4-polyfarnesene accounts for the remaining circa 88%. All other peaks in the spectrum follow the expected ratios, indicating that the resultant polymer is the expected 1,4-PF polymer.

To further characterize the product it was subjected to the following tests:

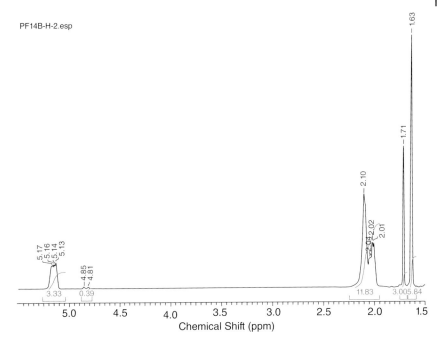

Figure 3.9 NMR of low molecular weight 1,4-PF.

3.5.1
Differential Scanning Calorimetry

A TA Q200 DSC (TA Instruments, New Castle, DE, USA) was utilized to determine the glass transition temperatures (T_gs) of the material produced. Samples were scanned between the temperatures of −175 and 75 °C at a rate of 10 °C min^{-1}. Figure 3.10 shows the corresponding curve.

The x-axis is the temperature and the y-axis is the heat flow. The T_g is identified as a step change in the heat flow, reported from the mid-point of the transition. The thermal traces indicate that the sample had a T_g of ∼ −76 °C. This value is well within the range of traditional elastomers [the T_g of *cis*-1,4-poly-(isoprene) is −73 °C, of *trans*-1,4- poly(isoprene) is −53 °C, of 3,4-polyisoprene is 5 °C, *cis*-1,4-poly-(butadiene) is −107 °C, and *trans*-1,4-poly(butadiene) is −107 °C]. The absence of any other thermal events indicates 1,4-PF does not crystallize within the studied temperature range.

3.5.2
Gel Permeation Chromatography

Gel permeation chromatography (GPC) was utilized to determine the molecular weight and polydispersity of the synthesized samples. Because of polydispersity, the molecular weight of a sample is generally reported as the number averaged

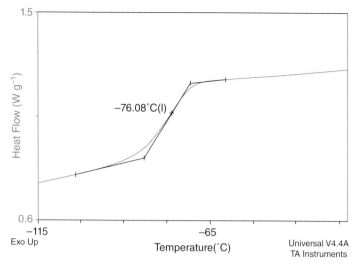

Figure 3.10 DSC of 1,4-PF $M_n = 100\,000$.

(M_n) or the weight average (M_w) molecular weight. In cases where several peaks overlap, impeding the determination of a unique polydispersity for each peak, the molecular weight indicated by the peak (M_p) is reported. In this case the GPC curves (not shown) indicated a calculated M_n of ~105 000 and M_w was ~147 000. This results in a polydispersity of 1.4. This is relatively high for anionic polymerization which can often achieve polydispersities of less than 1.2 (free radical polymerization results in polydispersities of greater than 1.5 and polycondensation results in polydispersities greater than 2.0, so a polydispersity of 1.4, while high for anionic polymerization, is not high for polymers in general).

3.5.3
Thermal Gravimetric Analysis

The degradation temperature of the samples in air and under nitrogen was determined utilizing thermal gravimetric analysis (TGA). The sample was heated from room temperature to 580 °C at 10 °C min^{-1}. The x-axis indicates the temperature and the y-axis the percent sample weight lost. The 1% and 5% weight temperatures are reported. The 1% weight loss in air occurred at 210 °C while the 5% weight loss occurred at 307 °C (Figure 3.11).

These values are sufficiently high for the material to be used in typical processing techniques, such as compression molding, injection molding and extrusion, as well as post-processing conditioning, such as vulcanization, which requires temperatures ~170 °C [the degradation of natural rubber begins at 207 °C and that of poly(isobutylene) generally begins at 300 °C, putting 1,4-PF in the same thermal degradation range as other commercially used elastomers]. Under nitrogen both values increased to 307 and 339 °C, respectively, indicating somewhat improved stability in an inert atmosphere.

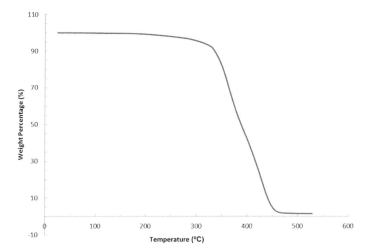

Figure 3.11 TGA of 1,4-PF M_n = 100 000 in air.

3.5.4
Tensile Strength

Because of the high molecular weight of the side chain in PF (204.35), a degree of polymerization of 490 was not sufficient to achieve a high enough entanglement molecular weight for the product to display measurable mechanical properties. At 100 000 Da 1,4-PF has a honey-like consistency. Despite the lack of mechanical strength, the material was observed to be very tacky, which suggests its possible use in adhesive applications.

In order to obtain mechanical strength data, we next sought to prepare a high molecular weight sample of 1,4-PF. From the corresponding GPC trace (not shown) the Mn of that sample was found to be 245 000 Da, with a M_w of 457 000 Da and peak molecular weight of 501 000 Da. This translates into an average of about 2000 repeat units per chain. The higher level of entanglement and consequent increase in relaxation time is evident in the consistency of the polymer, which no longer flows. The tensile strength of the samples was determined using an Instron tensile tester. The sample was cast into a film and cut to the appropriate dimensions. The thickness and width of the resulting samples were measured. A gage length of 2.54 cm was used with a crosshead speed 25 mm min^{-1}. The x-axis is strain or the percent elongation, and the y-axis is stress. The samples were soft and quickly yielded. The peak elongation was 6% with a maximum tensile strength of 19 psi. The sample did not strain harden, but continued to yield to about 40% elongation as the chains start to disentangle to relieve the stress on the sample. This stress–strain curve indicates that the sample is starting to exhibit rubbery behavior, while still retaining good tack.

A very interesting feature of the synthesis of the higher molecular weight polymer was the fact that it still occurred readily at 12% concentration, whereas to achieve high molecular weight poly(isoprene), the monomer concentrations have to be reduced to 5% or less, which hurts the process economics. Another notable feature was observed in the T_g determination of the high M_n 1,4-PF by TSC, namely that the T_g was determined also to be $-76\,°C$, that is, the same as that of the lower molecular weight sample. This is potentially beneficial because it indicates that any mechanical properties which are dependent on molecular weight can be improved without increasing the T_g of the material if so desired.

Finally, and again using standard conditions (n-BuLi/tetramethylethylenediamine (TMEDA), 12% monomer solution, cyclohexane as solvent), we sought to prepare the alternative 3,4-PF microstructure, targeting this time a M_n of 50 000. In this instance again complete monomer incorporation was seen and, using NMR, the ratio of 3,4-PF to 1,4-PF was determined to be 90:10. DSC thermal traces (not shown) indicated that the low molecular weight sample had a T_g of ~ $-76\,°C$, that is, the exact same temperature as the 1,4-PF microstructure. The fact that both microstructures have the same T_g is unusual. As mentioned above, the T_g 1,4-poly(isoprene) ranges between -73 and $-53\,°C$ for the cis and trans forms, respectively, while the T_g of 3,4-poly(isoprene) is $5\,°C$. This behavior of PF makes it more like poly(butadiene) whose cis and trans forms both have the same T_g. The implication is that one can blend the two structures in one chain to manipulate composite properties without detrimentally impacting the system T_g, a significant advantage over conventional elastomers. No other thermal events were detected in the -175 to $75\,°C$ temperature range, indicating that 3,4-PF does not crystallize either, even at very low temperatures. The TGA curves for 3,4-PF in air showed the 1% weight loss at $191\,°C$ and the the 5% weight loss at $265\,°C$, lower than those of the 1,4-PF but still sufficiently high for most standard processing techniques. Like the low molecular weight 1,4-PF, the even lower M_w 1,3-PF was a viscous liquid and no mechanical properties could be measured.

One interesting feature of this polymer was seen in its GPC data. As one can observe in Figure 3.12, the polymer presented two peaks, indicating the presence of two distinct weight fractions.

The M_n of the primary peak was calculated to be ~46 000 and the M_w was ~48 000. This gives a polydispersity of 1.04. This is relatively narrow, but not usual for anionic polymerization. The secondary peak has a peak molecular weight of ~97 000, or about two times that of the primary peak, which suggests that it is probably caused by coupling of the homopolymer rather than chain elongation.

Copolymerization of farnesene with a variety of co-monomers was also examined, but is not discussed here for the sake of brevity. Interested readers may consult the aforementioned and other more recent Amyris patents in this area [45, 50].

3.6
Lubricants

While several Diels–Alder reactions of β-farnesene are mentioned in the literature (see Section 3.4 and relevant references therein), we noted the absence of any mention of such reactions where this substance acted as both the diene and dienophile.

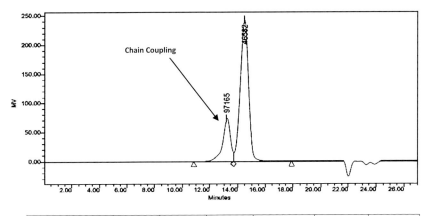

	Mn (Da)	Mw (Da)	MP (Da)	Mz (Da)	Mz+1 (Daltons)	Polydispersity	Mz/Mw	Mz+1/Mw
1			97165					
2	45818	47644	46582	49134	50527	1.039844	1.031269	1.060511

Figure 3.12 GPC of 3,4-PF.

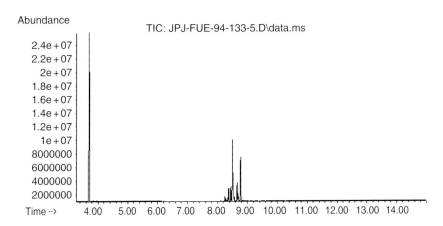

Figure 3.13 GC of hydrogenated β-farnesene.

Our initial interest in these compounds arose during the process development for scale up of the catalytic hydrogenation of β-farnesene to produce 2,6,10-trimethyldodecane, our renewable diesel fuel candidate. Under certain reaction conditions some minor impurities could be detected as a characteristic grouping of peaks in the 8–9 min C30 region of the gas chromatograms of hydrogenated β-farnesene (Figure 3.13).

The mass spectra of these peaks indicated molecular weights ranging from 414 to 420, consistent with a C30 β-farnesene dimer with 1–4 degrees of unsaturation. Hydrogenation under more forcing conditions depleted the lower molecular weight peaks but no peaks with molecular weights higher than 420 were observed. These results were largely independent of the type of catalyst used and blank

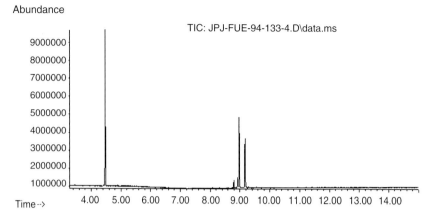

Figure 3.14 GC of β-farnesene after heat treatment.

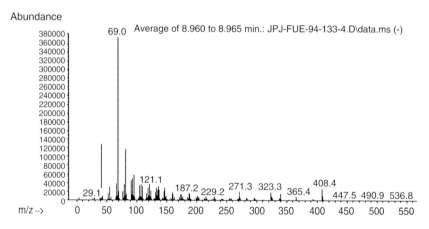

Figure 3.15 MS of the circa 8.9 min GC peak.

reactions using similar temperature profiles but without hydrogen or catalyst (indicating a purely thermal process) provided a simplified GC profile in which four major peaks grouped into two pairs could be observed (Figure 3.14).

Mass spectrometry of these peaks (Figure 3.15) gave a molecular weight of 408, consistent with a C30 molecule with 7 degrees of unsaturation, which could, in principle, be due to the presence of seven double bonds (i.e., a linear dimer with a dehydrosqualene-like structure) or six double bonds and a ring, but the hydrogenation results whereby one of the degrees of unsaturation resisted all attempts to saturate it, indicated we were probably dealing with the latter case, that is a group of dimers of β-farnesene containing a ring structure. The lower molecular weight peaks (414, 416, 418) and increased number of peaks upon hydrogenation could be explained by the presence of a series of isomeric partially hydrogenated intermediates.

Ultimately it was determined that all the evidence was consistent with a thermal Diels–Alder reaction between two β-farnesene molecules in which one of them acted as the diene and the other as the dienophile (Scheme 3.1).

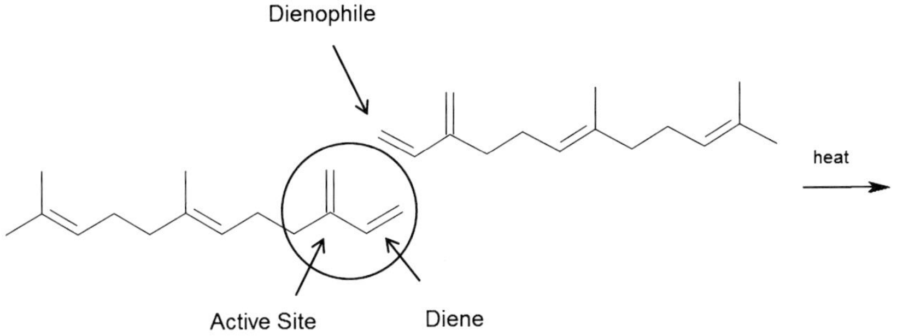

Scheme 3.1 Homo Diels–Alder reaction of β-farnesene.

Subsequent isolation of the impurities and their further characterization by extensive NMR studies confirmed the structures of the products to be those shown in Scheme 3.2 (the other isomers are omitted for clarity). The ratio of major to minor pairs is roughly 4:1.

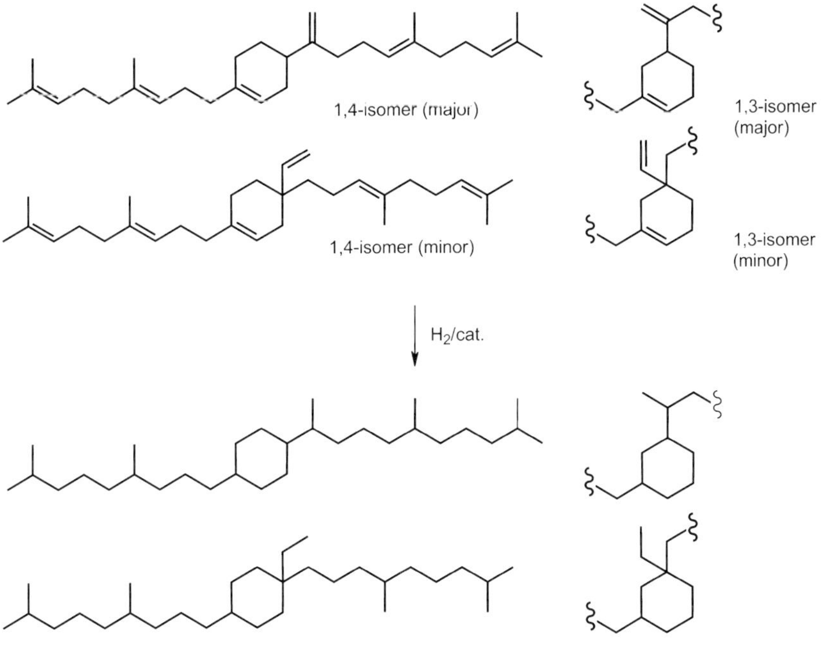

Scheme 3.2 Homo Diels–Alder reaction products of β-farnesene.

Similar products are known to be formed in the Diels–Alder reaction of isoprene (Scheme 3.3) [51].

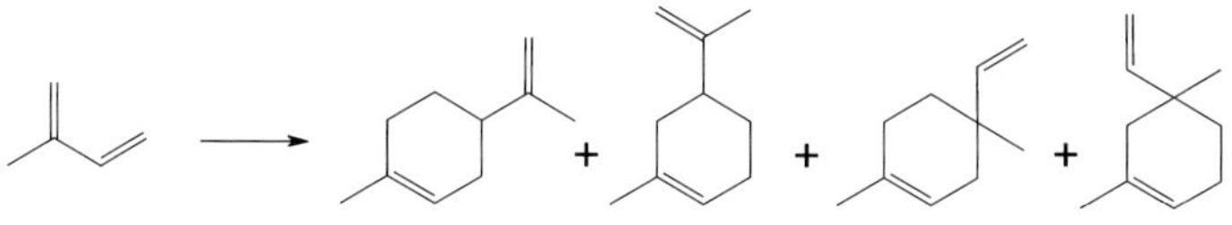

Scheme 3.3 Diels–Alder adducts of isoprene.

The ^{1}H-NMR of the unsaturated Diels–Alder adduct mixture exhibits signals for the side-chain trisubstituted olefinic protons at δ 5.11, the exo-methylene protons (δ 4.76) and the protons of the cyclohexene olefins (δ 5.35 and 5.42). Also present are ABX and AMX arrays (δ 5.69, X portion and 5.65, X portions) representing the vinyl protons of the 1,4- and 1,3- minor isomers, respectively. The ^{1}H-NMR spectra of the hydrogenated mixtures are too complex to provide much meaningful information, other than confirmation of a complete hydrogenation (i.e., the absence of alkene protons). This is because each 1,4-isomer has three pro-chiral centers (boxed carbons in Figure 3.16) and hence gives $2^3 = 8$ diasteromers upon hydrogenation, while the two 1,3-isomers, with five (major) and four (minor) pro-chiral centers can give $2^4 = 16$ and $2^5 = 32$ isomers each, respectively.

It is known that many unsaturated cyclic terpenes, when heated in the presence of a hydrogenation catalyst, but in the absence of hydrogen, can undergo a disproportionation whereby some of the molecules abstract hydrogens from other molecules and the molecules giving up the hydrogens are ultimately aromatized [52].

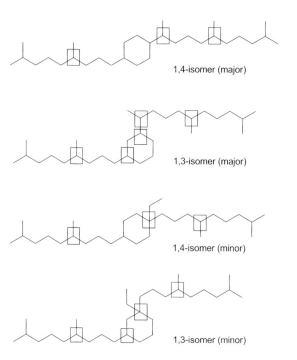

Figure 3.16 Pro-chiral centers in the different Diels–Alder adducts.

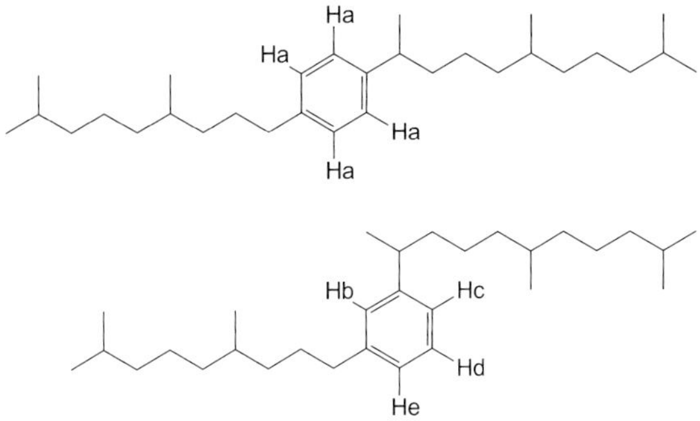

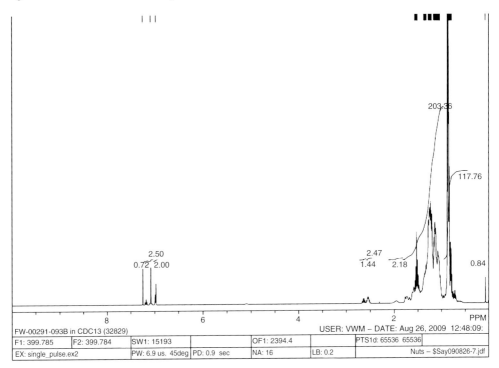

Figure 3.17 Aromatized Diels–Alder β-farnesene adducts.

Figure 3.18 ¹H-NMR of the aromatized Diels–Alder β-farnesene adducts.

Treatment of the thermal dimer mixture in heptane with 5% Pd/C at 100 °C in an autoclave for 60 h effects modest (5%) aromatization of the two major isomers, presumably to afford the products shown in Figure 3.17.

The minor components cannot aromatize under these conditions due to the presence of the quaternary carbons in the rings. After hydrogenation of the remaining olefinic linkages to declutter the downfield region, the corresponding ¹H-NMR spectrum (Figure 3.18) supports the structures proposed above for the aromatic compounds.

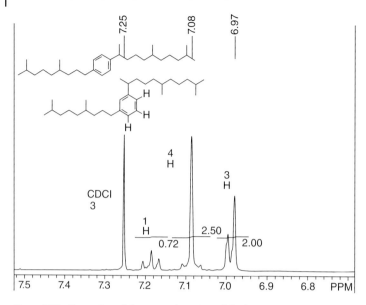

Figure 3.19 Expansion of the aromatic region of the ^{1}H-NMR spectrum of the aromatized Diels–Alder β-farnesene adducts.

The H_a protons of the 1,4-substituted ring appear as a 4H AB system at δ 7.09, which agrees well with the calculated value of 7.08–7.09 ppm. H_d appears as a triplet at δ 7.18 (calc. 7.13) which is actually an overlapping set of doublets due to the adjacent H_c (δ 6.94, calc. 6.95) and H_e (δ 6.99, calc. 6.94). The isolated proton, H_b, is superimposed on the signals for H_c and H_e and appears as a singlet at δ 6.98 (calc. 6.99) (Figure 3.19).

The observation that both the saturated and unsaturated Diels–Alder isomer mixtures were colorless viscous oils, with the unsaturated one being the noticeably more viscous of the two, led us on the one hand to seek ways to prepare these material in large amounts, and on the other, to test their properties as potential lubricants [53, 54].

Regarding the preparation, it was discovered that by simply heating neat β-farnesene high yields of the desired mixture of Diels–Alder adducts could be obtained. The reaction is exothermic (Figure 3.20), so for larger scale operation the use of a diluent is recommended. A fairly simple relationship between reaction temperature and conversion was observed (Figure 3.21).

GPC of the crude reaction mixtures (Figure 3.22a) shows the presence of two peaks, the major one corresponding to the thermal Diels–Alder adducts and a much smaller one corresponding to higher oligomers. Simple distillation isolates the pure Diels–Alder adducts (Figure 3.22b).

This purification process revealed an interesting property of these materials for their potential use as lubricant base oils, namely that by controlling the fraction of pure adduct present it was possible to control the viscosity. Since the higher oligomers have much higher viscosity than the distilled pure adducts, by blending crude adduct containing heavier oligomers with pure distilled adduct it was pos-

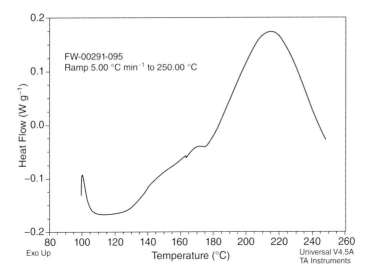

Figure 3.20 DSC curve of the thermal Diels–Alder reaction of β-farnesene.

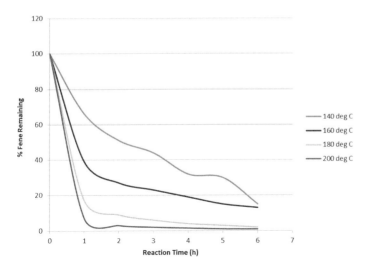

Figure 3.21 Formation of thermal Diels–Alder adducts of β-farnesene at different temperatures.

sible to modulate the viscosity of the mixtures, as indicated by the corresponding viscosity versus % pure (distilled) adduct curves (Figure 3.23). These curves show the dynamic and kinematic viscosities at 40 and 100 °C versus percentage of distilled hydrogenated adduct, indicating how the viscosities drop as more pure less viscous dimer is blended in. Since the amount of higher oligomer present depends to some extent on the reaction conditions (temperature and time, or the presence of suitable catalysts) this allows the production of mixtures of crude adduct with

74 | *3 The Development of Catalytic Processes from Terpenes to Chemicals*

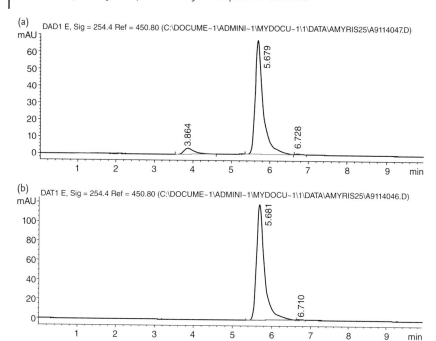

Figure 3.22 (a) Crude β-farnesene thermal Diels–Alder adducts; (b) distilled β-farnesene thermal Diels Alder adducts.

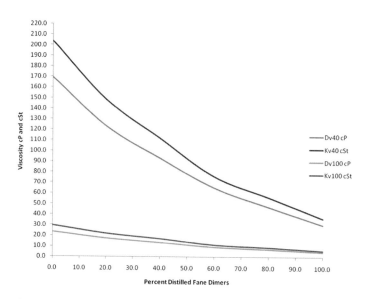

Figure 3.23 Viscosity versus distillation curves for hydrogenated β-farnesene Diels–Alder adducts.

Table 3.1 Mid-grade viscosity base oil comparison.

Property	Typical value		
	Amyris 24	PAO 25	PAO 40
Kinematic viscosity (cSt@212°F/100°C)	24.23	25.0	40.0
Kinematic viscosity (cSt@104°F/40°C)	180.3	208	410
Viscosity index	143.8	155	145
Pour point (°C)	−35	−45	−34
Flash point (°C)	228	518	561
Volatility, Noack (wt%)	13.4	1.1	–
Specific gravity (60°F/15.6°C)	0.848	0.843	0.851
Color, Saybolt	26	30	26
Biodegradable	Yes	No	No
Cost per gal	–	$15–25	>$20

different viscosities, and a wide range of lubricant properties can be spanned by this blending of products.

To identify potential applications in the lubricant field the Amyris renewable β-farnesene Diels–Alder adducts were subjected to a series of standard tests at third-party laboratories. Table 3.1 compares a crude hydrogenated Diels–Alder adduct (Amyris 24) with two commercial synthetic poly-α-olefin (PAO) base oils of different viscosity. One may see that, for the most part, the Amyris base oil compares favorably with the commercial products (with the exception perhaps of volatility and flash point, but these are parameters that can be adjusted through distillation and/or the use of suitable additives). A key differentiator is the biodegradability of the Amyris renewable feedstock based lubricant. It is also expected that, given the exceedingly simple production process, the renewable base oil will be cost competitive with the current high-end synthetics once produced on scale.

Subsequent work at Amyris has identified a variety of β-farnesene-derived molecules which present a range of properties that makes them suitable as base oils or blendstocks for many different lubricant applications. In 2011 Amyris announced the formation of Novvi S.A., a joint venture with the Brazilian company Cosan to develop, manufacture and sell worldwide the NovaSpec™ line of synthetic, renewable base oils for the lubricants market, leveraging Amyris' platform technology and Cosan's production capabilities [55]. Amyris is also planning to launch in the near future a family of finished high performance renewable lubricants under the tradename Evoshield™ [56].

3.7
Conclusions

To summarize, this article has offered a brief overview of the place of terpenes, and in particular sesquiterpenes, in the field of renewable chemical feedstocks.

These molecules, produced by engineering of microorganisms to produce them from fermentable sugars via large-scale industrial fermentation, represent novel building blocks that, in the hands of Amyris chemists, have led to the discovery and development of a wide range of high performance new products, exemplified in this article by new polymers and lubricants. We feel that in this work we have only begun to scratch the surface of the enormous potential of β-farnesene and the other renewable chemical feedstocks and look forward to finding novel derivatives and applications for these building block. We also feel very strongly that solutions to the challenges of achieving price parity with petroleum-derived chemicals exist or are imminent, and will only accelerate with contributions of others in the field. These solutions will come not only from improved yields on sugar and improved product recovery processes, in addition to new pathways and end products, but particularly from the successful development of cellulosic feedstocks as lower cost sources of fermentable sugars that shall benefit all companies like Amyris active in this area. Hopefully, also publications like this one will attract more bright minds from biology, chemistry and engineering to the field to help solve all these challenges for the benefit of all mankind.

References

1 Moray, S. (1826) An account of a new explosive engine. *Am. J. Sci. Arts*, **11**, 141–145.
2 Hale, W.J. (1934) *The Farm Chemurgic: Farmward the Star of Destiny Lights Our Way*, The Stratford Company, Boston, MA, USA, p. 201.
3 Benson Ford Research Center (2010) Popular Research Topics: Soybean Car, http://www.thehenryford.org/research/soybeancar.aspx (accessed 2 October 2012).
4 For a history of chemurgy, see: Finlay, M.R. (2003) Old efforts at new uses: a brief history of chemurgy and the american search for biobased materials. *J. Indust. Ecol.*, **7**, 33–46. For a comprehensive current overview of the state of the art in the use of renewable feedstocks for the production of chemicals, see Ulber, U., Sell, D., and Hirth, T. (eds.) (2011) *Renewable Raw Materials: New Feedstocks for the Chemical Industry*, 1st edn, Wiley-VCH Verlag GmbH: Weinheim, Germany and references therein.
5 For an overview of the artemisinin project see: Hale, V., Keasling, J.D., Renninger, N., and Diagana, T.T. (2007) Microbially derived artemisinin: a biotechnology solution to the global problem of access to affordable antimalarial drugs. *Am. J. Trop. Med. Hyg.*, **77**, 198–202.
6 OneWorldHealth (2008) Press Releases, http://www.oneworldhealth.org/press_releases/view/pr_1227057050 (accessed 1 October 2012).
7 http://www.amyris.com/ science/industrial-synthetic-biology-platform. Details are available in several Amyris patents, for example: (a)Renninger, N.S. (2007) Production of terpenes by recombinant microorganisms. WO 2007139924, A2, Dec 06, 2007; (b) Tsuruta, H., Lenihan, J.R., and Regentin, R. (2009) Production of isoprenoids using genetically engineered microorganisms. WO 2009042070 A2.
8 Amyris, Inc., N/ (2011) Markets, http://www.amyris.com/en/markets/fuels (accessed 3 October 2012).
9 Relevant patents are, *inter alia*, for diesel: (a) Renninger, N.S., and McPhee, D.J. (2008) Fuel compositions comprising farnesane and farnesane derivatives and method of making and using same. US

7399323 B2; (b) Renninger, N.S., and McPhee, D.J. (2010) Fuel compositions comprising farnesane and farnesane derivatives and method of making and using same. US 7846222 B2; f or gasoline; (c) Ryder, J.A., and Fisher, K. (2009) Fuel compositions comprising tetramethylcyclohexane. US 7540888 B2; for jet fuel; (d) Ryder, J.A. (2009) Jet fuel compositions. US 7589243 B2; (e) Ryder, J.A. (2010) Jet fuel compositions and methids of making and using same. US 7671245 B2.

10 New York Times (2007) First, Cure Malaria. Next, Global Warming, June 3, 2007, http://www.nytimes.com/ 2007/06/03/business/yourmoney/03stream.html?pagewanted=all (accessed 3 October 2012).

11 Lange, B.M., Rujan, T., Martin, W., and Croteau, R. (2000) Isoprenoid biosynthesis: the evolution of two ancient and distinct pathways across genomes. *Proc. Natl. Acad. Sci. U.S.A.*, **97**, 13172–13177.

12 Werpy, T., and Petersen, G. (2004) *Top Value Added Chemicals From Biomass. Volume I: Results of Screening for Potential Candidates from Sugars and Synthesis Gas.* U.S. Department of Energy, Office of Scientific and Technical Information: Oak Ridge, TN, USA.

13 Sell, C.S. (2006) *Terpenoids. Kirk-Othmer Encyclopedia of Chemical Technology [Online]*, Wiley & Sons, Posted September 15, 2006; 143 pp, http://onlinelibrary.wiley.com/doi/10.1002/ 0471238961.2005181602120504.a01.pub2/full (accessed 1 October 2012).

14 Summit Chemical PTE, Ltd. (Wuzhou, China) (2012) Statistics Data for Pine Chemicals Exports, http://www.pinechem.net/files/h_stat.htm (accessed 1 October 2012).

15 Sell, C.S. (ed.) (2006) *The Chemistry of Fragrances. From Perfumer to Consumer*, RSC Publishing, Cambridge, UK, pp. 81–88.

16 (a) Netscher, T. (2007) Synthesis of Vitamin E, in *Vitamins and Hormones* vol. 76 (ed. Litwack, G.), Elsevier Inc., Amsterdam, The Netherlands, pp. 156–202; (b) Baldenius, K.-U., von dem Bussche-Hünnefeld, L., Hilgemann, E., Hoppe, P., and Stürmer, R. (2011) Vitamins 4. Vitamin E (Tocopherols, Tocotrienols). *Ullman's Encyclopedia of Industrial Chemistry [Online]*, Wiley-VCH, Posted October 15, 2011, 11 pp, http://onlinelibrary.wiley.com/doi/10.1002/14356007.o27_o07/abstract (accessed 2 October 2012); (c) Weber, F., and Rüttimann, A. (2011) Vitamins, 5. Vitamin K. *Ullman's Encyclopedia of Industrial Chemistry [Online]*, Wiley-VCH, Posted October 15, 2011, 19 pp, http://onlinelibrary.wiley.com/doi/10.1002/14356007.o27_o08/abstract (accessed 3 October 2012).

17 Pickett, J.A., and Griffiths, D.C. (1980) Composition of aphid alarm pheromones. *J. Chem. Ecol.*, **6**, 349–360.

18 Kunert, G., Otto, S., Rose, U.S.R., Gershenzon, J., and Weisser, W.W. (2005) Alarm pheromone mediates production of winged dispersal morphs in aphids. *Ecol. Lett.*, **8**, 596–603.

19 Beale, M.H., Birkett, M.A., Bruce, T.J., Chamberlain, K., Field, L.M., Huttly, A.K., Martin, J.L., Parker, R., Phillips, A.L., Pickett, J.A., Prosser, I.M., Shewry, P.R., Smart, L.E., Wadhams, L.J., Woodcock, C.M., and Zhang, Y. (2006) Aphid alarm pheromone produced by transgenic plants affects aphid and parasitoid behavior. *Proc. Natl. Acad. Sci. U.S.A.*, **103**, 10509–10513.

20 Kunert, G., Reinhold, C., and Gershenzon, J. (2010) Constitutive emission of the aphid alarm pheromone, (E)-β-farnesene, from plants does not serve as a direct defense against aphids. *BMC Ecol. [Online]*, **10**, 23, http://www.biomedcentral.com/1472-6785/10/23 (accessed 5 October 2012).

21 Wohlers, P. (1981) Aphid avoidance of plants contaminated with alarm pheromone (E)-β-farnesene. *J. Appl. Entomol.*, **92**, 329–336.

22 Stökl, J., Brodmann, J., Dafni, A., Ayasse, M., and Hansson, B.S. (2010) "Smells like Aphids: Orchid Flowers Mimic Aphid Alarm Pheromones to Attract Hoverflies for Pollination." *Proc. R. Soc. B [Online]*, Posted October 13, 2010, http://rspb.royalsocietypublishing.org/content/early/2010/10/06/rspb.2010.1770.full (accessed 5 October 2012).

23 Sun, Y., Qiao, H., Ling, Y., Yang, S., Rui, C., Pelosi, P., and Yang, X. (2011) New analogues of (E)-β-Farnesene with insecticidal activity and binding affinity to aphid odorant-binding proteins. *J. Agric. Food Chem.*, **59**, 2456–2461.

24 The Good Scents Company (1980) Data: (E)-beta-farnesene, http://www.thegoodscentscompany.com/data/rw1053891.html (accessed 2 October 2012).

25 (a) Akutagawa, S., Taketomi, T., Kumobayashi, H., Takayama, K., Someya, T., and Otsuka, S. (1978) Metal-assisted terpenoid synthesis. V. The catalytic trimerization of isoprene to trans-β-farnesene and its synthetic applications for terpenoids. *Bull. Chem. Soc. Jpn.*, **51**, 1158–1162; (b) Morel, D. (1982) Selective addition of a compound having an activated carbon atom to a substituted conjugated diene. Eur. Pat. Appl. EP 44771.

26 (a) Arkoudis, E., and Stratakis, M. (2008) Synthesis of cordiaquinones B, C, J, and K on the basis of a bioinspired approach and the revision of the relative stereochemistry of cordiaquinone C. *J. Org. Chem.*, **73**, 4484–4490; (b) Dawson, G.W., Griffiths, D.C., Pickett, J.A., Smith, M.C., and Woodcock, C.M. (1982) Improved preparation of (E)-β-farnesene and its activity with economically important aphids. *J. Chem. Ecol.*, **8**, 1111–1117.

27 (a) Brieger, G. (1967) Convenient preparation of trans-β-farnesene. *J. Org. Chem.*, **32**, 3720; (b) Sun, D., Zhong, J., and Chen, J. (1996) Synthesis of (E)-β-farnesene from farnesol via treatment of farnesyl acetate with tetrakis(triphenylphosphine)palladium. *Huaxue Tongbao*, 37–38; (c) Torihara, M., and Tamai, H. (1988) Manufacture of farnesene compounds. Jpn. Kokai Tokkyo Koho JP 63290834 A.

28 (a) Cazes, B., Guittet, E., Julia, S., and Ruel, O. (1979) On a sulfurated Grignard reagent equivalent to that of 2-(chloromethyl)-1,3-butadiene. New syntheses of ipsenol (2-methyl-6-methylene-7-octen-4-ol) and of (E)-β-farnesene ((E)-7,11-dimethyl-3-methylene-1,6,10-dodecatriene. *J. Organomet. Chem.*, **177**, 67–74; (b) Oprean, I., Ciupe, H., and Taitan, L. (1982) Trans-β-Farnesene. Rom. Pat. RO 77594 A2; (c) Vig, O.P., Vig, A.K., and Kumar, S.D. (1975) New synthesis of β-myrcene and β-farnesene. *Indian J. Chem.*, **13**, 1244–1246.

29 Pan, Y., Xu, Z., Chen, J., and Sun, D. (1993) High order cuprate species(II), stereospecific carbocupration of terminal alkynes and its synthetic utility for (±)-ipsenol and (E)-β-farnesene. *Chem. Res. Chin. Univ.*, **9**, 104–109.

30 (a) Baeckstroem, P., Li, L., Wickramaratne, M., and Norin, T. (1990) A synthesis of trans-β-farnesene from myrcene which includes a modified work up method for DIBAH reductions of esters to aldehydes. *Synth. Commun.*, **20**, 423–429; (b) Mimura, T., Kimura, Y., and Nakai, T. (1979) Dithiocarbamates in organic synthesis. VIII. Synthetic utility of allylic thiolcarbamates. Stereoselective synthesis of simple natural products including optically active manicone and β-sinensal.. *Chem. Lett.*, 1361–1364.

31 Kuraray Co. Ltd. (2007) Products Flow Chart, http://siriuslink.com/client/kuraray/ fc/product_flowchart.pdf (accessed 5 October 2012).

32 Ploner, K.J., Wild, J., and Sigg-Gruetter, T. (1974) *trans*-1,3-Dienes for perfumes. Ger. Offen. DE 2422244.

33 Surburg, H., and Panten, J. (2006) *Common Fragrance and Flavor Materials*, 5th edn, Wiley-VCH Verlag GmbH, Weinheim, Germany, p. 27.

34 Dawson, G.W., Griffiths, D.C., and Pickett, J.A. (1982) Derivatives of (E)-β-farnesene useful in insect control. PCT Int. Appl. WO 8204249 A1.

35 Kisa, F., Morikawa, T., and Fujii, K. (2005) Preparation of high boiling point acrylate-monoterpene adducts by Diels-Alder reaction of hydroxylated (meth)acrylates with monoterpenes and hydrogenation. Jpn. Kokai Tokkyo Koho JP 2005298468 A.

36 Takada, S., and Kisa, F. (2007) Terpene dimethylol compounds as materials for polymers with improved water resistance and electric insulation. Jpn. Kokai Tokkyo Koho JP 2007332123 A.

37 Ishihara, K., Nakano, K., and Akakura, M. (2008) Organocatalytic enantioselective diels-alder reaction of

dienes with α-(N,N-diacylamino) acroleins. *Org. Lett.*, **10**, 2893–2896.

38 Veselovskii, V.V., Dragan, V.A., and Moiseenkov, A.M. (1987) Regiospecific ene-type reaction of benzenesulfinyl chloride with linear isoprenoids. *Izvest. Akad. Nauk SSSR, Ser. Khim.*, 2787–2790.

39 Matsui, M., and Yamamoto, T. (1996) Preparation of α-tocopherol and its derivatives. Jpn. Kokai Tokkyo Koho JP 08151376 A.

40 Mulhauser, M. (1986) Terpenoid allylic tertiary esters. Fr. Demande FR 2570374 A1.

41 Brimble, M.A., Rowan, D.D., and Spicer, J.A. (1996) Synthesis of chiral hydroxylated farnesene derivatives. *Synthesis*, 116–122.

42 Hamabura, K., Urawa, Y., Narabe, Y., Hisatake, Y., and Kijima, S. (1992) Condensation of terpenes with iminobutanoates. Eur. Pat. Appl. EP 503634 A1.

43 Courtney, J.L., and McDonald, S. (1969) Photochemistry of α- and β-farnesenes. *Aust. J. Chem.*, **22**, 2411–2416.

44 Price, K.N., Reilman, R.T., Scheibel, J.J., Shumate, R.E., and Urbin, S.A. (2011) Branched acyclic alcohol surfactants and use in cleaning compositions. US Pat. Appl. US 20110034363.

45 Newmark, R.A., and Majumdar, R.N. (1988) Carbon-13 NMR spectra of *cis*-polymyrcene and *cis*-poly-farnesene. *J. Polym. Sci. A*, **26**, 71–77.

46 Lybarger, H.M. (2000) Isoprene. *Kirk-Othmer Encyclopedia of Chemical Technology [Online]*, Wiley & Sons, Posted December 4, 2000, 18 pp, http://onlinelibrary.wiley.com/doi/10.1002/0471238961.0919151612250201.a01/full (accessed 3 October 2012).

47 Senyek, M.L. (2008) Isoprene Polymers. *Encyclopedia of Polymer Science and Technology [Online]*, Wiley & Sons, Posted March 14, 2008, 80 pp, http://onlinelibrary.wiley.com/doi/ 10.1002/0471440264.pst175/full (accessed 5 October 2012).

48 For a discussion review of this reaction see: Quirk, R.P. (2002) Anionic Polymerization. *Encyclopedia of Polymer Science and Technology [Online]*, Wiley & Sons, Posted March 15, 2002, 54 pp, http://onlinelibrary.wiley.com/doi/10.1002/0471440264.pst019/full (accessed 5 October 2012).

49 McPhee, D.J., and Graham, M.J. (2010) Adhesive compositions comprising polyfarnesene. US 7655739 B1.

50 McPhee, D.J. (2010) Farnesene interpolymers and their manufacture. US Patent Appl. US 20100056714, A1.

51 Walling, C., and Peisach, J. (1958) Organic reactions under high pressure. IV. The dimerization of isoprene. *J. Am. Chem. Soc.*, **80**, 5819–5824.

52 Thomas, A.F., and Bessiere, Y. (1989) Limonene. *Nat. Prod. Rep.*, 291–309.

53 Fisher, K., and Woolard, F.X. (2009) Preparation of farnesene dimers and/or farnesane dimers and lubricant compositions thereof. US 7592295 B1.

54 Fisher, K., and Woolard, F.X. (2010) Lubricant compositions based on farnesane dimers. US 7691792 B1.

55 http://www.novvi.com.br (accessed 2 October 2012).

56 http://www.amyris.com/en/markets/chemicals/renewable-synthetic-lubricants/evoshield (accessed 2 October 2012).

4
Furan-Based Building Blocks from Carbohydrates

Robert-Jan van Putten, Ana Sousa Dias, and Ed de Jong

4.1
Importance of Furans as Building Blocks

The world is more and more confronted with the reduction of fossil oil reserves, strong fluctuations of fossil fuel prices and the increase of CO_2 emissions with the ensuing problem of the greenhouse gas (GHG) effect. These environmental, social and economic challenges have created the need for sustainable alternatives to fossil fuels and chemicals [1, 2]. The use of plant biomass as starting material is one of the alternatives for reducing the dependence on fossil oil for transportation fuels and is the main alternative for replacing petrochemicals. The biomass can be transformed into energy, transportation fuels, various chemical compounds and materials such as natural fibers by applying biochemical, chemical, physical and thermal processes [1, 3–7]. However, the potential competition with food and feed applications and the consequent rise in prices is an important aspect to take into consideration. The fermentation and the chemical conversion of carbohydrates into value-added compounds has received increasing interest in the last decade, and in a biorefinery different advantages may be taken from both processes [2, 4, 5, 7–11]. Some of the most important chemical transformations of carbohydrates are arguably the hydrolysis and subsequent dehydration of polysaccharides into the furan platform products, furfural and 5-hydroxymethylfurfural [12, 13]. Furfural has a wide industrial application profile and is considered as one of the top 30 building blocks that can be produced from biomass [12, 14–17]. 5-Hydroxymethylfurfural (HMF) is promising as a versatile, renewable furan chemical for the production of chemicals, polymers and biofuels, similar to furfural [13, 15]. While furfural has been produced on an industrial scale for decades [12], the production of HMF has not yet reached industrial scale [13, 15]. In this chapter we will discuss occurrence and composition of carbohydrates in terrestrial and aquatic biomass and the subsequent chemistry involved to convert the monomeric carbohydrates into furfural and HMF.

4.2
Sources of Carbohydrates

Carbohydrates are among the most abundant organic compounds on Earth and represent the major portion of the world's annual production of renewable biomass. They are by far the most omnipresent component of biomass and are, therefore, often the preferred feedstock for the biobased economy. In fermentative processes there is sometimes more room for feedstock flexibility (proteins, triglycerides/fatty acids) but catalytic conversions, such as the transformation of biomass into furan molecules, are restricted to carbohydrates. Sources of carbohydrates include conventional forestry, wood processing by-products (e.g., wood chips, pulp and paper industrial residue), agricultural crops and surpluses (e.g., corn stover, wheat and rice straw), and so-called energy crops (e.g., switchgrass, Miscanthus, willow) grown on degraded soils, and aquatic biomass (algae, seaweeds). Typical carbohydrate compositions are shown in Table 4.1. The majority of terrestrial biomass consists of carbohydrates (60–80%), the other main component is lignin (20–25%), proteins are mainly found in fresh (i.e. green) plant material. Amounts of triglycerides, extractives and inorganic materials are very much dependent on species and harvest time. The bulk of the carbohydrates present in biomass is composed of poly/oligosaccharides, such as hemicelluloses, cellulose, starch, and inulin. Sucrose is an omnipresent disaccharide consisting of a glucose and fructose moiety, whereas monosaccharides such as glucose and fructose are present in far lesser amounts. In particular, lignocellulosic plant matter is available in large quantities and is relatively cheap, while aquatic biomass is given great potential for the future. In the following sections the three major classes of carbohyrates are discussed, that is, the storage carbohydrates, the structural carbohydrates and the carbohydrates of aquatic origin.

4.2.1
Storage Carbohydrates

Often the biological energy storage systems are based on carbohydrates like sucrose, starch and inulin, which will be discussed here.

4.2.1.1 Sucrose
Sucrose (table sugar, saccharose, α-D-glucopyranosyl-(1→2)-β-D-fructofuranoside) is a dimer of glucose and fructose, linked by a glycosidic bond between their anomeric positions. This also makes it a non-reducing sugar, as it lacks anomeric hydroxyl groups. It is found naturally in many food crops, sometimes together with the monosaccharides fructose and to a lesser extent glucose. In many fruits, such as pineapple and apricot, sucrose is the main saccharide. In others, such as grapes, apples and pears, fructose is the main saccharide. The two dominating sugar crops in the world are sugarcane (*Saccharum spp.*) and sugar beets (*Beta vulgaris*). Sugarcane makes up around 80% of the total sugar production with beet accounting for most of the remainder. Minor commercial sugar crops include

4.2 Sources of Carbohydrates | 83

Table 4.1 Carbohydrate composition of the main biomass types.

Origin	Species	Carbohydrate content (%d.m.)[c]	C6-Sugars[a]							C5-Sugars[a]			Oth[a]	Ref
			Glu	Fru	Man	Gal	Rha	Fuc	UrA	Xyl	Ara	Rib		
Hardwoods (average)	Mixed (stem)	38–50% cellulose 67–75% carb	43		0.4	0.9	0.5	0.1	0.2	16	1.3			[18, 19]
Softwoods (average)	Mixed (stem)	40–50% cellulose 67–75% carb	44		4.9	7.8	0.4	0.3		8.9	5.9			[18, 19]
Grasses	Sugarcane (bagasse + sugar)	18% sucrose	41	9	0.5	1.6				23	2			[20]
Tubers	Chicory	68% inulin, 14% sucrose, 5% cellulose	13	74										[21, 22]
Agricultural residues	Corn cobs	75	39						N.D.[c]	30	3.3		4[b]	[23]
	Wheat straw	57	32						N.D.[c]	20	2.8		2.6[b]	[23]
	Rice husks	49	30						N.D.[c]	17	2		1.1[b]	[23]
Green algae	Spirulina	13.6	7.3		1.2	0.4	3			1			0.5[e]	[24]
	Chlorophyceae	6–16	4.8		1.3	1.6	0.9	1.2		0.5	0.2	0.3		[25]
	Prasinophyceae	14–17	13		1	0.7	0.3	0.2		0.4	0.1	0.2		[25]
	Ulva	39	1.7				11[d]		24[d]	2.1				[26]

a) Glu = glucose; Fru = fructose; Man = mannose; Gal = galactose; Rha = rhamnose; Fuc = fucose; UrA = uronic acids; Xyl = xylose; Ara = arabinose; Rib = ribose; Oth = others.
b) Acetyl groups.
c) N.D. = not determined; d.m. = dry mass.
d) 21.7% Glucuronic acid and iduronic acid; rhamnose present as O-3-sulfate rhamnose.
e) 2-O-Methyl-L-rhamnose and 3-O-methyl-L-rhamnose.

sweet sorghum (*Sorghum vulgare*), date palm (*Phoenix dactylifera*) and sugar maple (*Acer saccharum*). Currently, worldwide production of sugar is around 170 million tonnes a year [27].

4.2.1.2 Starch

Pure starch is a white, tasteless and odorless powder that is insoluble in cold water or alcohol. It consists of two types of molecules: the linear and helical amylose consisting of glucose linked with α-(1→4)glycosidic bonds and the branched amylopectin which also contains glucose linked by α-(1→6) glycosidic bonds (Table 4.2). Depending on the plant, starch generally contains 20 to 25% amylose and 75 to 80% amylopectin by weight. The global production of starch-producing crops in 2008 was around 3 billion tons, representing around 1.7 billion tons of starch. However, in 2003 only around 55 million tons of starch was produced globally [28]. The main starch-producing crops are corn, wheat and rice, together accounting for more than 80% of the production. Other important starch-producing crops are cassava, barley and potato. Starch crops are grown globally but the specific crops are very much climate dependent.

4.2.1.3 Inulin

Inulin, a non-digestible carbohydrate, is a fructan commonly found in many plants as a storage carbohydrate [21]. It is present in many regularly consumed vegetables, fruits and cereals, including leek, onion, garlic, wheat, chicory, artichoke, and banana. Industrially, inulin is predominantly obtained from chicory roots (*Cichorium intybus*), and is currently used as a functional food ingredient that offers a unique combination of interesting nutritional properties and important technological benefits. Inulin has been defined as a polydisperse carbohydrate material consisting mainly, if not exclusively, of β-(2→1) fructosyl-fructose links. A starting glucose moiety can be present, but is not necessary. When referring to the definition of inulin, both GF_n and F_m compounds are considered to be included under this same nomenclature, where n or m represent the number of fructose units (F) linked with one terminal glucose (G). The degree of polymerization (DP) of plant inulin is rather low (maximally 200) and varies according to the plant species, weather conditions and the physiological age of the plant. In chicory inulin, the number of fructose units linked to a terminal glucose (n) can vary from two to 70 [21]. This also means that inulin is a mixture of oligomers and polymers. The DP of inulin, as well as the presence of branches, are important properties since they strongly influence the functionality of most inulins. Native (non-purified) inulin always contains glucose, fructose, sucrose, and small oligosaccharides [21, 22].

4.2.2
Structural Carbohydrates

Cellulose and hemicellulose can be found in the cell wall of all terrestrial plant cells. In terrestrial biomass the combined cellulose and hemicellulose fraction

Table 4.2 Main types of di-, oligo- and polysaccharides present in biomass.

Saccharide type	Biological origin	Abbreviation	Amount[a]	Units Backbone[c]	Side chains	Linkage	DP[b]
Sucrose	Sugar cane, sugar beet	GF	12–20	α-D-Glcp-β-D-Frcf		α-(1→2)-β	2
Starch (amylose)	Corn, wheat, cassava, potato, rice		20–30%	α-D-Glcp		α-(1→4)	300–3000
Starch (amylopectin)	Corn, wheat, cassava, potato, rice		70–80%	α-D-Glcp	α-D-Glcp	α-(1→4) α-(1→6)	2000–200 000
Inulin	Chicory, Agave, Jerusalem Artichoke, Dahlia	GF(F)$_n$	70%	α-D-Glcp-β-D-Frcf(n)		β-(2→1)	<200
Cellulose	All terrestrial plants		40–50%	β-D-Glcp		β-(1→4)	100–>10 000
Arabinogalactan	Softwoods (Larch*)	AG	1–3; 35[d]	β-D-Galp	β-D-Galp α-L-Araf β-L-Arap	β-(1→6) α-(1→3) β-(1→3)	100–600
Xyloglucan	Hardwoods, softwood, grasses	XG	2–25	β-D-Glcp β-D-Xylp	β-D-Xylp β-D-Galp α-L-Araf α-L-Fucp Acetyl	β-(1→4) α-(1→3) β-(1→2) α-(1→2) α-(1→2)	
Galactoglucomannan	Softwoods	GGM	10–25	β-D-Manp β-D-Glcp	β-D-Galp Acetyl	α-(1→6)	40–100
Glucomannan	Hardwoods and softwoods	GM	2–5	β-D-Manp β-D-Glcp		β-(1→4)	40–70

(*Continued*)

Table 4.2 (Continued)

Saccharide type	Biological origin	Abbreviation	Amount[a]	Backbone[c]	Side chains	Linkage	DP[b]
Glucuronoxylan	Hardwoods	GX	15–30	β-D-Xylp	4-O-Me-α-D-GlcpA Acetyl	α-(1→2)	100–200
Arabinoglucuronoxylan	Grasses and cereals, softwoods	AGX	5–10	β-D-Xylp	4-O-Me-α-D-GlcpAβ-L-Araf α-L-Araf-Feruloy	α-(1→2) α-(1→3) α-(1→2) α-(1→3)	50–185
Arabinoxylans	Cereals	AX	0.15–30	β-D-Xylp	α-L-Araf	α-(1→2) α-(1→3)	
Glucuronoarabino-xylans	Grasses and cereals	GAX	15–30	β-D-Xylp	4-O-Me-α-D-GlcpA Acetyl		
Algal polysaccharides							
Homoxylans	Algae, grasses	X		β-D-Xylp		β-(1→3); β-(1→4)	
Laminarin	Brown algae			B-D-Glup		α-(1→4)	
Ulvan	Ulvaceae		40	β-D-GlcAp- α-L-Rhap 3S		β-(1→4)	

a) %, dry biomass;
b) Degree of polymerization;
c) Monosaccharide present in the pyranose (p) or furanose (f) form
d) (up to) in the heartwood of larches.

almost always represents more than 50% of the total biomass, based on dry weight. Cellulose is a linear polymer composed of β-D-glucopyranose (glucose) units forming microfibrils that give strength and resistance to the cell wall. The hemicellulose consists of a wide variety of polysaccharides (composed of pentoses, hexoses, hexuronic acids) which are interspersed with the microfibrils of cellulose, conferring consistency and flexibility to the structure of the cell wall [8].

4.2.2.1 Cellulose

Cellulose is the basic structural component of plant cell walls and comprises about a third of all vegetable materials. Cellulose is a complex polysaccharide, consisting of 3000 or more β-(1→4) linked D-glucose units (Table 4.2). It is present in wood in quantities between 40 and 50% on a dry matter basis (Table 4.1). It is the most abundant of all naturally occurring organic compounds, comprising over 50% of all the carbon in vegetation. Cellulose is a straight-chain polymer where no coiling or branching occurs, and the molecule adopts an extended and rather stiff rod-like conformation. The chains can stack together to form larger microfibrils which make cellulose highly insoluble in water. Cellulose microfibrils may also associate with water and matrix polysaccharides, such as the (1→3, 1→4)-β-D-glucans, heteroxylans (arabinoxylans) and glucomannans [18, 29].

4.2.2.2 Hemicelluloses

Hemicelluloses are the world's second most abundant renewable polymers, after cellulose, in lignocellulosic materials. Hemicelluloses are a heterogeneous class of polymers representing, in general, 15–35% of plant biomass and may contain pentoses (β-D-xylose, α-L-arabinose), hexoses (β-D-mannose, β-D-glucose, α-D-galactose) and/or uronic acids (α-D-glucuronic, α-D-4-O-methylgalacturonic and α-D-galacturonic acids). Other sugars such as α-L-rhamnose and α-L-fucose may also be present in small amounts and the hydroxyl groups of sugars can be partially substituted with acetyl groups [20, 30, 31]. Composition and amounts depend strongly on plant source, plant tissue and geographical location. Hemicelluloses are usually bonded to other cell-wall components, such as cellulose, cell-wall proteins, lignin, and phenolic compounds by covalent and hydrogen bonds, and by ionic and hydrophobic interactions. The most relevant hemicelluloses are the xylans and the glucomannans, with xylans being the most abundant. Xylans are the main hemicellulose components of secondary cell walls constituting about 20–30% of the biomass of hardwoods (angiosperms) and herbaceous plants. In some tissues of grasses and cereals xylans can account for up to 50% [30]. Xylans are usually available in large amounts as by-products of forest, agriculture, agro-industries, wood and pulp and paper industries. Mannan-type hemicelluloses such as glucomannans and galactoglucomannans are the major hemicellulosic components of the secondary wall of softwoods (gymnosperms) whereas in hardwoods they occur in minor mounts. Depending on their biological origin, different hemicellulose structures can be found (Table 4.2). Upon hydrolysis, the hemicelluloses are converted into the corresponding monosaccharides (Table 4.1). The major hemicelluloses are discussed below.

Glucuronoxylans (GX) Hemicelluloses in various hardwood species differ from each other both quantitatively and qualitatively. The main hemicelluloses of hardwood are glucuronoxylans (O-acetyl-4-O-methylglucurono-β-(1,4)-D-xylan; GX), which can also contain small amounts of glucomannans (GM). In hardwoods, GX represent 15–30% of their dry mass and consist of a linear backbone of β-(1,4)-D-xylopyranosyl units. Some xylose units are acetylated at C2 and C3 and one in ten molecules has a uronic acid group (4-O-methylglucuronic acid) attached by α-(1,2) linkages (Table 4.2). The percentage of acetyl groups ranges between 8% and 17% of total xylan (3.5–7 acetyl residues per 10 xylose units). The xylosidic bonds between the xylose units are easily hydrolyzed by acids, but the linkages between the uronic acid groups and xylose are very resistant. Acetyl groups are easily cleaved by alkali, and the acetate formed during Kraft (alkaline) pulping of wood mainly originates from these groups. Besides these main structural units, GX may also contain small amounts of L-rhamnose and galacturonic acid. The latter increases the polymer resistance to alkaline agents. The average DP of GX is in the range of 100–200 [20, 31].

Glucomannan (GM) In addition to xylan, hardwoods contain 2–5% of a glucomannan, which is composed of β-glucopyranose and β-mannopyranose units linked by β-(1→4)-bonds (Table 4.2). However, the mannose/glucose monomer ratio may vary depending on the original source of GM. The ratio of glucose to mannose varies between 1:2 and 1:1. Galactose is not present in hardwood mannan. The mannosic bonds between the mannose units are more rapidly hydrolyzed by acid than the corresponding glycosidic bonds, and GM is easily depolymerized under acidic conditions. There may be certain short side branches at the C3 position of the mannoses and acetyl groups randomly present at the C6 position of a sugar unit. The acetyl groups frequently range from 1 per 9 to 1 per 20 sugar units [31].

Xyloglucans (XG) Besides xylan and glucomannan, xyloglucans are also present in the primary cell walls of some higher plants (mainly in hardwoods, and less in softwoods) [31]. They can also appear in small amounts (2–5%) in grasses. Xyloglucans consist of β-1,4-linked D-glucose (cellulosic) backbone with 75% of these residues substituted at O-6 with D-xylose. L-Arabinose and D-galactose residues can be attached to the xylose residues forming di-, or triglycosyl side chains. Also, L-fucose has been detected attached to galactose residues. In addition, xyloglucans can contain O-linked acetyl groups. Xyloglucans interact with cellulose microfibrils by the formation of hydrogen bonds, thus contributing to the structural integrity of the cellulose network [20].

Galactoglucomannans (GGM) The major hemicelluloses in softwoods (gymnosperms) are acetylated galactoglucomannans (GGM), accounting for up to 20–25% of their dry mass [20]. GGM consist of a linear backbone of β-D-glucopyranosyl and β-D-mannopyranosyl units, linked by β-(1,4) glycosidic bonds, partially acetylated at C2 or C3 and substituted by α-D-galactopyranosyl units attached to glucose and mannose by α-(1,6) bonds. GGM contain around 6% acetyl groups, corresponding to 1 acetyl group per 3–4 hexose units on average [20] (Table 4.2). Some GGM are water soluble, presenting higher galactose content than the insoluble GGMs. There are two main types of acetylgalactoglucomannans in softwoods, one being galactose-poor (5–8% of dry wood) and the other galactose-rich (10–15% of dry wood). The ratios of galactose:glucose:mannose are approximately 0.1:1:3 and 1:1:3 for the two woods, respectively [31]. GGM have an approximate DP between 100 and 150, which is equivalent to a molecular weight (Mw) around 16 000–24 000 Da. GGMs are easily depolymerized by acids, especially the bonds between galactose and the main chain. The acetyl groups are much more easily cleaved by alkali and acid [31]. Glucomannans (GM) occur in minor amounts in the secondary wall of hardwoods (<5% of the dry wood mass) [20]. Like GGM, they have a linear backbone of β-D-glucopyranosyl (Glc$p$) and β-D-mannopyranosyl (Man$p$) units but the ratio Glc$p$:Man$p$ is lower. In GGM and GM the extent of galactosylation governs the tendency of their association to the cellulose microfibrils and, hence, their extractability from the cell wall matrix [30].

Arabinoglucuronoxylans (AGX) Arabinoglucuronoxylans (arabino-4-O-methylglucuronoxylans) are the major components of non-woody materials (e.g., agricultural crops) and a minor component of softwoods (5–10% of dry mass). They consist of a linear β-(1,4)-D-xylopyranose backbone containing 4-O-methyl-D-glucuronic acid (MeGlcA) and α-L-arabinofuranosyl linked by α-(1,2) and α-(1,3) glycosidic bonds (Table 4.2) [20]. The xylopyranose backbone might be slightly acetylated [31]. The typical ratio arabinose:glucuronic acid:xylose is 1:2:8. Conversely to hardwood xylan, AGX can be less acetylated, but may contain low amounts of galacturonic acid and rhamnose. The average DP of AGX ranges between 50 and 185 [20]. In addition, because of their furanosidic structure, the arabinose side chains are easily hydrolyzed by acids [31].

Arabinogalactan (AG) The heartwood of larches contains exceptionally large amounts of water-soluble AG, which is only a minor constituent in other softwood species [31]. Its concentration and quality are not affected by seasonal variability. AGs are highly branched polysaccharides with molecular weights ranging from 10 000 to 120 000 Da. All Larch AG isolated from the Larix sp. is of the β-(3,6)-D-galactan type and consist of galactose and arabinose in a 6 to 1 ratio. Larch AG has a galactan backbone that features β-(1→3) linkages and galactose β-(1→6) and arabinose β-(1→6 and 1→3) side chains [31] (Table 4.2). The highly branched structure is responsible for the low viscosity and high solubility in water of this polysaccharide [31]. It has the ability to bind fat, retain liquid, and has dispersing properties. AG also possesses a high biological activity. Larch AG is currently used in a variety of food, beverage, nutraceutical, and medicine applications [31].

Arabinoxylan (AX) Arabinoxylans are the main hemicelluloses of the grasses (Gramineae). AXs have been generally present in a variety of tissues of the main cereals: wheat, rye, barley, oats, rice, corn, and sorghum, as well as other plants [31]. AXs are generally present in the starchy endosperm (flour) and outer layers (bran) of cereal grain. They are similar to hardwood xylan, but the amount of L-arabinose is higher. In AX, the linear β-(1→4)-D-Xyl$p$ backbone is substituted by α-L-Ara$f$ units in the positions 2-O and/or 3-O (Table 4.2). In addition, the AXs are also substituted by α-D-glucopyranosyl uronic unit or its 4-O-methyl derivative in the position 2-O, as can be found in wheat straw, bagasse and bamboo. O-acetyl substituents may also occur [31]. According to the amount of glucuronic acid and arabinose, the types of AX are classified as arabinoglucuronoxylan (AGX) and glucuronoarabinoxylan (GAX), respectively [30]. AGX are the dominant hemicelluloses in the cell walls of grasses and cereals, such as sisal, corncobs and straw. Compared to AGX, the GAX have an arabinoxylan backbone, which contains about ten times fewer uronic acid side chains than arabinose, and also contains xylan which is double-substituted by uronic acid and arabinose units. Ferulic acid and p-coumaric acid can occur esterified to the C-5 of arabinosyl units of GAXs [31]. The physical and/or covalent interaction with other cell wall constituents restricts xylan extractability [20].

β-(1→3, 1→4)-glucans (1314G) β-(1→3, 1→4)-Glucans consist of a linear chain of β-D-glucopyranosyl units linked by (1→3) and (1→4) bonds (Table 4.1). 1314G are present in Poaceae (grasses and cereals) as well as in Equisetum, liveworts and Charopytes. The mixed linkage glucans are dominated by cellotriosyl and cellotetrasyl units linked by β-(1→3) linkages, but longer β-(1→4)-linked segments also occur. Cellulose is also β-D-glucan, which is linked by (1→4)-glycosidic bonds, and thus cellulose has high stiffness (crystallinity) and is insoluble in most solvents. Contrary to cellulose, the β-(1→3) linkages existing in 1314G make glucans flexible and soluble [31].

Complex heteroxylans (CHXs) Structurally more complex heteroxylans are present in cereals, seeds, gum exudates and mucilages. In this case the β-(1,4)-D-xylopyranose backbone is decorated with single uronic acid and arabinosyl residues, and also various mono- and oligoglycosyl side chains [20].

4.2.3
Aquatic Carbohydrates

For aquatic biomass a completely different picture exists. The total content of carbohydrates on dry mass basis can be low (as low as 20%) with a much more diverse composition and a larger presence of further oxidized (uronic acids) as well as reduced (mannitol) carbohydrates. Homopolymers of xylose, so-called homoxylans only occur in seaweeds (red and green algae). Algae are the base of the aquatic food chain. They consist of a very extended and diverse group of organisms. Most of them have never been characterized for their chemical composition [32]. Their main components differ from those of terrestrial biomass (e.g., cellulose, hemi-cellulose and lignin). In this overview the macroalgae and microalgae are discussed and typical examples are given from representatives from the brown, red and green algal families in Table 4.2.

4.2.3.1 Macroalgae

Among macroalgae, the *Laminaria spp* and *Ulva spp* are the most important prospects from a biobased economy perspective. Currently, the vast majority of seaweed is collected for human consumption and for hydrocolloid production. Seaweed exploitation is still in its infancy. The majority of Asian seaweed resources are cultivated, in most other countries around the world natural stocks are harvested. Large-scale cultivation projects of up to 41 km^2 have been foreseen in the USA and Japan, but cost and engineering barriers have not been overcome. Macroalgae have about 80–85% moisture content and are therefore costly to transport. The presence of salt, polyphenols and sulfated polysaccharides also needs to be carefully managed in order to avoid negative effects on the subsequent conversion processes [32].

4.2.3.2 Brown Macroalgae

The main structural components of brown macroalgae (e.g., *Ascophyllum nodosum*, *Sargassum*, *Lamarinales*) are alginic acid, mannitol, laminarin and fucoidan. Alginic acid is the major polysaccharide in brown algae. It is a polymer of 6-carbon sugar acids, D-mannuronic and L-guluronic acid. The average length of these blocks is about 20 units with a different mixture of the two acids. The proportion of alginic acid in algae can be between 10 and 40% of the algae dry weight. The main energy (carbon) storage compounds present in brown algae are laminarin (polymer) and mannitol (monomer). Laminarin is composed of β-glucan with the main component being β-(1-3)-linked glucans, containing large amounts of sugars and a low fraction of uronic acids. The proportion of laminarin in brown algae ranges between 2 and 34% of the algal dry weight. Mannitol (sugar alcohol derived from mannose) can be found in a range of 5–25% of dry weight. Fucans or fucoidan are the sulfated polysaccharides present in brown algae. They are composed mainly of L-fucose (a six carbon sugar) with small proportions of other sugars, such as mannose, galactose, xylose and glucuronic acid. Fucan has an extremely complex structure and its presence in the cell wall of brown algae protects them from desiccation [33]. The proportion of fucans is 5–20% of alga dry weight [33].

4.2.3.3 Microalgae

There are at least 30 000 known species of microalgae. Only a handful is currently of commercial significance. These are generally cultivated for extraction of high-value components such as pigments or proteins. A few species are used for feeding shellfish or other aquaculture purposes. Key research tasks for commercialization of algae for biochemicals and energy purposes are to screen species for favorable composition and for ease of cultivation and processing, among other criteria. The main focus of screening is currently on lipid productivity, and subsequent esterification, but carbohydrate conversion options should not be ignored.

4.2.3.4 Green Algae

The green algal *Spirulina plutensis* contains 13.6% carbohydrate, the sugar composition of which is comprised principally of glucose along with rhamnose, mannose, xylose, galactose and two unusual sugars. The latter were identified by a combination of GC-MS, NMR and de-O-methylation as 2-*O*-methyl-L-rhamnose and 3-*O*-methyl-L-rhamnose (Table 4.2). Water soluble polysaccharides are complex and heterogeneous while the acid-soluble polysaccharide is a homogeneous glucan [24].

4.2.4
Conclusions on Carbohydrate Feedstocks

Storage carbohydrates are uniform in composition and relatively easy to isolate and purify. Therefore, many fermentative and catalytic processes have identified these feedstocks as their initial feedstock of choice. Because of costs and societal debates (food versus fuel and indirect land use debates) many researchers from both industry and academia are investigating the use of lignocellulose as feedstock. Pure cellulose has the same advantages as starch, that it is only built up from glucose and relatively easy to hydrolyze (although much more difficult than starch) when pure (and amorphous). However, to make use of lignocellulose economically the hemicellulose also needs to be used. This overview clearly shows that due to the heterogeneity of the monosaccharides incorporated and the large diversity in linkages and side groups, both an enzymatic hydrolysis system and a catalytic/fermentative conversion system need to be quite robust to make optimal use of the cellulose and hemicellulose fractions. Many barriers for development are still present regarding the cultivation and economical use of both macroalgae and microalgae. The unusual carbohydrate composition of most of the algae opens up possibilities to synthesize unique building blocks. However, only if this feature is fully utilized may there be potential for biochemicals and biofuels generation from the carbohydrates of both macroalgae and microalgae. The biochemicals and biofuels contribution from algae by 2020 is likely to be very modest. The interest for high added-value products, such as nutraceuticals, pigments, proteins, functional foods, and other chemical constituents is currently commercially more important than chemicals and biofuels applications [32].

4.3
Carbohydrate Dehydration

4.3.1
Introduction

The formation of furans from sugars has been known since the early nineteenth century [12, 13]. Furfural was discovered in 1821 by Döbereiner, by the distillation of bran with dilute sulfuric acid [2, 34]. The resulting compound was first named furfurol (the name comes from the Latin word *furfur* that means bran cereal, while the final *ol* means oil). The furfural molecule has an aldehyde group and a furan ring with aromatic character, and a characteristic smell of almonds. In the presence of oxygen, a colorless solution of furfural tends to become initially yellow, then brown, and finally black. This color is due to the formation of oligomers/polymers with conjugated double bonds formed by radical mechanisms and can be observed even at concentrations as low as 10^{-5} M [16]. Despite the fact that furfural has an LD_{50} between 50 and 2330 mg/kg for mice, rats, guinea pigs and dogs, man tolerates its presence in a wide variety of fruit juices, wine, coffee and tea [16, 17]. The highest concentrations of furfural are present in cocoa and coffee (55–255 ppm), in alcoholic beverages (1–33 ppm) and in brown bread (26 ppm) [16]. There is no commercially attractive route for the production of furfural from petrochemical resources [35]. The synthesis of 5-hydroxymethylfurfural (HMF) from biomass was already described in 1895 by Düll [36] and Kiermayer [37]. Due to their high potential as platform chemicals for a variety of applications, furfural and HMF were mentioned by Bozell in the "top 10 + 4" list of biobased chemicals [15], together with 2,5-furandicarboxylic acid (FDCA), which is formed by oxidation of HMF [13].

The formation of furans from sugars takes place through an acid-catalyzed dehydration of sugar molecules at elevated temperature. In general, furfural is formed from C-5 sugars and HMF is formed from C-6 sugars. It is, therefore, not surprising that furans, especially HMF, can be found in essentially all carbohydrate-containing heat-treated food. Furfural is known to have some toxic effects, whereas for HMF this is still unclear [13]. The hydrolysis of polysaccharides and subsequent dehydration into furfural and HMF may be promoted by Brönsted or Lewis acid catalysts [12, 13]. Furfural production through traditional processes is accompanied by acidic waste stream production and high energy consumption. Marcotullio and de Jong state that modern furfural production process concepts will have to consider environmental concerns and energy requirements besides economics, moreover they will have to be integrated within widened biorefinery concepts [38]. The industrial use of aqueous mineral acids as the catalysts, such as sulfuric acid for furfural production, poses serious operational (corrosion), safety and environmental problems (large amounts of toxic waste). Hence, it is seen as desirable to replace conventional aqueous mineral acids by "green" non-toxic catalysts for converting sugars into furfural and HMF. The use of solid acids as catalysts may have several advantages over liquid acids, such as easier separation and reuse of the solid catalyst, longer catalyst lifetimes, toleration of a wide range of temperatures and pressures, and easier/safer catalyst handling, storage and disposal.

4.3.2
Commercial Furfural Production and Applications

The industrial production of furfural was driven by the need of the USA to become self-sufficient during the First World War. Between 1914 and 1918, intensive exploration for converting agricultural wastes into industrially more valuable

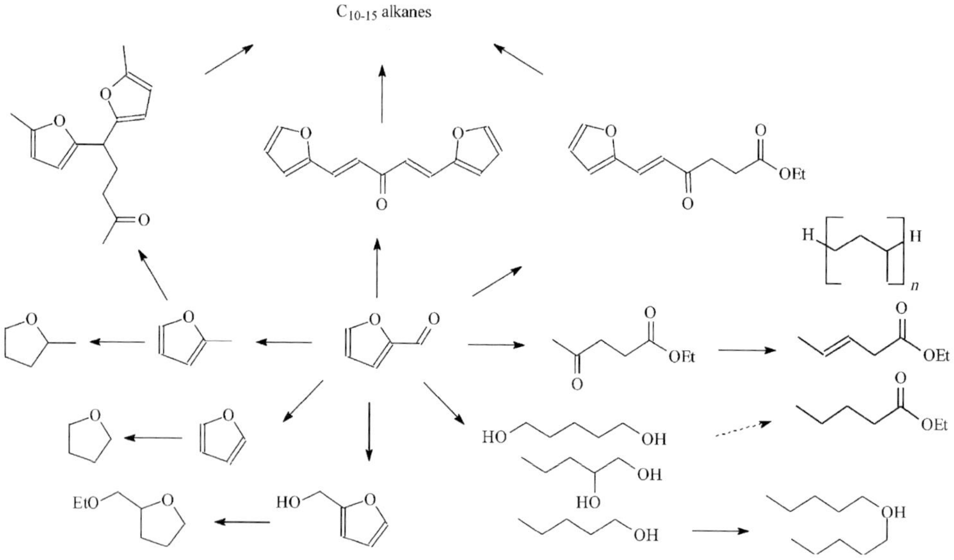

Figure 4.1 Some of the main outlets of furfural [12].

products was initiated. In 1921, the Quaker Oats company in Iowa initiated the production of furfural from oat hulls using "left over" reactors [16]. Over time, there was an increased industrial production of furfural and the discovery of new applications [34]. Presently, the annual world production of furfural is about 300 000 tons and, although there is industrial production in several countries, the main production units are located in China, the Dominican Republic and South Africa [2, 16, 17, 35].

Figure 4.1 gives an overview of some of the main outlets of furfural. Most of the furfural produced worldwide is converted through a hydrogenation process into furfuryl alcohol (FA), which is primarily used as foundry resin, but also increasingly applied as resin to improve wood durability and for the manufacture of polymers and plastics [12]. The aldehyde group and furan ring furnish the furfural molecule with outstanding properties for use as a selective solvent [16, 17, 39]. Furfural has the ability to form a conjugated double bond complex with molecules containing double bonds, and, therefore, is used industrially for the extraction of aromatics from lubricating oils and diesel fuels, or unsaturated compounds from vegetable oils. Furfural is used as a fungicide and nematocide in relatively low concentrations [16]. Additional advantages of furfural as an agrochemical are its low cost, safe and easy application, and its relatively low toxicity to humans. Nakagawa and Tomishige [40] have recently reviewed the catalyst system used to produce 1,5-pentanediol from tetrahydrofurfuryl alcohol. Other furan compounds obtained from furfural include levulinic acid [41] and tetrahydrofuran. Furfural and many of its derivatives can be used for the synthesis of new polymers based on the chemistry of the furan ring [17, 39, 42–44]. Furfural derivatives are also excellent starting points for fuel applications [14, 45, 46].

4.3.3
Furfural Formation from Pentose Feedstock

Commercially, the pentosans (mainly xylan) present in the hemicellulose fraction of agricultural streams such as corn cobs and sugarcane bagasse are hydrolyzed, using homogeneous acid catalysts in water, giving rise to pentose (xylose), which, by dehydration and cyclization reactions, leads to furfural with a theoretical mass yield of approximately 73% (Scheme 4.1). Nowadays, other feedstocks are also considered. Huber and his group developed a new process to produce furfural from waste aqueous hemicellulose solutions from the pulp and paper and cellulosic ethanol industries using a continuous two-zone biphasic reactor [47]. A two-stage hybrid fractionation process was investigated to produce cellulosic ethanol and furfural from corn stover. In the first stage, zinc chloride ($ZnCl_2$) was used to selectively solubilize hemicellulose. During the second stage, the remaining solids were converted into ethanol using commercial cellulase and fermentative microorganisms. This hybrid fractionation process recovered 94% of glucan, 90% of xylan, 71% of arabinan, and 75% of lignin under optimal reaction conditions (1st stage: 5% acidified $ZnCl_2$, 7.5 ml min^{-1}, 150 °C (10 min) and 170 °C (10 min); 2nd stage: simultaneous saccharification and fermentation (SSF) using *S. cerevisiae*). Yoo *et al.* found that the furfural yield from the hemicellulose hydrolysates could be up to 58% based on carbon [48]. Yemis and Mazza researched the potential of a microwave-assisted process which provided a highly efficient conversion of wheat straw, triticale straw, and flax shives: obtained furfural yields based on carbon were 48%, 46%, and 72%, respectively [49, 50]. Sahu and Dhepe also presented a solid acid-catalyzed one-pot method for the selective conversion of solid hemicellulose without its separation from other lignocellulosic components, such as cellulose and lignin, resulting in 56% furfural yields in biphasic systems [51]. An interesting approach was disclosed by Vom Stein and coworkers [52] by working with "real samples". They prepared aqueous solutions of $FeCl_3$-NaCl (or seawater) to evaluate the dehydration of xylose into furfural, which can be extracted *in situ* into 2-methyltetrahydrofuran (2-MTHF) as second phase. Furfural was also successfully obtained when aqueous non-purified xylose effluents directly from lignocellulose fractionation are tested [52]. Also Marcotullio and De Jong observed good results with $FeCl_3$ [38].

The hydrolysis of pentosans to pentoses in the presence of H_2SO_4 is faster than the dehydration of the pentose monomers to furfural [16, 17]. Hence, kinetic studies are generally focused on the rate limiting process, which is the dehydration

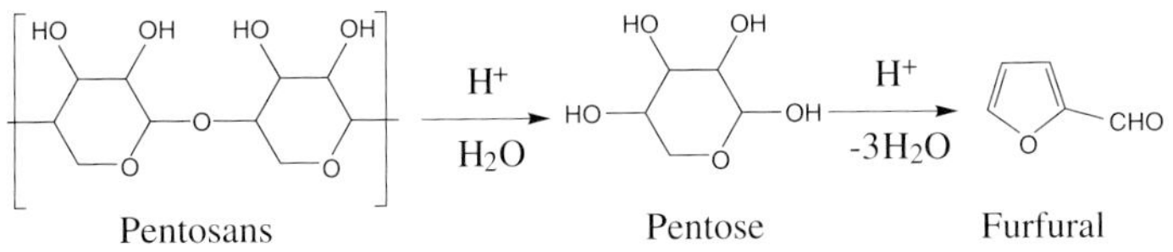

Scheme 4.1 Net conversion of pentosans to furfural.

of pentoses. Xylose and arabinose are monomers found in pentosans, which can be converted into furfural, and some studies have shown that the dehydration of arabinose is slower than that of xylose [16, 53]. The concentration of xylose in the various raw materials is almost always much higher than that of arabinose. Considering these factors, it seems reasonable to investigate the kinetics of the dehydration process using xylose as substrate [16, 39, 42–44, 54–56]. In the dehydration and cyclization of xylose to furfural, three molecules of water are released per molecule of furfural produced. It is generally accepted that the xylose to furfural conversion involves a complex reaction mechanism consisting of a series of elementary steps. The two mechanisms presented have in common the issue that the furfural is formed from the xylopyranose ring and not from its open-chain aldehyde isomer (Schemes 4.2 and 4.3). Considering the mechanism proposed by Zeitsch [16], the transformation of the pentose into furfural involves two eliminations in the positions 1,2 and one elimination in the position 1,4 (Scheme 4.2). The 1,2-eliminations imply the involvement of two neighboring carbon atoms and the formation of a double bond between them, while the 1,4-elimination involves two carbon atoms separated by two carbon atoms and the formation of the furan ring. Zeitsch summarizes the mechanism of the acid-catalyzed conversion of

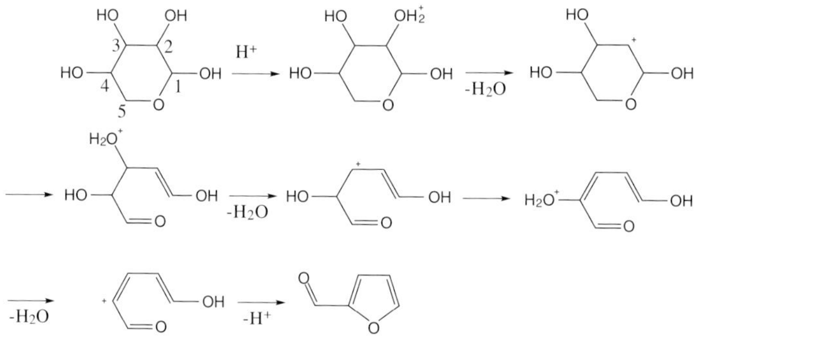

Scheme 4.2 Mechanism of the dehydration of pentoses to furfural proposed by Zeitsch [41].

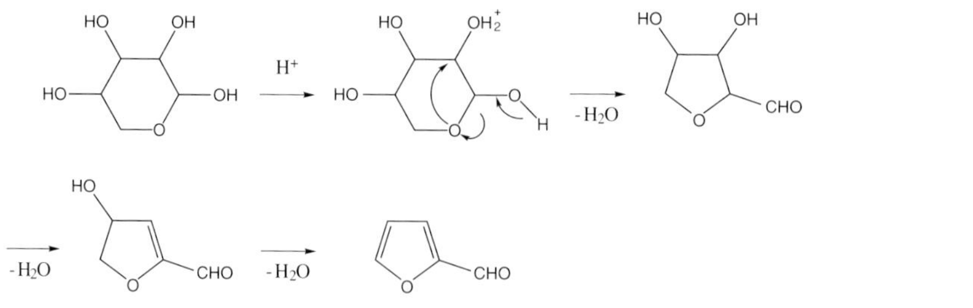

Scheme 4.3 Reaction mechanism proposed by Antal et al. involving the protonation of the hydroxyl group in position C2 [55].

pentose into furfural as a series of protonations of the pentose hydroxy groups, leading to the formation of carbocations through the elimination of water molecules [16].

According to Antal et al. [55], there are two possible mechanisms to obtain furfural from D-xylose, originating from the initial protonation of the hydroxy group at either position C1 or C2 (Scheme 4.3, only the mechanism resulting from the protonation of the hydroxyl group at the C2 position is shown). Both mechanisms involve the xylopyranose isomers, which lead to the formation of furfural by the loss of three molecules of water. A recent study of the xylose degradation using quantum mechanics modeling showed that the protonation of the hydroxyl group at C2 is favored (requires less energy) over that of C1 [57–59]. In acidic medium, the open chain xylose undergoes isomerization to lyxose, which may be further dehydrated to furfural, albeit at a lower rate than that observed for the dehydration of xylose to furfural [55]. Theander et al. have shown that keto-pentoses (e.g., xylulose) are dehydrated to furfural much faster than aldo-pentoses (e.g., xylose, arabinose) [60].

Huber and his group developed a new process to produce furfural from aqueous hemicellulose waste solutions from the pulp and paper and cellulosic ethanol industries using a continuous two-zone, biphasic reactor [47]. Besides furfural, substantial amounts of formic and acetic acid are produced, probably formed from the acid hydrolysis of formylated and acetylated xylose oligomers, respectively. Formic acid is also a direct product from xylose modification [60]. It was estimated that this approach uses 67 to 80% less energy than the current industrial processes to produce furfural. Under optimum conditions, a furfural yield of 90% can be achieved from the hot water extract containing 11 wt% xylose [47]. In another paper [61] a kinetic model for the dehydration of xylose to furfural in a biphasic batch reactor with microwave heating was presented. There are four key steps in their kinetic model: (i) xylose dehydration to form furfural; (ii) furfural reaction to form degradation products; (iii) furfural reaction with xylose to form degradation products, and (iv) mass transfer of furfural from the aqueous phase into the organic phase (methyl isobutyl ketone – MIBK). It was estimated that furfural yields in a biphasic system can reach 85%, whereas under these same conditions in a monophase system furfural yields of only 30% are obtained [61]. A kinetic model for the homogeneous conversion of D-xylose in high temperature water has also been developed [62]. Experimental testing evaluated the effects of operating conditions on xylose conversion and furfural selectivity, with furfural yields of up to 60% observed. The kinetics of formic acid-catalyzed xylose dehydration to furfural and furfural decomposition was also investigated using batch experiments within a temperature range of 130–200 °C [63]. The study showed that the modeling must account for other reactions from xylose besides dehydration to furfural. Moreover, the reactions between xylose intermediates and furfural play only a minor role and furfural decomposition reactions must take the uncatalyzed decomposition in water into account [63]. By-products formed in the xylose reaction may also derive from the fragmentation of xylose, such as glyceraldehyde, glycolaldehyde, formic acid, lactic acid, acetol [55, 60].

Similar to the dehydration of C6 sugars [13], it is much easier to dehydrate C5-ketoses than C5-aldoses, potentially giving high furfural yields. This was nicely demonstrated by the ^{1}H-NMR study by Theander and coworkers [60]. This approach can be achieved by combining solid acid and base catalysts in one pot via an aldose/ketose isomerization of xylose to xylulose by solid base and successive dehydration of xylulose to furfural by the solid acid catalyst. Good results were obtained in polar aprotic solvents such as N,N-dimethylformamide using Amberlyst-15 and hydrotalcite under moderate conditions [64]. Also the one-pot synthesis of furfural from arabinose using a combination of solid acid (Amberlyst-15) and base catalysts (hydrotalcite) was successfully demonstrated. Moreover, this approach of combined acid–base catalysts displayed good activity for the transformation from mixed sources of C5 and C6 sugars to the corresponding furans [65]. The use of chromium catalysts follows the same principle [66]. Recently, xylulose could be detected in the dehydration reaction of xylose in a water–tetrahydrofuran biphasic medium containing $AlCl_3 \cdot 6H_2O$ and NaCl under microwave heating at 140 °C [67].

On the other hand, as furfural is formed it can be transformed into higher molecular weight products by (i) condensation reactions between furfural and intermediates of the conversion of xylose to furfural (and not directly with xylose), and (ii) furfural polymerization [16]. Aldol condensation between two molecules of furfural does not occur due to the absence of a carbon atom in the Hα position in relation to the carbonyl group [68]. The side reactions (i) and (ii) lead to oligomers and polymers with (i) considered to be more relevant than (ii), although published characterization studies of the by-products formed are scarce [16]. The extent of these side reactions can be minimized by reducing the residence time of furfural in the reaction mixture and by increasing the reaction temperature [16, 56, 69]. If furfural is kept in the gas phase during the aqueous phase reaction it will not react with intermediates, which are "non-volatile". Agirrezabal-Telleria *et al.* [70] developed new approaches for the production of furfural from xylose. They propose to combine relatively cheap heterogeneous catalysts (Amberlyst 70) with simultaneous furfural stripping using nitrogen under semi-batch conditions. Nitrogen, compared to steam, does not dilute the vapor phase stream when condensed. This system allowed stripping 65% of the furfural converted from xylose and almost 100% selectivity in the condensate. Moreover, high initial xylose loadings led to the formation of two water–furfural phases, which could further reduce purification costs. Constant liquid–vapor equilibrium during stripping could be maintained for different xylose loadings. The modeling of the experimental data was carried out in order to obtain a liquid–vapor mass-transfer coefficient. This value could be used for future studies under steady-state continuous conditions in similar reaction systems [70]. Formic acid, a by-product of the furfural process [56], can be an effective catalyst for dehydration of xylose to furfural. There is a growing interest in the use of formic acid as catalyst because it has low corrosiveness and can be easily separated and reused. Using response surface methodology the optimal process parameters (xylose concentration $40\,g\,l^{-1}$, formic acid concentration $10\,g\,l^{-1}$, and a reaction temperature 180 °C) were determined to obtain high furfural yield and selectivity. Under these conditions, a maximum furfural yield of 74% and selectivity of 78% were achieved [67]. Extraction using supercritical CO_2 also enhances furfural yields [62, 71, 72].

The above mechanistic considerations for the homogeneous conversion of xylose to furfural using H_2SO_4 as catalyst may also be considered for solid acid catalysts. Nevertheless, differences in product selectivity between homogeneous and heterogeneous catalytic processes are expected due to effects such as shape/size selectivity, competitive adsorption (related to hydrophilic/hydrophobic properties), and the strength of the acid sites.

4.3.4
Production Systems of Furfural

Industrially, furfural is directly produced from the lignocellulosic biomass in the presence of mineral acids, mainly sulfuric acid, under batch or continuous mode operation (Table 4.3). Attempts to improve furfural yields have been made by process innovation, although the use of mineral acids remains a drawback [1, 68, 69]. The cost and inefficiency of separating these homogeneous catalysts from the products makes their recovery impractical, resulting in large volumes of acid waste, which must be neutralized and disposed of. Other drawbacks include corrosion and safety issues. The production of furfural is therefore one of many industrial processes where the reduction or replacement of the "toxic liquid" acid catalysts by alternative "green" catalysts is of high priority. Recently, Marcotullio and De Jong [38, 73] shed new light on some particular aspects of the chemistry of D-xylose reaction to furfural. Their aim was to clarify the reaction mechanism leading to furfural and to define new green catalytic pathways for its production. Specifically, their objective was to reduce the use of mineral acids by the introduction of alternative catalysts, for example, halides, in dilute acidic solutions at temperatures between 170 and 200 °C [19]. Results indicate that the Cl^- ions promote the formation of the 1,2-enediol from the acyclic form of xylose, and thus the subsequent acid-catalyzed dehydration to furfural (Scheme 4.4). For this reason the presence of Cl^- ions led to significant improvements for H_2SO_4 catalyzed reactions. The addition of NaCl to a 50 mM HCl aqueous solution gave 90%

Table 4.3 Industrial processes of furfural production.

Industrial process	Catalyst	Reaction type	Temperature (°C)
Quaker Oats	H_2SO_4	Batch	153
Chinese	H_2SO_4	Batch	160
Agrifurane	H_2SO_4	Batch	177–161
Quaker Oats	H_2SO_4	Continuous	184
Escher Wyss	H_2SO_4	Continuous	170
Rosenlew	Acids formed from the raw material	Continuous	180

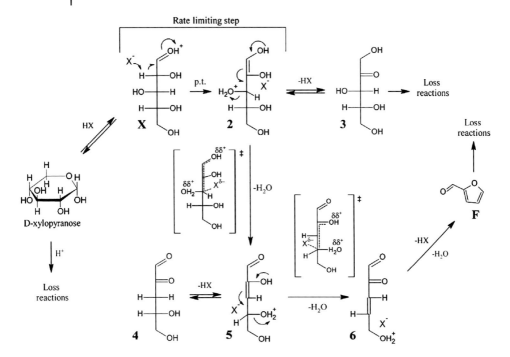

Scheme 4.4 Effect of aqueous halides on the reaction mechanism leading from D-xylose to furfural in acidic solutions according to Marcotullio and de Jong [38]. Aqueous halides are indicated as X⁻.

selectivity to furfural. Follow up experimental results by the same group show the halides to influence at least two distinct steps in the reaction leading from D-xylose to furfural under acidic conditions, via different mechanisms. The nucleophilicity of the halides appears to be critical for the dehydration, but not for the initial enolization reaction. By combining different halides synergic effects become evident, resulting in very high selectivities and furfural yields [73]. Also Rong et al. [74] found that the addition of inorganic salts (e.g., NaCl, FeCl$_3$) promoted the yield of furfural from xylose. Another approach to reduce the inorganic waste streams is to perform the reaction at high temperatures. It was shown that the reaction pathway for the xylose decomposition in high temperature liquid water can be changed by manipulating the temperature and pressure without any catalyst, with a maximum furfural yield of 50% [75].

The extensive studies carried out by the group of Dumesic and coworkers on the use of different solvent mixtures for the dehydration of sugars to HMF and furfural using mineral acids as catalysts at high temperatures give valuable insights on the solvent effects [41, 76]. Preferably, the solvent should have an excellent extracting capacity for the furan compound and should be used in minimal amounts, avoiding high dilution and long heating times.

4.3.5
Heterogeneous Catalysts

Many attempts have been made to develop heterogeneous catalytic processes for furfural production that offer environmental and economic benefits, but to the best of our knowledge none has been commercialized. The acid properties of solid acids may be negatively affected by the presence of water in the reaction medium. However, the dehydration of pentoses generates 3 molecules of water per molecule of furfural so the presence of water in the reaction broth is inevitable. Hence, one of the critical parameters in the choice of a stable, active heterogeneous catalyst is its tolerance toward water [77–83]. Several water-tolerant solid acids have been investigated in the conversion of saccharides to furan derivatives, including inorganic oxides and resins. Inorganic oxides have led to important improvements with respect to catalyst stability, recyclability, activity and selectivity in comparison to conventional mineral acids and commercial acid ion-exchange resins.

Conventional microporous zeolites, such as Faujasite HY and H-Mordenite, seem quite promising, achieving selectivities to furfural in the range of 90 to 95% at xylose conversions between 30 and 40%, with water as solvent and in the presence of toluene as co-solvent, at 170 °C [54, 84]. However, xylose conversion has to be kept low in order to avoid significant drops in the furfural selectivity [12, 44, 85]. Microporous AM-11 crystalline niobium silicates were studied as solid acid catalysts in the dehydration of xylose in water/toluene biphasic conditions. After 6 h at 160 °C, xylose conversions of up to 90% and furfural yields of up to 50% were achieved, and the thermally regenerated catalysts could be reused without loss of activity or selectivity [86]. Microporous silicoaluminophosphates (SAPO) have also been tested in the dehydration of xylose to furfural, under biphasic aqueous-organic conditions at 170 °C. At complete xylose conversion (reached after 16–24 h), furfural yields were up to 65% [87]. Zeolites, AM-11 and SAPO materials are sufficiently stable to be used at elevated temperatures and to be regenerated by thermal treatments under air. This constitutes an important advantage in comparison to ion-exchange resins as solid acid catalysts. The catalytic results obtained with the crystalline solid acids may be further optimized by, for example, using different solvent mixtures and compositions.

Heterogeneous heteropolyacids (HPAs) are promising candidates as green catalysts and are already used in several industrial processes, such as the hydration of olefins [12]. The advantages of HPAs in homogeneous liquid phase catalysis are their low volatility, low corrosiveness, high flexibility, safety in handling and generally high activity and selectivity compared to conventional mineral acids. Furthermore, side reactions, such as sulfonation, chlorination and nitration, and so on, which normally occur in the presence of mineral acids, are absent in the reactions catalyzed by HPAs. The Keggin-type HPAs are typically represented by the formula $H_{8-x}[XM_{12}O_{40}]$, where X is the heteroatom, x is its oxidation state and M is the addenda atom (Mo^{6+} or W^{6+}). HPAs were investigated in the liquid phase dehydration of D-xylose to furfural [88]. The catalytic results depend on the reaction temperature, type of solvent and HPA composition. The most promising systems were the tungsten-containing HPAs used with either DMSO or toluene/water as solvent. Furfural yields achieved within 8 h at 140 °C were below 70%. Kinetic studies showed that the initial reaction rate exhibits a first-order dependence on the initial concentration of xylose and a nonlinear dependence on the initial concentration of HPA.

The most important and common HPAs for catalysis are the Keggin acids, since they are the most stable and readily available. In particular, heteropolytungstate (PW) possesses the highest acid strength and thermal stability [12, 89]. Catalysts based on PW supported on silica have been used in the dehydration of xylose [90]. Zhang and coworkers also evaluated the benefits of a solid acid catalyst, the mesoporous molecular sieve MCM-41 with 1-butanol as a renewable extraction solvent in a biphasic system. At a reaction temperature of 170 °C, the optimal furfural yield was obtained within 3 h, and more than 97% D-xylose conversion and 44% furfural yield were achieved. Notably, a higher furfural yield was achieved in the presence of sodium chloride when using MCM-41 as a solid acid catalyst. These results also showed that 1-butanol was a good, as well as green and renewable, extraction solvent for furfural [91].

In order to increase the stability of the catalysts towards leaching, sulfonic acid functionalized mesoporous silicas were prepared taking into account the need for a covalent link between the active acid site and the support. Sulfonic acid functionalized mesoporous silicas are active and selective catalysts for a number of reactions. The active sulfonic group is obtained post-synthetically by sulfonation reactions, or by the oxidation of thiol-functionalized silicas previously synthesized by a one-step sol–gel or post-modification grafting route. The immobilization of (3-mercaptopropyl)-trimethoxysilane (MPTS) in toluene onto MCM-41 with controlled water content resulted in a "coated" material (MCM-41-SHc) with a monolayer of MPTS moieties (7.1 wt% S), and a less covered "silylated" material (MCM-41-SHs) was obtained in dry conditions (4.5 wt% S). The oxidation of the mercaptopropyl groups by hydrogen peroxide in a water–methanol solution was complete, but resulted in a reduction of the sulfur contents and in the formation of disulfide and partially oxidized disulfide species. The performance of these materials is summarized in Table 4.4. For the synthesized materials, the highest yield was obtained with MCM-41-SO$_3$Hc as a catalyst (70% yield after 24 h, at 140 °C, DMSO or water–toluene), which is higher than that attainable with commercial zeolites. The high furfural selectivity observed at high conversions may be explained by the presence of large uni-dimensional mesopores in the MCM-41 materials that promote the dehydration of xylose to furfural by allowing fast diffusion of furfural out of the catalyst as soon it is formed. This diffusion effect avoids the extensive consecutive degradation reactions of furfural. Even so, these materials deactivate with long residence times, accompanied by the appearance of a brownish color during the reaction (coke (humins) adsorption). In fact, the progressive catalyst deactivation between recycling runs might be connected with the inefficient removal of the adsorbed by-products, which load the catalyst surface and lead to the decrease in the sites that are active for the xylose dehydration by passivation effects. For better catalyst regeneration these by-products must be

Table 4.4 Catalytic performance of sulfonic acid-functionalized materials in the dehydration of D-xylose.

Catalyst	S_{BET} ($m^2 g^{-1}$)	V_p ($cm^3 g^{-1}$)	H^+ (meq g^{-1})[b]	TOF[c] (mmol·g_{cat}^{-1} h^{-1})	Conv. (%)	Select. (%)	Ref
None[a]	–	–	–	–	34/84	2/27	[12]
MCM-41[a]	833	0.59	–	0.8	30/86	4/52	[12]
MCM-41-SO$_3$Hs[a]	493	0.28	0.4	2.0 (5)	81/90	49/77	[12]
MCM-41-SO$_3$Hc[a]	438	0.24	0.7	2.1 (3)	84/91	65/82	[12]
Hybrid-SO$_3$H[a]	278	0.13	0.1	1.4 (14)	57/88	11/61	[12]
Amberlyst-15[a]	–	–	4.6	2.2 (0.5)	87/90	68/70	[12]
Amberlyst-70[d]	–	–	2.6	–	26/81	83/38	[70]
Amberlyst-70[g]	–	–	2.55	1.2	65	49	[92]
MSHS-SO$_3$H[e]	432	0.38	–	–	64	68	[93]
MCM-41-SO$_3$H[f]	686	0.68	–	–	37	65	[93]
SBA-15[g]	1126	1.85	1.1	0.74	67	82	[92]
SBA-15[h]	1126	1.85	1.1	–	96	85	[92]

a) Reaction conditions: 1 ml DMSO, 30 mg xylose, 20 mg catalyst, 140 °C, conversion after 4/24 h, selectivity to furfural after 4/24 h.
b) Measured by titrating the solid with NaOH.
c) Turnover frequency calculated after 4 h. In brackets the TOF values are expressed as mmol·(meq H^+)$^{-1}$ h^{-1}.
d) 1 ml DMSO, 30 mg xylose, 20 mg catalyst, 140 °C, conversion after 4/24 h, selectivity to furfural after 4/24 h.
e) Reaction conditions: 30 ml H_2O, 900 mg xylose, 100 mg catalyst, 190 °C, conversion and selectivity after 1 h.
f) Reaction conditions: 30 ml H_2O, 900 mg xylose, 100 mg catalyst, 170 °C, conversion and selectivity after 1 h.
g) Batch in 200 ml toluene/water, 2 wt.% xylose loading and catalyst P100-0.10 60 wt.% loading with respect to initial [xylose], 140 °C, conversion and selectivity after 20 h.
h) At 170 °C, otherwise identical to [g].

removed with a thermal treatment (350 °C is needed for the coke removal), but the sulfonic acid groups are only stable up to 250 °C. This constitutes a major drawback for practical application. One approach that allows the thermal regeneration of the catalysts is to prepare materials without an organic component.

Early studies showed that furfural yields of about 60% at circa 90% conversion could be achieved using conventional sulfated zirconia (SZ) or titania as solid acids and supercritical carbon dioxide as an extracting solvent, at 180 °C and 200 bar [94]. These results could be further improved by dispersing (per)sulfated zirconia ((P)SZ) on a mesoporous support with high specific surface area. Conventional SZ has a specific surface area usually in the range 80–100 $m^2 g^{-1}$ and a lack of ordered mesoporosity and textural homogeneity, making it suitable for traditional vapor-phase reactions involving small molecules, but less amenable to liquid-phase reactions. On the other hand, the use of supercritical CO_2 as extracting solvent is greener than the use of organic solvents, but it may have economic drawbacks associated with process requirements and energy consumption.

Conventional (per)sulfated bulk zirconia, mesoporous sulfated zirconia and (per)sulfated zirconia supported on an ordered mesoporous silica, MCM-41, with or without aluminum incorporation, were examined as acid catalysts for the dehydrocyclization of xylose to furfural in W/T, at 160 °C [90]. Furfural yields of up to 50% could be achieved at >90% conversion with the mesostructured bulk and silica-supported zirconia catalysts, which was better than that achievable with H_2SO_4 (using approximately the same equivalent amount of sulfur). While these materials were stable towards zirconium leaching, loss of sulfur was observed in recycling runs, but in some cases no decrease in catalytic activity was observed. Of all the investigated materials, MCM-41-supported sulfated zirconia containing Al seemed to be the most attractive catalyst for aqueous phase conversion of xylose, since it was the most stable to sulfur leaching and exhibited increasing activity and no significant loss of selectivity to furfural in three runs.

Niobium-containing materials, such as hydrated niobium oxide, a water-tolerant solid acid catalyst, exhibit unique activity, selectivity and stability for many different catalytic reactions [95, 96]. Ordered mesoporous MCM-41-type niobium silicates prepared by the doping of niobium in the micelle-templated silica, with Si/Nb molar ratios of either 25 or 50 (in the H^+-form), were found to be active catalysts for xylose dehydration in W/T and gave furfural yields consistently in the range of 34 to 39% (after 6 h reaction at 160 °C) [86]. The niobium-containing mesoporous MCM-41-type catalysts exhibited higher activities than crystalline AM-11 materials (for the first catalytic runs), but were less selective to furfural at conversions above 80% [86]. Partial loss of activity in recycling runs and leaching of Nb from MCM-41 occurred during the reaction. Aluminum-containing MCM-41 catalysts gave similar results to those obtained with AM-11 at 160 °C in W/T at 6 h reaction, and could be reused several times without loss of catalytic activity and selectivity, and no metal leaching was detected [97]. Another interesting group of heterogeneous catalysts is the transition metal oxide nanosheets. Crystalline layered metal oxide cation exchangers, such as titanates, niobates and titanoniobates, are potentially strong solid acids when in the H^+-form. However, the high charge density of the anionic sheets in these materials hinders the access of bulky substrate molecules to the acid sites. This problem has recently been addressed by exfoliating the layered metal oxides to give aggregates of nanosheets, where the two-dimensional sheet structure remains [98]. The composites have much higher specific surface areas than the acid-exchanged layered precursors and function as strong solid acid catalysts, rivaling or even beating niobic acid, which is a rare water-tolerant solid acid. It was found that they can be more active and somewhat more selective catalysts (4 h reaction, furfural yields of up to 55%) for the conversion of xylose to furfural than the microporous AM-11 crystalline niobium silicates, which in turn yielded more furfural than zeolites such as HY (the protonic form of Y-zeolite, with Si/Al = 5) and mordenite (Si/Al = 6), under similar reaction conditions (at 160 °C, in water/toluene) [86]. The acid-catalyzed, aqueous phase dehydration of xylose to furfural was investigated using vanadium phosphates (VPO) as catalysts. At low concentrations of $(VO)_2P_2O_7$ (5 mM) the catalytic reaction of xylose dehydration (0.67 M xylose in water, and toluene as solvent for the *in situ* extraction of furfural) gave circa 56% furfural yield, at 170 °C and 6 h reaction time [99]. Various sulfated metal oxides were also tested as solid acid catalysts for the dehydration of xylose to furfural under milder conditions. It was found that sulfated tin oxide exhibited the highest catalytic activity, the effects of the content of SO_4^{2-} group and the calcination temperature on the structural properties and catalytic performance were investigated, and it was shown that the sulfated tin oxide catalyst could be reused through a re-sulfation process after the reaction [100]. Lam et al. [101] evaluated graphene, graphene oxide and their sulfonated equivalents for the dehydration of xylose to furfural in water. In particular, sulfonated graphene oxide was demonstrated to be a rapid and water-tolerant solid acid catalyst, even at very low catalyst loadings down to 0.5 wt % versus xylose. The catalyst maintained its initial activity after 12 tested repetitions at 200 °C, with an average yield of 61%, in comparison to 44% for the autocatalyzed system. Spectroscopic and surface area analysis suggested that the aryl sulfonic acid groups were the key active sites for high temperature production of furfural in water [101]. The one-pot acid-catalyzed conversion of mono/di/polysaccharides (xylose, xylan, fructose, inulin, glucose, sucrose, cellobiose) to furfural or HMF in the presence of aluminum-containing mesoporous TUD-1 catalyst (denoted as Al-TUD-1, Si/Al = 21), at 170 °C was studied [102]. Xylose gave 60% furfural yield after 6 h reaction; hexose-based mono/disaccharides gave less than 20% HMF yield; polysaccharides gave less than 20 wt % furfural or HMF yields after 6 h [102].

4.3.6
5-Hydroxymethylfurfural Formation from Hexose Feedstock

HMF is formed through the acid-catalyzed dehydration of a hexose, as described in Scheme 4.5. Initially the synthesis of HMF from hexoses was performed in aqueous systems, catalyzed by homogeneous acids.

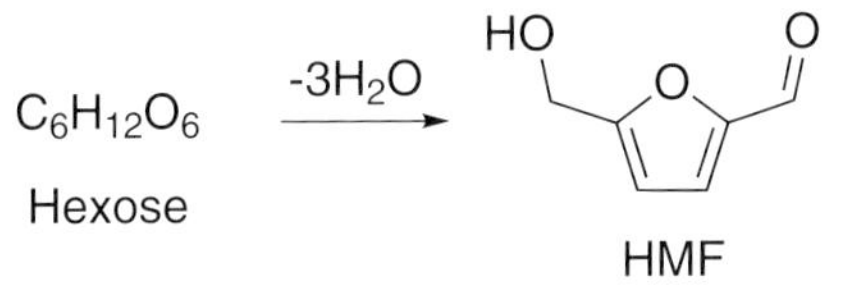

Scheme 4.5 The acid-catalyzed dehydration of hexose to HMF.

A number of mechanistic pathways have been proposed for this reaction, which can generally be divided into two groups. The first group is based on a pathway through acyclic intermediates and the second group is based on a pathway through cyclic intermediates.

Although there are differences between the various acyclic pathways proposed for the aqueous dehydration of hexoses, they generally propose the formation of the 1,2-enediol intermediate in the Lobry De Bruijn-Alberda Van Ekenstein transformation [103] between fructose and glucose as the key intermediate [104–107]. This is proposed to dehydrate to a 3-deoxyglucosone, followed by further dehydration and ring-closure to form HMF. A schematic representation is provided in Scheme 4.6.

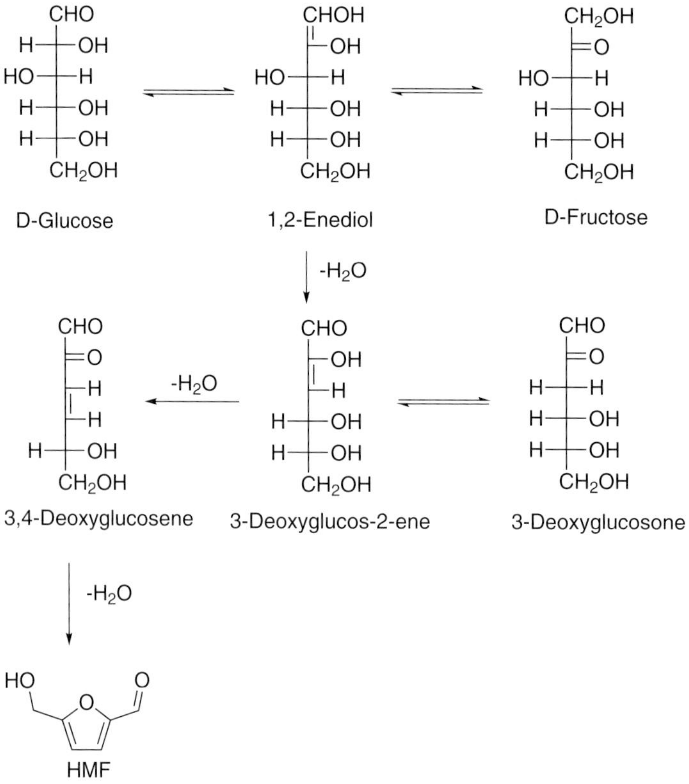

4 Furan-Based Building Blocks from Carbohydrates

Scheme 4.6 The dehydration of glucose and fructose through acyclic intermediates.

The proposed aqueous hexose dehydration pathways through cyclic intermediates generally assume dehydration to start at the C2 hydroxyl position of fructose (Scheme 4.7), leading to the formation of a tertiary carbocation [13, 105, 107]. This is then followed by consecutive dehydrations at C3 and C4 to form HMF. It is clear that in this proposed mechanism, glucose dehydration requires glucose to first isomerize to fructose before it can dehydrate to HMF. Under acidic reaction conditions, though, this is unfavorable as the isomerization is base-catalyzed.

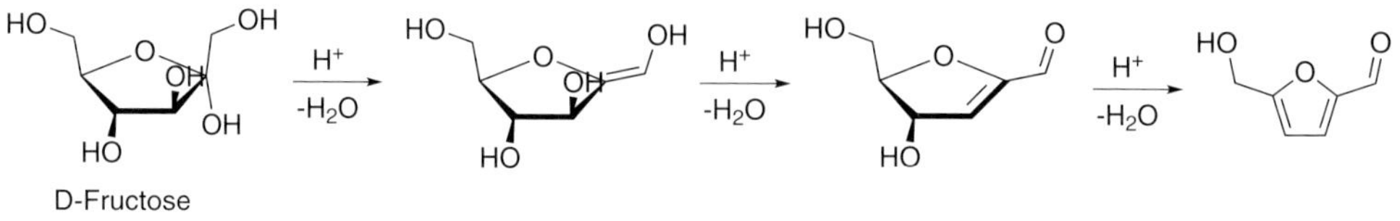

Scheme 4.7 The dehydration of fructose through cyclic intermediates.

The HMF yields and selectivities from the dehydration of fructose, a ketose, are generally much higher than those obtained from the dehydration of glucose, which is an aldose [13]. The HMF yields for homogeneous acid-catalyzed fructose dehydration in water are limited to around 60% at full conversion, whereas for glucose this is only around 10% at full conversion. Fructose is known to be significantly less stable than glucose, which shows in the required reaction conditions for dehydration. Fructose dehydrates to HMF at temperatures around 100 °C in the presence of acid, whereas glucose requires much more severe conditions of at least 140 °C in the presence of catalyst to form only small amounts of HMF (less than 10% yield). Quite large variations are seen in the reaction conditions applied by different groups. In some cases relatively high catalyst concentrations in the order of 0.1–1 M mineral acid are applied in fructose dehydration at relatively low temperatures between 100 °C and 150 °C with reaction times in the order of minutes. Others applied lower acid concentrations, but at either longer reaction times or higher temperatures [13]. Also a significant amount of work has been done with heterogeneous acid catalysts, like ion exchange resins and zeolites, showing comparable selectivities and yields to the homogeneous catalysts [13].

The HMF yield is limited by its inherent instability under aqueous acidic conditions. In the presence of acid, HMF reacts with water (so-called HMF hydration reaction) to form levulinic acid and formic acid, as described in Scheme 4.8 [106]. Other undesirable side-reactions are the formation of polymeric material, often referred to as humins [106], and retro-aldol reactions of sugars [108].

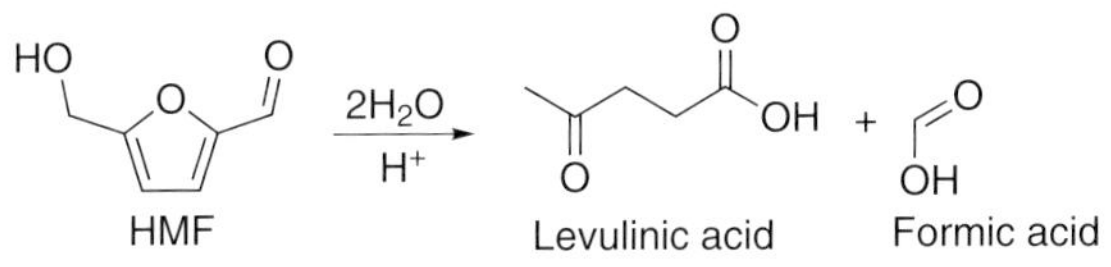

Scheme 4.8 The acid-catalyzed hydration of HMF to levulinic acid and formic acid.

In order to minimize side-reactions and HMF hydration, biphasic systems have been researched in which the HMF is extracted to the organic phase [76, 109–112]. The major extraction solvents used are methylisobutylketone, 1-butanol and 2-butanol. The *in situ* extraction has improved HMF yields from fructose dehydration in some cases to around 70% at full conversion. Due to the high solubility of HMF in water relatively large amounts of solvent are needed, generally at least 2 equivalents, in order to extract sufficient amounts of HMF [13].

In the early 1980s a number of researchers started performing HMF synthesis in organic solvents [113–115]. The biggest initial challenge here is that, apart from high-boiling coordinating solvents like dimethylsulfoxide (DMSO), N,N-dimethylformamide (DMF) and N-methylpyrrolidinone (NMP), most organic solvents do not dissolve sugars very well. The focus was mainly on DMSO and DMF, showing significant improvements in yield and selectivity due to the apparently high stability of HMF in these solvents [113–116]. In DMSO reaction temperatures of 100–120 °C are generally applied and the solvent shows catalytic activity as yields over 90% have been reported in the absence of catalyst [115, 116]. An important issue here is the known decomposition of DMSO at temperatures over 100 °C.

Since 2003, ionic liquids have been extensively researched as solvents for HMF synthesis by many research groups, however 20 years before that HMF synthesis in pyridinium salts was already performed by Fayet and Gelas, resulting in 70% yield starting from fructose [117]. Certain ionic liquids are known to dissolve sugars in high concentrations. The vast majority of this research has been done in imidazolium-based ionic liquids, a number of which are described in Scheme 4.9. As is the case for the coordinating organic solvents, the HMF yields for fructose dehydration to HMF in ionic liquids, in which the ionic liquid is often also the catalyst, are generally high (70–90%) and levulinic acid formation is, in most cases, not mentioned [13, 118]. In the work on ionic liquids some conflicting results have been published with the same or comparable ionic liquids. The exact composition of these ionic liquids is not always known and the purity applied is often less than 98%, which immediately shows that impurities in the solvent can be present in higher than catalytic amounts and, therefore, could influence the results [13, 119].

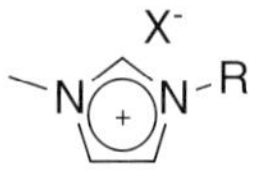

R = H, Ethyl, ^{n}Butyl, Hexyl
X = Cl$^-$, HSO$_4^-$

Scheme 4.9 Imidazolium-based ionic liquids for HMF synthesis.

As was already mentioned, HMF synthesis from glucose is much more challenging than from fructose. In 2007 Zhang and coworkers published a breakthrough in glucose dehydration to HMF by using $CrCl_2$ as a catalyst in an imidazolium-type ionic liquid [120]. They achieved an HMF yield of around 70%, essentially equal to the yield obtained from fructose in the same system. It is believed that $CrCl_2$ behaves as an isomerization catalyst that forms fructose, which can be dehydrated readily to HMF. $CrCl_3$ was later shown by other groups also to be effective and it is possible that Cr(II) is oxidized to Cr(III) *in situ* [13, 121, 122].

Since an *in situ* isomerization from glucose to fructose is now widely accepted to be necessary in order to produce HMF from glucose efficiently, more work has been published with bifunctional catalysts. Some groups continued in the direction started by Zhang, using ionic liquid systems in combination with an apparent isomerization catalyst, like $CrCl_n$, $SnCl_4$ and boric acid [13, 123, 124]. In addition, heterogeneous bifunctional catalysis in traditional solvents, using either a catalyst with both acidic and basic sites or a mixture of heterogeneous acid and base, was researched [13, 125–127].

Earlier research on HMF synthesis focused mainly on fructose and polymers thereof as substrates. Recent years have seen an enormous increase in interest in the development of biobased platform chemicals as a replacement for fossil oil based feedstock. For this reason it is preferable to use cheap feedstocks that do not compete with food. This has led to increased interest in cellulose, a, for humans, non-digestible polymer of glucose, as a feedstock. Cellulose is present in large amounts in plant waste material. Application in HMF synthesis will require both hydrolysis and dehydration of the cellulose, either in one reactor or in two separate steps. Recent years have shown a dramatic increase in research on HMF synthesis from cellulose. The main focus has been in line with the work on glucose, applying bifunctional catalyst systems, especially chromium salts in combination with a Brønsted acid. Especially in ionic liquids, the yields approach those obtained with glucose. The substrate concentration is mostly significantly lower, due to the much lower solubility of cellulose. The reaction times are also typically much longer for cellulose compared to glucose, likely due to the required hydrolysis prior to dehydration to HMF [13].

Although sugar dehydration to furans is a hot topic in academia, a lot of research has yet to be done in upscaling these processes to pilot plant, and ultimately industrial scale. This holds true especially for hexose dehydration to HMF. Only two pilot-scale processes are known for the production of HMF: a process from Süddeutsche Zucker-Aktiengesellschaft and a process from Roquette Frères. The first process concerns HMF production at around the 5 kg scale from fructose and inulin, a polymer of mainly fructose, catalyzed by oxalic acid at around 140 °C in water, in which the purification of HMF is done by chromatographic separation [128]. The second process concerns fructose dehydration in a water-MIBK (1:9 v/v) biphasic system in the presence of cationic resins at temperatures between 70 and 95 °C [129]. In both processes the fructose concentration in water was 20–25 wt% and the HMF yields are in the 40–50% range. The work-up procedures for HMF mentioned in these patents appear unfavorable for a large-scale plant as large-scale chromatographic separation is expensive and a very high solvent to water ratio requires a lot of energy for evaporation of the solvent from the product.

In order to produce HMF or a derivative thereof in a cost effective way, some challenges must be overcome. HMF is unstable under the reaction conditions in the presence of water, leading to the formation of levulinic acid, formic acid and polymeric materials. For this reason contact with water should be minimized. This can be achieved by performing the reaction using other solvents, or by continuously extracting the HMF from the aqueous phase.

The distribution of HMF over water and extraction solvents is generally not highly favorable towards the solvent, demanding large excess of extraction solvents and therefore energy-consuming work-up [130].

Performing HMF synthesis in solvents other than water is an appealing option. Here the choice has to be made between solvents that have lower boiling points, but exhibit low sugar solubility, and solvents that dissolve high concentrations of sugar, like DMSO and ionic liquids, though from which product separation is difficult due to the high affinity of HMF for these solvents.

Two processes focus on the production of derivatives of HMF in order to produce the furanic product more effectively. Mascal and coworkers have focussed their efforts on the production of 5-chloromethylfurfural (CMF) in a biphasic system of concentrated hydrochloric acid and 1,2-dichloroethane [131]. Avantium Chemicals opened their pilot plant in December of 2011 on alcohol-based production of HMF ethers, which will be used for the production of furan-based polymers [132].

4.4
Conclusions and Further Perspectives

Attempts have been made to convert (poly)saccharides into basic, versatile furan compounds, furfural and HMF, using heterogeneous catalytic routes and the published results obtained at the lab-scale seem quite promising and encouraging. The use of porous solid acids as catalysts instead of mineral acids for this reaction system may have several advantages, such as an easier separation of the catalyst from the products (e.g., by simple filtration), convenient regeneration (e.g., after thermal removal of coke) and the possibility to reuse the catalyst consecutively (avoiding treatments of effluent streams). Comparable or higher yields of the target product may also be reachable. Disadvantages of heterogeneous catalysts are the costs of the catalysts, the often more expensive capital expenditures involved, and the long-term selectivity and stability of the process. The results may be further optimized by fine-tuning the catalyst properties, such as acid–base (poor selectivity has been correlated to strong Brönsted acidity, and enhanced Lewis acidity seems favorable) and hydrophobic/hydrophilic properties (e.g., via dealumination functionalization using organosilanes), and pore size distribution (e.g., by appropriate choice of template). Together with the adjustment of the reaction conditions, such as the composition of sugar, catalyst and solvent mixtures (types of solvents), temperature and residence times, and reactor design, these developments could open up valuable perspectives in the application of solid acid catalysts to the conversion of saccharides to basic furan derivatives. The use of water instead of organic solvents, such as DMSO and dimethylformamide, to dissolve saccharides can be a more convenient, greener and cheaper approach. These issues lead to stricter requirements in terms of catalytic properties: (i) the solid acid catalysts must be sufficiently tolerant toward water and impurities present in the raw materials, (hydro)thermally stable and preferably readily prepared; (ii) high selectivity at high conversion for high substrate concentration is important for enhanced productivity. The use of extracting organic solvents to increase selectivity poses environmental concerns and future work in this direction must be considered carefully.

The use of heterogeneous catalysts for the production of furfural and HMF has not yet been implemented industrially, to the best of our knowledge. One critical factor is the choice of the biomass raw materials and the management/processing inputs in order to obtain liquid feed streams rich in saccharides for the heterogeneous catalytic hydrolysis/dehydration processes: the transformation of solid biomass using a solid catalyst would be subject to severe mass transfer limitations. Using di/oligo/polysaccharides as starting materials instead of the monosaccharides themselves to produce HMF and furfural in a one-pot process, thereby eliminating the separate hydrolysis step before the dehydration reaction, seems quite attractive. The design of versatile catalysts for mixed feed (fractions of cellulose, hemicellulose, starch) processing may make the process more cost-competitive. Another important aspect that must be considered is the transport of biomass to the industrial plant.

Collaborative efforts between academia and industry will be crucial in developing competitive technologies for the production of HMF and furfural using a suitable reaction medium with readily prepared (with optimized composition) and sufficiently robust solid acids.

References

1 Brown, R.C. (2003) *Biorenewable Resources – Engineering New Products from Agriculture*, 1st edn, Blackwell Publishing, USA.
2 Kamm, B., Gruber, P.R., and Kamm, M. (eds) (2006) *Biorefineries – Industrial Processes and Products, Status Quo and Future Directions*, vol. 1, vol. 2, Wiley-VCH Verlag GmbH, Weinheim.
3 Huber, G.W., Iborra, S., and Corma, A. (2006) Synthesis of transportation fuels from biomass: chemistry, catalysts, and engineering. *Chem. Rev.*, **106**, 4044–4098.
4 Gallezot, P. (2012) Conversion of biomass to selected chemical products. *Chem. Soc. Rev.*, **41**, 1538–1558.
5 Climent, M.J., Corma, A., and Iborra, S. (2011) Heterogeneous catalysts for the one-pot synthesis of chemicals and fine chemicals. *Chem. Rev.*, **111**, 1072–1133.
6 Climent, M.J., Corma, A., and Iborra, S. (2011) Converting carbohydrates to bulk chemicals and fine chemicals over heterogeneous catalysts. *Green Chem.*, **13**, 520–540.
7 Lichtenthaler, F.W., and Peters, S. (2004) Carbohydrates as green raw materials for the chemical industry. *C. R. Chim.*, **7**, 65–90.
8 Spiridon, I., and Popa, V.I. (2008) Hemicelluloses: major sources, properties and applications, in *Monomers, Polymers and Composites from Renewable Resources* (eds N.M. Belgacem and A. Gandini), Elsevier, Amsterdam, pp. 289–304.
9 Lin, Y.-C., and Huber, G. (2009) The critical role of heterogeneous catalysis in lignocellulosic biomass conversion. *Energy Environ. Sci.*, **2**, 68–80.

10 Stöcker, M. (2008) Biofuels and biomass-to-liquid fuels in the biorefinery: catalytic conversion of lignocellulosic biomass using porous materials. *Angew. Chem. Int. Ed.*, **47**, 9200–9211.
11 Dhepe, P.L., and Fukuoka, A. (2008) Cellulose conversion under heterogeneous catalysis. *ChemSusChem*, **1**, 969–975.
12 Dias, A.S., Lima, S., Pillinger, M., and Valente, A.A. (2010) Furfural and furfural-based industrial chemicals, in *Ideas in Chemistry and Molecular Sciences: Advances in Synthetic Chemistry*, vol. 8 (ed. B. Pignataro), Wiley-VCH Verlag GmbH, Weinheim, pp. 167–186.
13 Van Putten, R.-J., van der Waal, J.C., de Jong, E., Rasrendra, C.B., Heeres, H.J., and de Vries, J.G. (2012) Hydroxymethylfurfural, a versatile platform chemical made from renewable resources. *Chem. Rev.*, in press, http://dx.doi.org/10.1021/cr300182k.
14 Lange, J.-P., van der Heide, E., van Buijtenen, J., and Price, R. (2012) Furfural – a promising platform for lignocellulosic biofuels. *ChemSusChem*, **5**, 150–166.
15 Bozell, J.J., and Petersen, G.R. (2010) Technology development for the production of biobased products from biorefinery carbohydrates – the US Department of Energy's "Top 10" revisited. *Green Chem.*, **12**, 539–554.
16 Zeitsch, K.J. (2000) *The Chemistry and Technology of Furfural and Its Many by-Products, Sugar Series*, vol. 13, Elsevier, The Netherlands.
17 Hoydonckx, H.E., van Rhijn, W.M., van Rhijn, W., de Vos, D.E., and Jacobs, P.A. (2007) Furfural and derivatives. in

Ullmann's Encyclopedia of Industrial Chemistry, Wiley, 1–29.
18 Fengel, D., and Wegener, G. (1984) Wood: Chemistry, Ultrastructure, Reactions, Kessel Verlag, Munich, Germany, 613 pp, ISBN 3-935638-39-6.
19 Schädel, C., Blöchl, A., Richter, A., and Hoch, G. (2010) Quantification and monosaccharide composition of hemicelluloses from different plant functional types. Plant Physiol. Biochem., 48, 1–8.
20 Girio, F.M., Fonseca, C., Carvalheiro, F., Duarte, L.C., Marques, S., and Bogel-Łukasik, R. (2010) Hemicelluloses for fuel ethanol: a review. Bioresour. Technol., 101, 4775–4800.
21 Franck, A., and de Leenheer, L. (2002) Inulin, in Biopolymers, Polysaccharides II: Polysaccharides from Eukaryotes vol. 6 (eds E.J. Vandamme, S. De Baets, and A. Steinbüchel), Wiley-VCH Verlag GmbH, Weinheim, pp. 439–479.
22 Chi, Z.-M., Zhang, T., Cao, T.-S., Liu, X.-Y., Cui, W., and Zhao, C.-H. (2011) Biotechnological potential of inulin for bioprocesses. Bioresour. Technol., 102, 4295–4303.
23 Nabarlatz, D., Ebringerová, A., and Montané, D. (2007) Autohydrolysis of agricultural by-products for the production of xylo-oligosaccharides. Carbohydr. Pol., 69, 20–28.
24 Shekharam, K.M., Venkataraman, L.V., and Salimath, P.V. (1987) Carbohydrate composition and characterization of two unusual sugars from the blue green alga, Spirulina platensis. Phytochemistry, 26, 2267–2269.
25 Brown, M.R., and Jeffrey, S.W. (1992) Biochemical composition of microalgae from the green algal classes Chlorophyceae and Prasinophyceae. 1. Amino acids, sugars and pigments. J. Exp. Mar. Biol. Ecol., 161, 91–113.
26 Costa, C., Alves, A., Pinto, P.R., Sousa, R.A., Borges da Silva, E.A., Reis, R.L., and Rodrigues, A.E. (2012) Characterization of ulvan extracts to assess the effect of different steps in the extraction procedure. Carbohydr. Polym., 88, 537–546.
27 OECD-FAO Agricultural Outlook 2011–2020 (2011) Chapter 6 Sugar, http://www.oecd.org/site/oecd-faoagriculturaloutlook/48184295.pdf (accessed 20 November 2012).
28 Messias de Bragança, R., and Fowler, P. (2004) Industrial Markets for Starch, http://www.bc.bangor.ac.uk/_includes/docs/pdf/indsutrial%20markets%20for%20starch.pdf (accessed 19 November 2012).
29 Sinha, A.K., Kumar, V., Makkar, H.P.S., De Boeck, G., and Becker, K. (2011) Non-starch polysaccharides and their role in fish nutrition – A review. Food Chem., 127, 1409–1426.
30 Ebringerova, A., Hromadkova, Z., and Heinze, T. (2005) Hemicellulose. Adv. Polym. Sci., 186, 1–67.
31 Peng, F., Peng, P., Xu, F., and Sun, R.-C. (2012) Fractional purification and bioconversion of hemicelluloses. Biotechnol. Adv.. doi: 10.1016/j.biotechadv.2012.01.018
32 Bruton, T., Lyons, H., Lerat, Y., Stanley, M., and Rasmussen, M.B. (2009) A review of the potential of marine algae as a source of biofuel in Ireland. SEI Technical Report; Sustainable Energy Ireland: Dublin, Ireland, 88pp.
33 Anastasakis, K., Ross, A.B., and Jones, J.M. (2011) Pyrolysis behavior of the main carbohydrates of brown macro-algae. Fuel, 90, 598–607.
34 IFC, Historical overview and industrial development. (30 August 2012) Discovery of Furfural, http://www.furan.com/furfural_historical_overview.html (accessed 13 January 2013).
35 Mamman, A.S., Lee, J.-M., Kim, Y.-C., Hwang, I.T., Park, N.-J., Hwang, Y.K., Chang, J.-S., and Hwang, J.-S. (2008) Furfural: hemicellulose/xylose derived biochemical. Biofuels Bioprod. Bioref., 2, 438–454.
36 Düll, G. (1895) Über die einwirkung von oxalsäure auf inulin. Chem. Ztg., 19, 216–220.
37 Kiermayer, J. (1895) Über ein furfurolderivat aus lävulose. Chem. Ztg., 19, 1003–1006.
38 Marcotullio, G., and de Jong, W. (2010) Chloride ions enhance furfural formation from D-xylose in dilute aqueous acidic solutions. Green Chem., 12, 1739–1746.

39 Sain, B., Chaudhuri, A., Borgohain, J.N., Baruah, B.P., and Ghose, J.L. (1982) Furfural and furfural-based industrial chemicals. *J. Sci. Ind. Res.*, **41**, 431–438.

40 Nakagawa, Y., and Tomishige, K. (2012) Production of 1,5-pentanediol from biomass via furfural and tetrahydrofurfuryl alcohol. *Catal. Today*, **195**, 136–143.

41 Gürbüz, E.I., Wettstein, S.G., and Dumesic, J.A. (2012) Conversion of hemicellulose to furfural and levulinic acid using biphasic reactors with alkylphenol solvents. *ChemSusChem*, **5**, 383–387.

42 Win, D.T. (2005) Furfural – gold from garbage. *AU J. Technol.*, **8**, 185–190.

43 Gandini, A., and Belgacem, M.N. (1997) Furans in polymer chemistry. *Prog. Polym. Sci.*, **22**, 1203–1379.

44 Moreau, C., Belgacem, M.N., and Gandini, A. (2004) Recent catalytic advances in the chemistry of substituted furans from carbohydrates and in the ensuing polymers. *Top. Catal.*, **27**, 11–30.

45 Gruter, G.-J., and de Jong, E. (2009) Furanics: novel fuel options from carbohydrates. *Biofuels Technol.*, **1**, 11–17.

46 de Jong, E., Vijlbrief, T., Hijkoop, R., Gruter, G.-J.M., and van der Waal, J.C. (2012) Promising results with YXY Diesel components in an ESC testcycle using a PACCAR Diesel engine. *Biomass Bioenergy*, **36**, 151–159.

47 Xing, R., Qi, W., and Huber, G.W. (2011) Production of furfural and carboxylic acids from waste aqueous hemicellulose solutions from the pulp and paper and cellulosic ethanol industries. *Affiliation Information Energy Environ. Sci.*, **4**, 2193–2205.

48 Yoo, C.G., Kuo, M., and Kim, T.H. (2012) Ethanol and furfural production from corn stover using a hybrid fractionation process with zinc chloride and simultaneous saccharification and fermentation (SSF). *Process Biochem.*, **47**, 319–326.

49 Yemis, O., and Mazza, G. (2011) Acid-catalyzed conversion of xylose, xylan and straw into furfural by microwave-assisted reaction. *Bioresour. Technol*, **102**, 7371–7378.

50 Yemis, O., and Mazza, G. (2012) Optimization of furfural and 5-hydroxymethylfurfural production from wheat straw by a microwave-assisted process. *Bioresour. Technol.*, **109**, 215–223.

51 Sahu, R., and Dhepe, P.L. (2012) A one-pot method for the selective conversion of hemicellulose from crop waste into C5 sugars and furfural by using solid acid catalysts. *ChemSusChem*, **5**, 751–761.

52 vom Stein, T., Grande, P.M., Leitner, W., and Domínguez de María, P. (2011) Iron-catalyzed furfural production in biobased biphasic systems: from pure sugars to direct use of crude xylose effluents as feedstock. *ChemSusChem*, **4**, 1592–1594.

53 Kootstra, A.M.J., Mosier, N.S., Scott, E.L., Beeftink, H.H., and Sanders, J.P.M. (2009) Differential effects of mineral and organic acids on the kinetics of arabinose degradation under lignocellulose pretreatment conditions. *Biochem. Eng. J.*, **43**, 92–97.

54 Moreau, C., Durand, R., Peyron, D., Duhamet, J., and Rivalier, P. (1998) Selective preparation of furfural from xylose over microporous solid acid catalysts. *Indus. Crops Prod.*, **7**, 95–99.

55 Antal, M.J., Leesomboon, J.T., Mok, W.S., and Richards, G.N. (1991) Mechanism of formation of 2-furaldehyde from D-xylose. *Carbohydr. Res.*, **217**, 71–86.

56 Root, D.F., Saeman, J.F., Harris, J.F., and Neill, W.K. (1959) Chemical conversion of wood residues, part II: kintetics of the acid-catalyzed conversion of xylose to furfural. *Forest Prod. J.*, **9**, 158–165.

57 Nimlos, M.R., Qian, X., Davis, M., Himmel, M.E., and Nimlos, D.K.J. (2006) Energetics of xylose decomposition as determined using quantum mechanics modeling. *J. Phys. Chem. A*, **110**, 11824–11838.

58 Qian, X., Nimlos, M.R., Davis, M., Johnson, D.K., and Himmel, M.E. (2005) *Ab initio* molecular dynamics simulations of β-D-glucose and β-D-xylose degradation mechanisms in acidic aqueous solution. *Carbohydr. Res.*, **340**, 2319–2327.

59 Qian, X., Johnson, D.K., Himmel, M.E., and Nimlos, M.R. (2010) The role of hydrogen-bonding interactions in acidic sugar reaction pathways. *Carbohydr. Res.*, **345**, 1945–1951.

60 Ahmad, T., Kenne, L., Olsson, K., and Theander, O. (1995) The formation of 2-furaldehyde and formic acid from pentoses in slightly acidic deuterium oxide studied by 1H-NMR spectroscopy. *Carbohydr. Res.*, **276**, 309–320.

61 Weingarten, R., Cho, J., Conner, J.W.C., and Huber, G.W. (2010)) Kinetics of furfural production by dehydration of xylose in a biphasic reactor with microwave heating. *Green Chem.*, **12**, 1423–1429.

62 Kim, S.B., Lee, M.R., Park, E.D., Lee, S.M., Lee, H.K., Park, K.H., and Park, M.J. (2011) Kinetic study of the dehydration of d-xylose in high temperature water. *React. Kin. Mech. Cat.*, **103**, 2267–2277.

63 Lamminpää, K., Ahola, J., and Tanskanen, J. (2012) Kinetics of xylose dehydration into furfural in formic acid. *Ind. Eng. Chem. Res.*, **51**, 6297–6303.

64 Takagaki, A., Ohara, M., Nishimura, S., and Ebitani, K. (2010) One-pot formation of furfural from xylose via isomerization and successive dehydration reactions over heterogeneous acid and base catalysts. *Chem. Lett.*, **39**, 838–840.

65 Tuteja, J., Nishimura, S., and Ebitani, K. (2012) One-pot synthesis of furans from various saccharides using a combination of solid acid and base catalysts. *Bull. Chem. Soc. Jpn.*, **85**, 275–281.

66 Binder, J.B., Blank, J.J., Cefali, A.V., and Raines, R.T. (2010) Synthesis of furfural from xylose and xylan. *ChemSusChem*, **3**, 1268–1272.

67 Yang, W., Li, P., Bo, D., and Chang, H. (2012) The optimization of formic acid hydrolysis of xylose in furfural production. *Carbohydr. Res.*, **357**, 53–61.

68 Chheda, J.N., and Dumesic, J.A. (2007) An overview of dehydration, aldol-condensation and hydrogenation processes for production of liquid alkanes from biomass-derived carbohydrates. *Catal. Today*, **123** (1–4), 59–70.

69 Zeitsch, K.J. (2000) Furfural production needs chemical innovation. *Chem. Innov.*, **30**, 29–32.

70 Agirrezabal-Telleria, I., Larreategui, A., Requies, J., Güemez, M.B., and Arias, P.L. (2011) Furfural production from xylose using sulfonic ion-exchange resins (Amberlyst) and simultaneous stripping with nitrogen. *Bioresour. Technol.*, **102**, 7478–7485.

71 Sako, T., Sugeta, T., Nakazawa, N., Okubo, T., Sato, M., Taguchi, T., and Hiaki, T. (1991) Phase equilibrium study of extraction and concentration of furfural produced in a reactor using supercritical carbon dioxide extraction. *J. Chem. Eng. Jpn.*, **24**, 449–455.

72 Sako, T., Sugeta, T., Nakazawa, N., Okubo, T., Sato, M., Taguchi, T., and Hiaki, T. (1992) Kinetic study of furfural formation accompanying supercritical carbon dioxide extraction. *J. Chem. Eng. Jpn.*, **25**, 372–377.

73 Marcotullio, G., and de Jong, W. (2011) Furfural formation from D-xylose: the use of different halides in dilute aqueous acidic solutions allows for exceptionally high yields. *Carbohydr. Res.*, **346**, 1291–1293.

74 Rong, C., Ding, X., Zhu, Y., Li, Y., Wang, L., Qu, Y., Ma, X., and Wang, Z. (2012) Production of furfural from xylose at atmospheric pressure by dilute sulfuric acid and inorganic salts. *Carbohydr. Res.*, **350**, 77–80.

75 Jing, Q., and Lu, X.-Y. (2007) Kinetics of non-catalyzed decomposition of D-xylose in high temperature liquid water. *Chin. J. Chem. Eng.*, **15**, 666–669.

76 Román-Leshkov, Y., Chheda, J.N., and Dumesic, J.A. (2006) Phase modifiers promote efficient production of hydroxymethylfurfural from fructose. *Science*, **312**, 1933–1937.

77 Okuhara, T. (2002) Water-tolerant solid acid catalysts. *Chem. Rev.*, **102**, 3641–3666.

78 Li, L., Yoshinaga, Y., and Okuhara, T. (1999) Water-tolerant catalysis by Mo–Zr mixed oxides calcined at high temperatures. *Phys. Chem. Chem. Phys.*, **1**, 4913–4918.

79 Izumi, Y. (1997) Hydration/hydrolysis by solid acids. *Catal. Today*, **33**, 371–411.

80 Namba, S., Hosonuma, N., and Yashima, T. (1981) Catalytic application of hydrophobic properties of high-silica zeolites: I. Hydrolysis of ethyl acetate in aqueous solution. *J. Catal.*, **72**, 16–20.

81 Csicsery, S.M. (1986) Catalysis by shape selective zeolites-science and technology. *Pure Appl. Chem.*, **58**, 841–856.

82 Ishida, H. (1997) Liquid-phase hydration process of cyclohexene with zeolites. *Catal. Surv. Jpn.*, **1**, 241–245.

83 Kobayashi, S. (1998) New types of Lewis acids used in organic synthesis. *Pure Appl. Chem.*, **70**, 1019–1126.

84 Moreau, C. (2002) Zeolites and related materials for the food and non food transformation of carbohydrates. *Agro-Food-Industry Hi-Tech*, **13**, 17–26.

85 Moreau, C. (2006) Micro- and mesoporous catalysts for the transformation of carbohydrates, in *Catalysis for Chemical Synthesis* (ed. E. Derouane), John Wiley & Sons, Ltd, pp. 141–156.

86 Dias, A.S., Lima, S., Brandão, P., Pillinger, M., Rocha, J., and Valente, A.A. (2006) Liquid-phase dehydration of D-xylose over microporous and mesoporous niobium silicates. *Catal. Lett.*, **108**, 179–186.

87 Lima, S., Fernandes, A., Antunes, M.M., Pillinger, M., Ribeiro, F., and Valente, A.A. (2010) Dehydration of xylose into furfural in the presence of crystalline microporous silicoaluminophosphates. *Catal. Lett.*, **135**, 41–47.

88 Benvenuti, F., Carlini, C., Patrono, P., Raspolli Galleti, A.M., Sbrana, G., Massucci, M.A., and Galli, P. (2000) Heterogeneous zirconium and titanium catalysts for the selective synthesis of 5-hydroxymethyl-2-furaldehyde from carbohydrates. *Appl. Catal. A: Gen.*, **193**, 147–153.

89 Mizuno, N., and Misono, M. (1998) Heterogeneous Catalysis. *Chem. Rev.*, **98**, 199–217.

90 Lansalot-Matras, C., and Moreau, C. (2003) Dehydration of fructose into 5-hydroxymethylfurfural in the presence of ionic liquids. *Catal. Commun.*, **4**, 517–520.

91 Zhang, J., Zhuang, J., Lin, L., Liu, S., and Zhang, Z. (2012) Conversion of D-xylose into furfural with mesoporous molecular sieve MCM-41 as catalyst and butanol as the extraction phase. *Biomass Bioenergy*, **39**, 73–77.

92 Agirrezabal-Telleria, I., Requies, J., Güemez, M.B., and Arias, P.L. (2012) Pore size tuning of functionalized SBA-15 catalysts for the selective production of furfural from xylose. *Appl. Catal. B: Environ.*, **115–116**, 169–178.

93 Jeong, G.H., Kim, E.G., Kim, S.B., Park, E.D., and Kim, S.W. (2011) Fabrication of sulfonic acid modified mesoporous silica shells and their catalytic performance with dehydration reaction of D-xylose into furfural. *Micropor. Mesopor. Mater.*, **144**, 134–139.

94 Kim, Y.-C., and Lee, H.S. (2001) Selective synthesis of furfural from xylose with supercritical carbon dioxide and solid acid catalyst. *J. Ind. Eng. Chem.*, **7**, 424–429.

95 Carlini, C., Patrono, P., Raspolli Galletti, A.M., and Sbrana, G. (2004) Heterogeneous catalysts based on vanadyl phosphate for fructose dehydration to 5-hydroxymethyl-2-furaldehyde. *Appl. Catal. A: Gen.*, **275**, 111–118.

96 Moreau, C., Finiels, A., and Vanoye, L. (2006) Dehydration of fructose and sucrose into 5-hydroxymethylfurfural in the presence of 1-H-3-methyl imidazolium chloride acting both as solvent and catalyst. *J. Mol. Catal. A: Chem.*, **253**, 165–169.

97 Valente, A.A., Dias, A.S., Lima, S., Brandão, P., Pillinger, M., Plácido, H., and Rocha, J. (2006) Catalytic performance of microporous Nb and mesoporous Nb or Al silicates in the dehydration of D-xylose to furfural. *IX Congreso Nacional de Materiales*, Vigo, 1203–1206.

98 Dias, A.S., Lima, S., Carriazo, D., Rives, V., Pillinger, M., and Valente, A.A. (2006) Exfoliated titanate, niobate and titanoniobate nanosheets as solid acid catalysts for the liquid-phase dehydration of D-xylose into furfural. *J. Catal.*, **224**, 230–237.

99 Sádaba, I., Lima, S., Valente, A.A., and López Granados, M. (2011) Catalytic dehydration of xylose to furfural: vanadyl

pyrophosphate as source of active soluble species. *Carbohydr. Res.*, **346**, 2785–2791.
100. Suzuki, T., Yokoi, T., Otomo, R., Kondo, J.N., and Tatsumi, T. (2011) Dehydration of xylose over sulfated tin oxide catalyst: influences of the preparation conditions on the structural properties and catalytic performance. *Appl. Catal. A: Gen.*, **408**, 117–124.
101. Lam, E., Chong, J.H., Majid, E., Liu, Y., Hrapovic, S., Leung, A.C.W., and Luong, J.H.T. (2012) Carbocatalytic dehydration of xylose to furfural in water. *Carbon*, **50**, 1033–1043.
102. Lima, S., Antunes, M.M., Fernandes, A., Pillinger, M., Ribeiro, M.F., and Valente, A.A. (2010) Acid-catalysed conversion of saccharides into furanic aldehydes in the presence of three-dimensional mesoporous Al-TUD-1. *Molecules*, **15**, 3863–3877.
103. Speck, J.C., Jr. (1958) The Lobry de Bruyn- Alberda van Ekenstein transformation. *Adv. Carbohydr. Chem.*, **13**, 63–103.
104. Anet, E.F.L.J. (1964) 3-Deoxyglycosuloses (3-deoxyglycosones) and the degradation of carbohydrates. *Adv. Carbohydr. Chem.*, **19**, 181–218.
105. Feather, M.S., and Harris, J.F. (1973) Dehydration reactions of carbohydrates. *Adv. Carbohydr. Chem.*, **28**, 161–224.
106. Kuster, B.F.M. (1990) 5-Hydroxymethylfurfural. A review focusing on its manufacture. *Starch/Stärke*, **43**, 314–321.
107. Newt, F.H. (1951) The formation of furan compounds from hexoses. *Adv. Carbohydr. Chem.*, **6**, 83–106.
108. Aida, T.M., Tajima, K., Watanabe, M., Saito, Y., Kuroda, K., Nonaka, T., Hattori, H., Smith, J.R.L., and Arai, K. (2007) Dehydration of D-glucose in high temperature water at pressures up to 80MPa. *J. Supercritical Fluids*, **42**, 110–119.
109. Cope, A.C. (1959) Production and recovery of furans. US2917520.
110. Kuster, B.F.M., and van der Steen, H.J.C. (1977) Preparation of 5-hydroxymethylfurfural part I. dehydration of fructose in a continuous stirred tank reactor. *Starch/Stärke*, **29**, 99–103.
111. Kuster, B.F.M., and Laurens, J. (1977) Preparation of 5-hydroxymethylfurfural part II. dehydration of fructose in a tube reactor using polyethyleneglycol as solvent. *Starch/Stärke*, **29**, 172–176.
112. Moreau, C., Durand, R., Razigade, S., Duhamet, J., Faugeras, P., Rivalier, P., Ros, P., and Avignon, G. (1996) Dehydration of fructose to 5-hydroxymethylfurfural over H-mordenites. *Appl. Catal. A: Gen.*, **145**, 211–224.
113. Nakamura, Y., and Morikawa, S. (1980) The dehydration of D-fructose to 5-hydroxymethyl-2-furaldehyde. *Bull. Chem. Soc. Jpn.*, **53**, 3705–3706.
114. Szmant, H.H., and Chundury, D.D. (1981) The preparation of 5-Hydroxymethylfurfuraldehyde from high fructose corn syrup and other carbohydrates. *J. Chem. Tech. Biotechnol.*, **31**, 135–145.
115. Brown, D.W., Floyd, A.J., Kinsman, R.G., and Roshan-Ali, Y. (1982) Dehydration reactions of fructose in nonaqueous media. *J. Chem. Tech. Biotechnol.*, **32**, 920–924.
116. Musau, R.M., and Munavu, R.M. (1987) The preparation of 5-hydroxymethyl-2-furaldehyde (HMF) from D-fructose in the presence of DMSO. *Biomass*, **13**, 67–75.
117. Fayet, C., and Gelas, J. (1983) Nouvelle méthode de préparation du 5-hydroxyméthyl-2-furaldéhyde par action de sels d'ammonium ou d'immonium sur les mono-, oligo- et poly-saccharides. Accès direct aux 5-halogénométhyl-2-furaldéhydes. *Carbohydr. Res.*, **122**, 59–68.
118. Zakrzewska, M.E., Bogel-Łukasik, E., and Bogel-Łukasik, R. (2010) Ionic liquid-mediated formation of 5-hydroxymethylfurfurals. A promising biomass-derived building block. *Chem. Rev.*, **111**, 397–417.
119. Zhao, H., Brown, H.M., Holladay, J.E., and Zhang, Z.C. (2008) Prominent roles of impurities in ionic liquid for catalytic conversion of carbohydrates. *Top. Catal.*, **55**, 33–37.
120. Zhao, H., Holladay, J.E., Brown, H., and Zhang, Z.C. (2007) Metal chlorides in ionic liquid solvents convert sugars to

5-hydroxymethylfurfural. *Science*, **316**, 1597–1600.

121 Cao, Q., Guo, X., Yao, S., Guan, J., Wang, X., Mu, X., and Zhang, D. (2011) Conversion of hexose into 5-hydroxymethylfulfural in imidazolium ionic liquids with and without a catalyst. *Carbohydr. Res.*, **346**, 956–959.

122 Qi, X., Watanabe, M., Aida, T.M., and Smith, J.R.L. (2010) Fast transformation of glucose and di-/polysaccharides into 5-hydroxymethylfurfural by microwave heating in an ionic liquid/catalyst system. *ChemSusChem*, **3**, 1071–1077.

123 Hu, S., Zhang, Z., Song, J., Zhou, Y., and Han, B. (2009) Efficient conversion of glucose into 5-hydroxymethylfurfural catalyzed by a common Lewis acid $SnCl_4$ in an ionic liquid. *Green Chem.*, **11**, 1746–1749.

124 Ståhlberg, T., Rodriguez-Rodriguez, S., Fristrup, P., and Riisager, A. (2011) Metal-free dehydration of glucose to 5-(Hydroxymethyl)furfural in ionic liquids with boric acid as a promoter. *Chem. Eur. J.*, **17**, 1456–1464.

125 Nikolla, E., Román-Leshkov, Y., Moliner, M., and Davis, M.E. (2011) "One-pot" synthesis of 5-(hydroxymethyl)furfural from carbohydrates using tin-beta zeolites. *ACS Catalysis*, **1**, 408–410.

126 Ohara, M., Takagaki, A., Nishimura, S., and Ebitani, K. (2010) Syntheses of 5-hydroxymethylfurfural and levoglucosan by selective dehydration of glucose using solid acid and base catalysts. *Appl. Catal. A: Gen.*, **383**, 149–155.

127 Chareonlimkun, A., Champreda, V., Shotipruk, A., and Laosiripojana, N. (2010) Reactions of C_5 and C_6-sugars, cellulose, and lignocellulose under hot compressed water (HCW) in the presence of heterogeneous acid catalysts. *Fuel*, **89**, 2873–2880.

128 Rapp, K.M. (1987) Process for preparing pure 5-hydroxymethylfurfuraldehyde. Süddeutsche Zucker-Aktiengesellschaft. US4740605A1.

129 Fleche, G., Gaset, A., Gorrichon, J.-P., Truchot, E., and Sicard, P. (1982) Procedé de fabrication du 5-hydroxymethylfurfural. US4339387.

130 Román-Leshkov, Y., and Dumesic, J.A. (2009) Solvent effects on fructose dehydration to 5-hydroxymethylfurfural in biphasic systems saturated with inorganic salts. *Top. Catal.*, **52**, 297–303.

131 Mascal, M., and Nikitin, E.B. (2008) Direct, high-yield conversion of cellulose into biofuel. *Angew. Chem.*, **120**, 1–4.

132 Gruter, G.-J.M., and Dautzenberg, F. (2007) Method for the synthesis of 5-alkoxymethylfurfural ethers and their use. Furanix Technologies B.V.: The Netherlands WO2007104514.

5
A Workflow for Process Design – Using Parallel Reactor Equipment Beyond Screening

Erik-Jan Ras

5.1
Introduction

In 2004, a team of researchers from NREL, PNNL and EERE published a report assessing the top value-added chemicals accessible from sugar-based biomass conversion [1]. The classification strategy used distinguishes between drop-in replacements of existing molecules, for example, biobased acrylic acid, novel products, for example, poly-lactic acid and novel building blocks, for example, levulinic acid. In 2007 a follow-up report [2] extended the list of accessible building blocks with aromatics derived from lignin materials.

Many of the chemicals from Scheme 5.1 are under investigation by several academic and industrial groups, targeting microbial, enzymatic, fermentation or

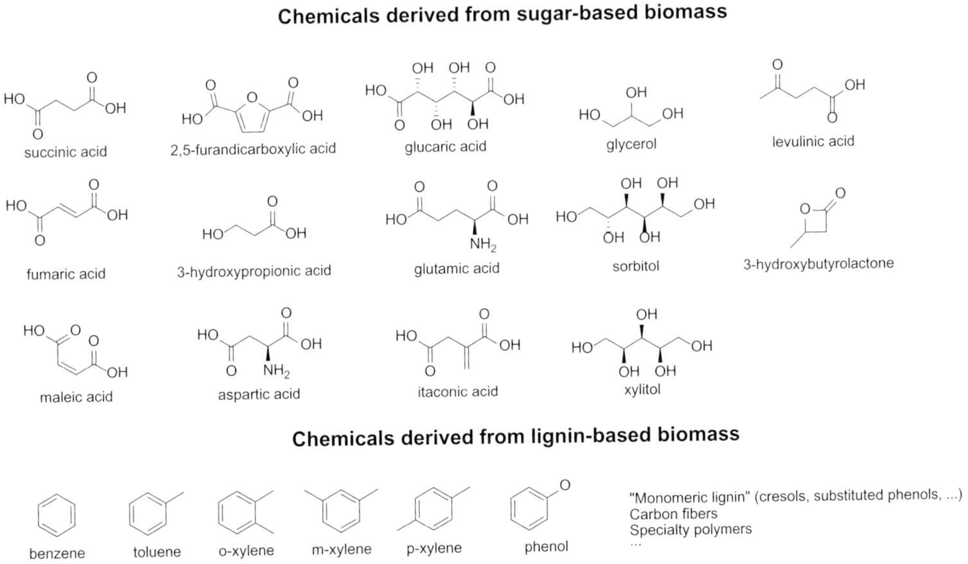

Scheme 5.1 Top listed value-added chemicals in the 2004 and 2007 reports [1, 2].

Catalytic Process Development for Renewable Materials, First Edition. Edited by Pieter Imhof and Jan Cornelis van der Waal.
© 2013 Wiley-VCH Verlag GmbH & Co. KGaA. Published 2013 by Wiley-VCH Verlag GmbH & Co. KGaA.

thermochemical-based processes. Even more groups are studying the further conversion of these chemicals to derived intermediates, products and polymers. It is here where a large amount of chemo-catalytic transformations are sought. In contrast to the early days of petrochemical research, the current state of the art in catalytic research offers the opportunity to test not only many catalysts in parallel but also many different process options in parallel.

5.2
The Evolution of Parallel Reactor Equipment

The high-throughput approach to materials discovery found its origin in the pharmaceutical industry. In particular, the discovery of new drug molecules is often accelerated by testing immense component libraries against biological targets. Perhaps inspired by the near infinite amount of possible catalysts, scientists in heterogeneous catalysis were inspired by these techniques. From the late 1990s the number of scientific and patent publications related to high-throughput catalysis increased significantly (Figure 5.1). In the early days of high-throughput experimentation for heterogeneous catalysis, parallel reactor equipment was often marketed as a means to rapidly discover new catalysts using many experiments of low information density. Many academic and industrial groups developed proprietary equipment using a plurality of reactor configurations and analytical techniques [3, 4]. The applications studied were often limited to relatively simple, gas-phase reactions, like the oxidation of H_2 [5]. Later, more and more real world applications were translated to high-throughput equipment.

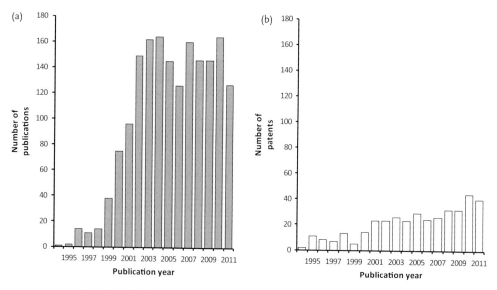

Figure 5.1 Number of scientific publications (a) and patent publications (b) related to high-throughput catalysis for the period 1994 to 2011 based on a search in Scifinder™.

Nowadays, complex applications like Fischer–Tropsch synthesis [6], hydrodesulfurization and biomass conversion [7–9] can be studied making use of modern, commercially available parallel reactor equipment. In fact, one could consider the high-throughput approach mature and "just" a part of the toolbox available to the modern chemist. Although classically associated with early catalyst screening, there are many other stages of research where parallel reactor equipment has added value over conventional bench-scale equipment.

5.3
The Evolution of Research Methodology – Conceptual Process Design

In current chemical research, finding a catalyst and a combination of process conditions is only one part, albeit an important one, of the challenges encountered in developing a new chemical process. The overall economics of the process that is considered will determine whether the process will find its way into the market place. This has become more and more apparent in the development of processes based on sustainable feedstock, where the entire process has to be developed, from feedstock selection to process steps. It is best practice to carry out the techno-economic evaluation of a process in parallel with its development. In this manner those elements of the process having a strong negative impact on the overall production cost can be identified early on. The typical elements of a techno-economic analysis, although generalized, are shown in Figure 5.2. Although catalyst and conditions are at the heart of any process, there are many other elements to consider. As a chemist, we have a natural tendency to focus on maximizing activity and selectivity of our catalyst. This is accomplished by varying catalyst composition, feedstock composition, and process conditions. In reality, unfortunately, there are quite a lot of processes where the main contributions to overall cost are

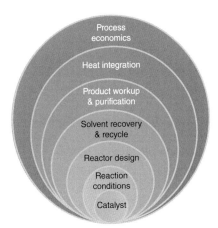

Figure 5.2 Different layers of information involved in a typical integrated conceptual process design study.

not, or only very loosely, related to the catalytic performance obtained. If this is the case, it would be beneficial to identify the main factors determining cost early on in the process development cycle.

The estimation of production cost will become more accurate as research progresses and more data become available. More important is the fact that, in some cases, the estimated cost is identified to be so high that alternative process steps should be considered. When summarizing the research trajectory, from idea to pilot plant, different levels of detail are required at each stage. From an experimental data perspective, the information content of each experiment will increase as research progresses. Consequently, the understanding of the type of process layout required will also be improved along the same lines. As more detailed information becomes available, the estimates of required investments (both capital and operational) will be narrowed down (Figure 5.3).

To illustrate the typical conceptual process design approach, an example from literature [10] is used where two processes for the conversion of fructose to chemicals are compared. The first process is the conversion of fructose to 5-hydroxymethylfurfural (HMF). The second is the conversion of fructose to 2,5-methylfuran (DMF). Both processes also yield levulinic acid (LA) as a major side product (Scheme 5.2). HMF is a building block that has many outlets in applications like transportation fuels, (fine) chemicals and polymers [1]. DMF has been suggested as an alternative biobased fuel, with favorable energy density compared to ethanol [11]. The main side product obtained in both processes, LA, can be applied in subsequent conversions to fuel additives, monomers and solvents [12]. The performance of both processes has been summarized in Table 5.1. Since LA is produced in large quantities (33.5 vs. 38.0 metric ton per day) in both processes it should be considered as a product with value instead of as a waste

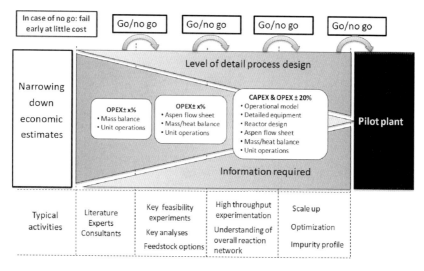

Figure 5.3 Increasing accuracy of process economic parameters as a process study progresses.

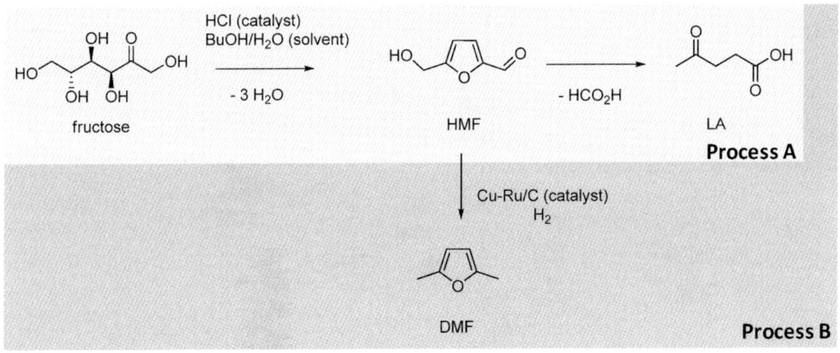

Scheme 5.2 Major products obtained from the conversion of fructose in the two processes evaluated in [10].

Table 5.1 Output of the two processes (A and B) in terms of the main products obtained for a fixed 300 metric ton per day feed of fructose (this corresponds to 1.67E + 06 moles per day) for both processes. The data is derived from Tables 2 and 6 in reference [10].

Process A – Fructose to HMF and LA			Process B – Fructose to DMF and LA		
HMF	174.2	Metric ton/day	DMF	96.6	Metric ton/day
	1.38E + 06	Moles/day		1.00E + 06	Moles/day
	83.0	Selectivity (%)		60.3	Selectivity (%)
LA	33.5	Metric ton/day	LA	38.0	Metric ton/day
	2.88E + 05	Moles/day		3.27E + 05	Moles/day
	17.3	Selectivity (%)		19.7	Selectivity (%)

stream. Note that failing to do so may render the process uneconomic due to the low carbon utilization.

Considering the process layout that has been used to simulate the HMF process (Figure 5.4) it becomes clear that the actual reactor, a continuous stirred tank reactor (CSTR) in this case, actually represents only a relatively small portion of the equipment needed for the process. Most of the equipment involved is related to the purification of products (HMF and LA), the recycle of unconverted feedstock (fructose) and co-solvent (butanol). This partition between reactor- and purification-related equipment is common. This is reflected in the analysis from the authors of the installed equipment cost for the process. Over 90% of the equipment cost (i.e., the capital investment needed to build a plant) is attributed to these sections of the process (Areas 200 and 400 in Figure 5.4). The other cost aspect that needs to be evaluated is the operating cost. The analysis of the authors shows that the recycle of fructose and the purification of levulinic acid (Area 400 in Figure 5.4) is a major contribution to the operating cost. The most significant contribution to operating cost is feedstock price, accounting for nearly 50% of the total cost. Although the authors state that increasing the yield of HMF is the easiest way to decrease the overall cost of the process, the current selectivity of 83% may be difficult to

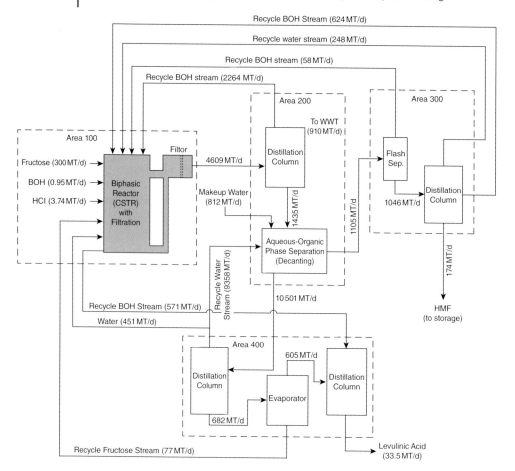

Figure 5.4 HMF production process block diagram. BOH: butanol; WWT: wastewater treatment (reproduced from [10] with permission).

improve on. A more detailed analysis shows that the limiting steps are (i) the separation of HMF to the butanol phase and (ii) the fructose conversion of 75%. This is in line with the cost impact of fructose recycle on both installed equipment and operating cost. To summarize the findings, in this case it would be best to focus research on identifying a catalyst with greater activity and maintained or improved selectivity compared to the currently used hydrochloric acid, reducing the cost impact of fructose recycle. If this catalyst could operate in a non-aqueous system it would reduce the impact of the overall separation process.

The DMF process outline (Figure 5.5) shows a large overlap with the HMF process outline (Figure 5.4). The largest difference is the additional reactor (Area 300 in Figure 5.5), a packed bed reactor for the hydrogenation and hydrogenolysis of HMF to DMF. This means the proportion of installed equipment cost related to reactors is larger for the DMF process than for the HMF process discussed

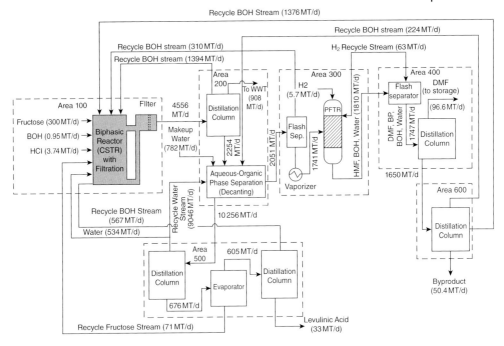

Figure 5.5 DMF production process block diagram. BOH: butanol; WWT: wastewater treatment (reproduced from [10] with permission).

before. Still, the overall installed equipment cost is dominated by the various purification steps. In terms of operating cost, the conclusions are quite similar to those for the HMF process.

Again, the feed stock cost makes up nearly 50% of the overall operating cost, and a large contribution comes from the recycle of unconverted fructose and the purification of levulinic acid (Area 500 in Figure 5.5). The levulinic acid purity obtained, 87% in the DMF process versus 98% in the HMF process, is considerably lower. The decrease in value for this lower purity fraction could be countered by adding an additional purification step at additional cost, both installed equipment and operational. The authors' analysis indicates that the cost impact factors for the HMF-part of the DMF process are similar to those identified in the analysis of the HMF process. The performance of the combined hydrogenation and hydrogenolysis of HMF to DMF has additional cost implications. For example, the catalyst cost (Cu-Ru /C) represents a large part of the fixed cost of the process. This implies that identifying a cheaper catalyst with similar performance would reduce the production cost. The performance of the hydrogenation reaction in general also offers a handle, since the obtained selectivity to DMF is only 60%. Increasing this selectivity by using a different catalyst and/or process conditions would result in an instant reduction in production cost. Also, the separation of HMF produced from the aqueous phase relies on large quantities of NaCl being added. This additive could have a negative impact on the performance of the hydrogenation reaction that is performed downstream. For both processes, the value that can be obtained for the produced LA is of concern. The market price and potential market demand for this product are described by the authors as uncertain.

From the two scenarios outlined it is clear that factors determining production cost for a given process are not always obvious. In fact, the complexity is such that one may find that just optimizing catalyst performance may not transform a non-economic process into an economic one. This means that the evaluation of the overall process performance and cost throughout the research and development cycle should be an integrated approach.

Conceptual process design (CPD) addresses this need by identifying problematic steps early on. Also, by considering multiple process outlines (reactor types, separation methods, recycle streams) one can maximize the chances of focusing on the most significant research topics in relation to process cost. This may entail optimizing catalyst performance, but it could also be the reduction of a particular side product or impurity, or the investigation of different purification methods. Of course, many other potential research topics could be identified. By integrating this economic evaluation in the actual research trajectory it is even possible that processes will be identified to be non-economic. Such a conclusion is ideally reached before many years of research effort have been invested.

The analysis as described above is ideally revised at each stage of process development. In a typical approach, multiple reactor types and separation methods are compared at an early stage. Parallel reactor equipment can be used in many stages, but near the pilot plant stage often additional data obtained from bench scale equipment is also obtained. The following sections will highlight typical experimental approaches, hardware and strategy, for making optimal use of the advantages of parallel reactor equipment[1].

5.4
Essential Workflow Elements

5.4.1
Catalyst Testing Equipment

Parallel reactor equipment is nowadays readily available. This holds true for batch and tubular reactors, and to a lesser extent also for continuous stirred tank reactors. Many companies provide both more conventional bench scale reactors and

---

1) Although these sections are written largely from the perspective of the equipment used and developed by Avantium, it is safe to assume that equipment from other vendors has similar capabilities. Since detailed information is typically not readily available in the public domain from the various vendors, this document is biased towards the information available within Avantium. For reasons of confidentiality details on catalysts, feedstocks and processes are obfuscated where needed.

modest size parallel systems based on similar technology [13–17]. Companies providing both fee-for-service research as well as equipment for sale [18–20] are not as abundant. Generally, batch reactors are more readily available in the marketplace than continuous reactors. The reason for this is the greater inherent complexity of continuous reactors. Although more complex compared to batch reactors, continuous reactors are still amenable to application in a parallel set-up. One needs to consider carefully the design principles from a chemical engineering perspective, but the same is true when constructing a well designed single bench-scale reactor. For industrial applications, since most large scale processes are operated in a continuous fashion, parallelization and miniaturization of continuous reactors is often preferred. As an example, the typical properties of the Avantium Flowrence* reactor are highlighted below [18]. Other vendors supply reactors of similar design and capabilities [13–17, 19, 20].

In order to use catalytic test data for scale-up activities it is essential to measure intrinsic kinetics. The two most important aspects that need to be addressed to ensure this are heat transfer and mass transfer. Performing a catalytic test under isothermal conditions is the first prerequisite. Our flow reactors, with a typical internal diameter of 2.0 mm, are designed to ensure this. When a reactor set-up is configured, the isothermal zone length is determined by measuring the temperature inside the reactors at various heights. An example outcome of this procedure is shown in Figure 5.6. In this case an isothermal zone length of 65 mm[2] is found where, at a set point of 400 °C, the measured temperature lies within 1 °C of the set point. In terms of bed volume this means that the maximum isothermal bed volume equals 200 µl. Assuming the smallest bed volume that can be loaded accurately equals 25 µl, a wide space velocity window is accessible. Based on a constant flow, the difference between the highest and lowest accessible space velocity is a factor 8.[3]

The second prerequisite that needs to be addressed is the requirement to obtain data in the absence of mass transfer limitations. This is ensured by testing the catalyst in powder form when kinetic experiments are being performed. Generally speaking, the requirements for achieving plug flow conditions can be summarized by the following criteria [21, 22][4]:

$$\frac{d_{reactor}}{d_{particle}} > \sim 8-10 \quad \text{and} \quad \frac{L_{reactor}}{d_{particle}} > \sim 10-50$$

In a typical set-up, we use particles with diameters in the range 50–200 µm, easily meeting the criteria above (for a 100 µm particle, a 65 mm bed height and a reactor

2) The isothermal zone length varies with desired temperature, for lower temperatures a longer isothermal zone is accessible. For cases where larger bed volume or length is required to achieve the desired conversion, longer reactors are used.

3) Note that to achieve more representative results it may be required to fill up the catalyst bed for different reactors to the same volume using an inert diluent material.

4) When running near the extremes of the conversion curve somewhat more sophisticated rules need to be observed to ensure plug flow behavior, taking into account conversion level and linear velocity.

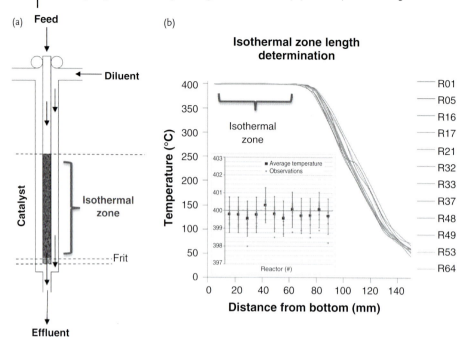

Figure 5.6 Results of an isothermal zone length determination for a temperature set point of 400 °C (b). For reference purposes, a schematic representation of an individual reactor is given (a).

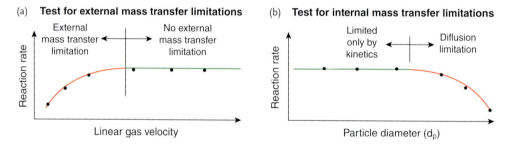

Figure 5.7 Typical diagnostic tests for (a) external and (b) internal mass transfer limitations.

diameter of 2.6 mm: $d_r/d_p = 26$ and $L/d_p = 65$). A more extended set of criteria for ensuring ideal testing conditions can be found in the literature (for example, [21, 22]).

When transferring results to scale-up and engineering it is important to make sure that, for the most relevant catalysts, some additional data are generated using different catalyst particle sizes and using different gas velocities (Figure 5.7). This will greatly enhance the understanding of mass transfer effects and thus facilitate the scale-up of the catalyst from powder to extrudate.

The most robust method for mapping internal and external mass transfer limitations is an experiment where the linear gas velocity and the particle diameter are varied in a designed manner. This can be achieved by performing an isothermal experiment where a series of different catalyst mass loadings, each across a range of different particle sizes, is loaded into the parallel set-up. During the run recipe, the linear gas velocity is increased. By extracting the rates of reaction at the (for the process) relevant space velocity, a two-dimensional map can be created linking internal and external mass transfer effects and thus effectively identifying the window of operability. Since this method requires a large number of reactors to be used per catalyst it is typically used near the end of a catalyst development project.

5.4.2
Kinetics and Pseudo-Kinetics

In order to demonstrate the validity of microscale data to develop a process to the level of a pilot plant this section outlines the results obtained by one of our customers[5]. The application studied here is the hydrotreatment of vacuum gas oil (VGO), in particular the hydrodesulfurization (HDS). VGO hydrotreating is generally perceived as one of the most complicated applications to scale down to laboratory scale. A large contribution to this complexity is that the application is performed in trickle phase, which is inherently more complex than gas-phase or full liquid-phase.

The data presented here are part of a catalyst qualification study, where we compared the performance of various catalyst powders and extrudates on microscale. The customer subsequently tested the most successful catalysts (in extrudate form) in their pilot plant and obtained the same ranking and trends. The feed used in this study was an industrial VGO containing 22.000 ppm sulfur. The weight-hourly space velocity used in our reactors was 1.2 h 1 using an H_2 to liquid ratio of 800. The temperature window explored was 355 to 385 °C in 10 °C increments. The pressure used was 80 barg. For evaluating the performance of catalysts a general model describing the rate constant for HDS is used:

$$K_s = \frac{WHSV}{n-1} \cdot \left(\frac{1}{S_{sample}^{n-1}} - \frac{1}{S_{feed}^{n-1}} \right) \quad \text{with} \quad n = 1.6 \text{ and } S_{feed} = 2.2 \cdot 10^4 \text{ ppm}$$

Comparing the resulting apparent activation energies (350 kJ mol^{-1} for powder and 289 kJ mol^{-1} for extrudates) it is clear that, as expected, the extrudates show more internal mass transfer limitations than the powder. Comparing K_0 values, the lower value found for the extrudates is explained by the increased chance of bypassing. This being said, the performance obtained for extrudates in a small scale test is remarkably good. Obviously the powder catalysts are more suitable for obtaining intrinsic kinetic data (Figure 5.8).

---

5) For confidentiality reasons we cannot disclose the name of the customer and the nature of the catalysts used.

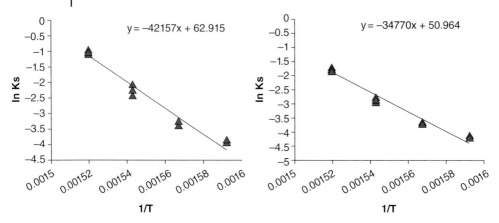

Figure 5.8 Arrhenius plot for an identical catalyst tested in (a) powder and (b) extrudate form. Each catalyst was tested in four reactors.

The ultimate objective of our customer was to scale the best performing catalysts on microscale to pilot plant scale directly. Figure 5.9 shows the performance data for two representative catalysts on microscale and pilot scale[6]. As expected the sulfur content, more specifically the HDS reaction rate, is lower in the pilot plant than on the microscale. This is explained simply by the additional mass transfer and heat transfer issues associated with the larger scale of operation. Since the temperature and mass transfer effects on rate of reactions were well studied at small scale, reaching the same performance level at pilot scale was a simple matter of slightly adjusting the operating temperature and space velocity. The key message here is that a detailed study of the factors determining performance at the small scale can directly be translated to handles for fine-tuning large scale performance. Please note that at two different scales it is unlikely that the same conditions result in exactly the same performance, so understanding the key factors is of critical importance. Obviously, the same is true for making the step from pilot plant to full commercial scale.

The example outlined here demonstrates a case where the additional scale-up step was not required. In some cases, for example, in the case of large exothermic effects, an additional step may still be advisable. In this case, the small-scale data should be carefully examined in order to design a number of appropriate experiments using a bench-scale reactor.

5.4.3
Statistical Design of Experiments

With the use of parallel reactor equipment larger numbers of experiments, and thus larger numbers of variables, become accessible. When considering catalyst

---

6) The smaller difference in microscale to pilot performance for catalyst B compared to catalyst A can be explained by the fact that catalyst B has a greater pore volume and higher surface area.

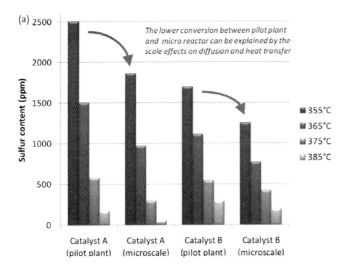

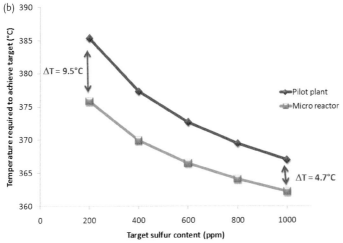

Figure 5.9 Direct comparison of catalyst extrudates A and B tested on microscale and pilot plant scale (a) and difference in temperature required to reach a certain target level of sulfur for catalyst A (b).

synthesis, catalyst pretreatment and process conditions alone, the number of possible experiments quickly becomes incomprehensible. Statistical experimental design techniques are ideal for structuring experimental design campaigns. From early screening to final optimization and process robustness studies, experimental design methods are well established. Many textbooks are available [23–26] as well as ready-to-use software [27, 28]. At the most basic level, experimental design methods can be classified as either "screening" or "optimization" methods. Screening methods are predominantly used to reduce the number of variables that need to be considered for optimization. Optimization methods are used to identify the "sweet spot", the most ideal combination of catalyst composition and operating conditions. The cost in terms of experiments needed is the most important practical difference. Screening methods require relatively few experiments but provide an answer with less information content. Optimization methods on the other hand require more experiments but also provide much more information. In a typical usage scenario, a screening design is first used to reduce the number of variables considered to the most relevant subset. This screening design is then followed by an optimization design based on the most influential variables. Figure 5.10 gives an overview of typical design types and the required number of experiments associated.

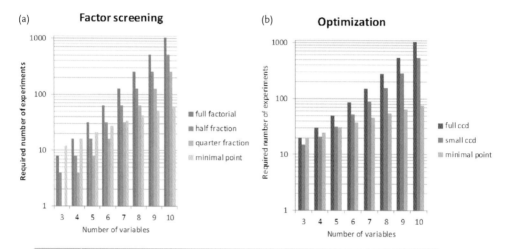

number of variables	Factor screening				Optimization		
	full factorial	half fraction	quarter fraction	minimal point	full ccd	small ccd	minimal point
3	8	4	n.a.	12	20	15	20
4	16	8	4	16	30	21	25
5	32	16	8	21	50	32	31
6	64	32	16	27	86	52	38
7	128	64	32	34	152	88	46
8	256	128	64	42	282	154	55
9	512	256	128	51	540	284	65
10	1024	512	256	61	1042	542	76

Figure 5.10 Number of experiments required per number of variables under investigation for a number of typical screening (a) and optimization (b) designs.

Although historically microreactors are predominantly positioned in the screening stage of catalyst development, there are numerous applications of the same technology in later stage catalyst development. In particular, the optimization of process conditions, catalyst composition and feed composition is readily optimized using parallel reactor technology. The benefit of using a parallel reactor set-up is twofold: (i) using multiple reactors decreases the timeline, and (ii) using multiple reactors reduces the risk by allowing multiple catalyst systems to be optimized simultaneously. This being said, for conducting these optimization-type experiments, a different experimental strategy is typically required compared to "simple" screening experiments. The remainder of this section outlines a possible strategy[7] for a combined optimization of feed composition (three-component feed) and process window (temperature and space velocity).

In the collaboration using this strategy we were faced with, from screening, a relatively large number of lead catalysts that could serve as candidates for optimization. Optimizing these leads one at a time using a single reactor would compromise the project timeline significantly. By making use of a parallel reactor set-up, in our case 64 reactors in parallel, the various leads can be optimized in a much more efficient manner. The first things to consider are the physical constraints of the parallel reactor set-up used. In the case of our set-up we have 64 reactors divided over 4 blocks of 16 reactors each. Each of these blocks can have its own temperature set point and each reactor can contain a different catalyst or catalyst loading. In a typical set-up, the pressure and feed composition are fixed for the total of 4 reactor blocks[8]. Within the run recipe, pressure and feed composition can be varied as desired. Given these constraints a logical experimental strategy would be the variation of space velocity by means of catalyst loading, and using a fixed temperature per block of 16 reactors. In this collaboration we used 5 different loadings per catalyst, effectively spanning an order of magnitude in terms of space velocity and 4 temperatures spaced 10 °C apart. Using this strategy, 3 catalysts could be treated per run. The loading and temperature strategy used is outlined in Figure 5.11.

Given the fact that a stable catalyst system is used, the feed composition can be varied within the run recipe. By including reference feed composition points at regular intervals in the run recipe, stability of the catalyst can be probed by comparing key performance indicators at the various instances of this reference feed composition. This approach should be used if the catalyst stability towards changes in the feed is in question or unknown. If stability issues are known a priori, a different strategy with shorter and less variable runs should be used.

---

7) For confidentiality reasons the chemistry used is obfuscated. The strategy and outcome were obtained in a customer collaboration which was successfully scaled to pilot plant within weeks after the results were transferred to the customer. During the implementation phase at the customer facilities frequent discussions were held to cross-reference results and to discuss strategy.

8) Technically speaking the feed section and pressure regulation can also be set up for each individual block. In reality, the extra effort required for this, as well as the additional complexity, outweigh the benefits in most cases.

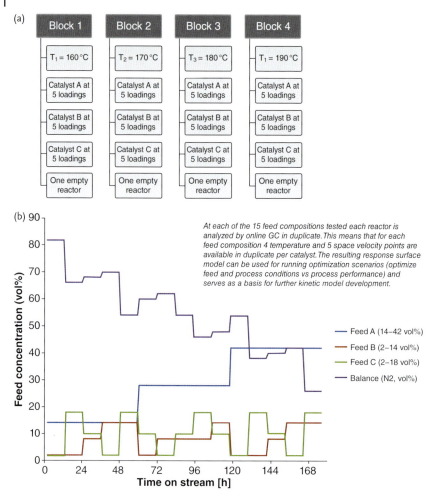

Figure 5.11 Reactor loading strategy used (a) and various feed compositions used during the run recipe (b) for the optimization scenario described in the text.

If experiments are set up and conducted in the manner outlined above, the resulting data nicely fit into the context of a response surface model design. In these models, the responses are described as a mathematical function of the inputs (in this case: temperature, space velocity and concentrations of the three feed components). A model is fitted to each response (in this case: conversions, space-time yield and selectivity to side products) describing the explored parameter space in a mathematical model. Using these models, optimization scenarios can be explored, for example: "What conditions and feed composition do I need in order to maximize STY and minimize side product selectivity?" or "What is the impact

Table 5.2 Summary of the parameter space studied, responses obtained and overall process objectives.

Variables	Responses	Objectives
Feed composition (A, B, C) Temperature Space velocity	Conversions (A, B, C) Side-product selectivity Impurity content Space-time yield	Maximize space-time yield Minimize side-product selectivity to a value less than or equal to 10% Minimize impurity content to a value less than or equal to 1000 ppm

of increasing the feed concentration on selectivity?". The design space explored and the process objectives are outlined in Table 5.2.

Since each of the responses is described as a mathematical model based on the variables, realizing the objectives within the explored parameter space can be achieved using a multi-objective optimization. In such an optimization routine, objectives and constraints can be applied to both variables and responses. It is important to realize that the responses in a typical catalytic process are coupled, and that in the end often a combination of variable settings is chosen that meet the objectives on all responses, rather than the combination of variable settings that meets just one objective. The basis of selecting the right operating window is process economics. The economics should take into account operating cost, recycle and separation considerations, and productivity. To clarify, please consider the example below where the three objectives stated in Table 5.2 are treated on an individual basis (Table 5.3).

Clearly, the process windows obtained when optimizing for the individual objectives are not acceptable. As an alternative the objectives can be combined. After review of the economics it was decided that conversion of C was most critical to recycle and separation and a target value of 50% conversion was assigned. Moreover, it was decided that side-product selectivity and impurity content were more important than space-time yield. The fact that we have models describing each of these responses means that this scenario (and any other scenario) can be explored *in silico* without the need to first perform additional experiments.

Figure 5.12 shows the obtained response surfaces for this scenario. The settings required to obtain this performance are high inlet concentrations of feeds A and B, an intermediate inlet concentration of feed C and high values of space velocity and temperature. The impurity level for these settings is 800 ppm, well below target.

Optimization of process conditions is only one of the many applications for the use of response surface methodology. The key message is that parallel reactor equipment is an important tool to facilitate this type of research. First, because of the high data quality and consistency obtained, but also because more prospective catalyst candidates can be explored simultaneously, reducing the risk of late stage failures.

Table 5.3 Results of optimizing individual responses.

Scenario 1 – Maximize space-time yield		
Required variable settings	Obtained responses	Interpretation
Feed A – high Feed B – high Feed C – high Temperature – high Space velocity – high	Conversion A – intermediate Conversion B – low 😟 Conversion C – intermediate Side product – low 👍 Impurity content – intermediate STY – high 👍	Good solution for primary objectives. Generally low conversions, increasing separation and recycle cost. High temperature could have catalyst lifetime impact.

Scenario 2 – Minimize side-product selectivity		
Required variable settings	Obtained responses	Interpretation
Feed A – intermediate Feed B – intermediate Feed C – low Temperature – low Space velocity – high	Conversion A0000000 – low 😟 Conversion B – low 😟 Conversion C – low 😟 Side product – low 👍 Impurity content – low 👍 STY – low 😟	Space-time yield is at an unacceptably low level. Conversions generally speaking too low.

Scenario 3 – Minimize impurity content		
Required variable settings	Obtained responses	Interpretation
Feed A – intermediate Feed B – high Feed C – low Temperature – high Space velocity – low	Conversion A – high 👍 Conversion B – low 😟 Conversion C – low 😟 Side product – low 👍 Impurity content – 👍 STY – low 😟	Space-time yield is at an unacceptably low level.

5.4.4
Data Analysis

With the use of parallel reactor equipment the amount of data available to a researcher increases dramatically. Not only can more catalysts be explored, but also more process conditions per catalyst become accessible. This results in 100s or 1000s of data points being created in a relatively short timeframe. When confronted with such a data set one quickly learns that typical spreadsheet software does not offer sufficient flexibility to explore the data efficiently in a graphical manner. Specialized software packages are available, both commercial [29–31] and open source [32, 33].

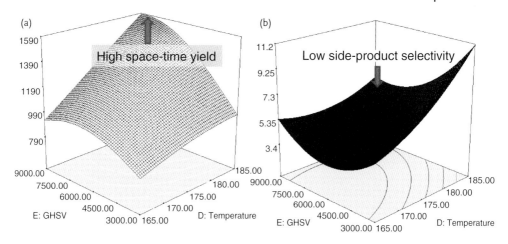

Figure 5.12 Response surfaces for space-time yield (a) and side-product selectivity (b) as a function of space velocity and temperature.

After the first visual exploration of the data one may choose to apply modeling methods. Methods like principal component analysis (PCA) [34] and partial least squares regression (PLS) [35] are commonplace in chemometrics, yet in catalysis are used less frequently than one would expect. PCA can be used to quickly identify the main trends in a data set and to identify outlying behaviors. It relies on reducing the dimensionality of a dataset by replacing the original variables by linear combinations thereof, so-called principal components or latent variables. This method is demonstrated in the remainder of this section by means of an example probing the reaction network observed in the hydrogenation of 5-ethoxymethylfurfural (EMF) [7, 9]. This work has been presented at the ORCS conference in 2010 in Monterey, CA. Where PCA "merely" aims to highlight and simplify the structure of a data set, PLS aims to correlate principal components identified to one or more responses. The resulting regression model can be used to make predictions for future experiments and to quantify the contribution of variables to a response. This method is demonstrated using a set of near-infrared spectra of various diesel samples that are correlated with the boiling behavior of the samples [36].

5.4.5
Example of PCA Applied to Catalysis

PCA, in essence, is a technique aiming to reduce the dimensionality of a data set to facilitate interpretation. It achieves this goal by replacing the original variables by linear combinations thereof using singular value decomposition. The result is twofold. First, a new data set is obtained in which each of the observations is expressed in terms of these linear combinations (the latent variables or principal components). The value of an observation in terms of a latent variable is termed

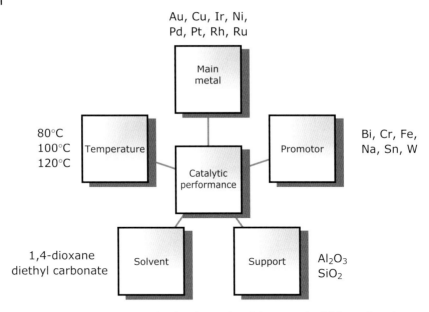

Figure 5.13 Parameter space explored in the combined data sets. The full factorial combination of these parameters results in 576 combinations which have all been tested. Additionally, monometallic catalysts and blanks were also tested.

"score". Secondly, per latent variable, a vector is obtained quantifying the importance of each of the underlying raw variables in a linear combination. These weights are termed "loadings". We demonstrate the method here by analyzing the yields of various intermediates, and side products obtained in the hydrogenation of 5-ethoxymethyl furfural (EMF). The total data set comprises 576 data points, including variations in catalyst composition, support type, solvent and temperature using a full factorial experimental design (Figure 5.13).

PCA analysis reduces the yields in the original parameter space to two PCs, together explaining 68% of the overall variance in the original data. Practically this means we can describe almost 70% of what is going on in our experiments by looking at two factors rather than at the original eight (Figure 5.14).

In Figure 5.15a, we see that the loadings for all yields are positive in the first PC. This simply means that this first PC is directly correlated with catalyst activity. Every observation with a high score on the first PC will have a high conversion, and every observation with a low score on the first PC will have a low conversion. Translating this observation to the scores plot (Figure 5.14) implies that, generally, conversion increases from left to right. This means we can effectively explain a large portion of the differences in conversion using the first PC. Looking at the loadings on the second PC (Figure 5.15b), we see large differences in effect size (magnitude of the bar). Moreover, the effects have different directions. This gives information about the relationship between the variables. For example, the yield

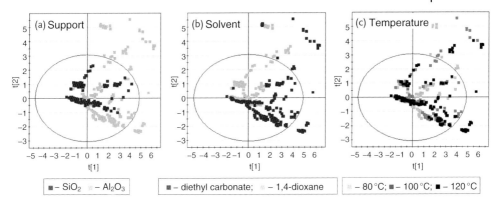

Figure 5.14 Scores plot of the two PC model colored by (a) support, (b) solvent and (c) temperature. In all cases, the x-axis corresponds to the first PC t[1], explaining 49% of variance in the data and the y-axis corresponds to the second PC t[2], explaining 19% of variance in the data.

of the unsaturated aldehyde **3** is anti-correlated with the yield of the etherification product **5**. This means that if a high yield of **3** is found, a low yield of **5** is expected and vice versa. From a chemistry point of view this makes sense, because if the furan ring has been hydrogenated to produce **3**, the formation of the unsaturated component **5** is no longer possible. More generally, the second PC gives information about the selectivity of each observation.

Combining these two PCs allows us to quickly find the maximum yield of a given component in the scores plot. For example, in our case a high yield of the saturated aldehyde **3** is expected at a moderately positive score on the first PC based on the loading for **y(3)** in Figure 5.15a and a strongly positive score on the second PC based on the loading for **y(3)** in Figure 5.15b. Following the same reasoning, we can expect a high yield of the etherification product **5** at a strongly positive score on the first PC, and a strongly negative score on the second PC. Subtle trends can also be identified. An example of this is the saturated alcohol **4**, which only occurs in small amounts (a maximum of 5% yield and an average of only 0.1% yield across all data points). This component may easily be overlooked by conventional visual data analysis. In our case, the PCA model shows (Figure 5.15) that a "high" yield of **4** always coincides with a high yield of etherification product **5**. This is explained by the fact that high amounts of **5** can only be formed if substantial quantities of its direct precursor **2** are first formed. Instead of etherification to **5**, a small portion of the unsaturated alcohol **2** could be further hydrogenated to the saturated alcohol **4**. The reason why only a minor amount of the unsaturated alcohol follows this path is because it requires a change in adsorption mode (ring vs. oxygen).

Turning our attention to the effect of the metals used, the first thing that should be noted is that the two most successful silica-supported catalysts in the

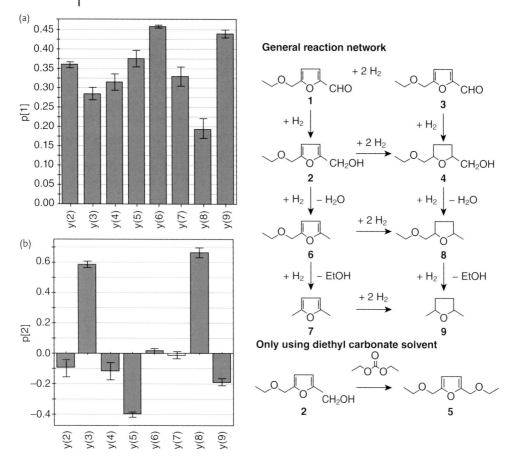

Figure 5.15 Loadings plots for the first PC (a) and second PC (b), accompanied by the reaction scheme for indicating the identity of components. The error bars show the 95% confidence interval on the loadings. Only y(7) and possibly y(6) could be considered insignificant since the confidence interval either includes or is very close to zero.

data set are not explicitly identified by the model at hand. This is not an error, but simply a confirmation of the most important attribute of PCA models: only trends valid across the majority of the data are identified. Less systematic trends, like two odd catalysts outperforming the other catalysts, will not be identified. Clearly, this implies that PCA analysis should only be used together with conventional data analysis and not as its replacement. One of the major trends that can easily be identified is the high selectivity of Pd catalysts towards the unsaturated alcohol **3**. From Figure 5.15 we learn that a high yield **y(3)** coincides with a high score on both the first and second PC. This translates to the top-right quadrant in the scores plot (Figure 5.15), which is almost exclusively populated with Pd catalysts.

5.4.6
Example of PLS Applied to Diesel Properties

PLS extends the dimension reduction achieved by PCA with a regression model. Here the latent variables (the "scores") are used to derive a model, correlating the original variable to one or more responses. Using this method, a relatively simple model can be obtained, even for large data sets. Also, compared to classical experimental design methods, PLS is more forgiving towards less structured data sets as typically encountered in early catalysis research. It is an especially useful method when dealing with a large number of variables that could potentially relate to the responses, without too much prior knowledge on such correlations being available. The method is demonstrated briefly here by means of a model correlating near-infrared spectra of diesel samples with the boiling behavior of these samples.

NIR spectroscopy is often used as a rapid analytical method in the petrochemical industry. By making use of chemometric tools like PLS, a multivariate calibration, taking the entire spectrum into account rather than the recorded intensities at a few specific wavelengths, is easily constructed. Using this method, even if to the bare eye hardly significant, small and subtle differences between spectra that correlate with the response of interest will be identified and used.

To get a feel for this feature, consider the spectra in Figure 5.16. Although there is well over 100 °C difference in the 50% mass loss temperature between

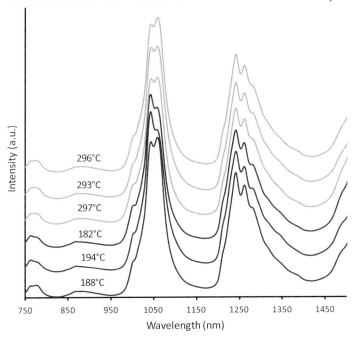

Figure 5.16 NIR spectra of 6 diesel samples and the associated temperatures corresponding to 50% mass loss obtained using simulated distillation.

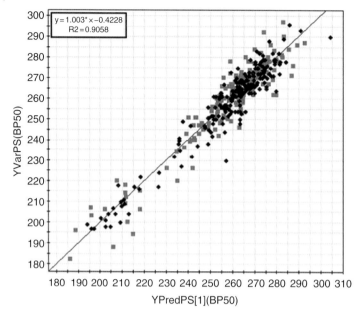

Figure 5.17 Measured boiling point (vertical) versus predicted boiling point (horizontal) using PLS. The observations used to construct the model ($n = 210$) are indicated in black and the observations that have been used to validate the model ($n = 185$) are indicated in gray.

the least and most volatile sample, the differences in the obtained NIR spectra are subtle in nature in the regions around 1050 and 1250 nm. One would think that such small differences would provide difficulties in establishing a correlation between spectra and boiling behavior but this is far from true. Figure 5.17 shows the prediction performance of a PLS model relating the NIR spectra of 395 diesel samples against the 50% mass loss temperatures from simulated distillation. The quality of the model ($R^2 = 0.91$, $Q^2 = 0.90$, RMSEE = 5.9 °C, RMSEP = 6.9 °C) is excellent, indicating a solid relationship between the NIR spectra and the boiling points.

Looking more clearly into the inner workings of the model, it becomes clear that indeed relatively large changes in sample volatility correspond to relatively small differences in NIR spectra. This is best evaluated by comparing the loadings of the PLS model (Figure 5.18). Indeed, the regions identified by observing the spectra correspond to the regions having the greatest impact on boiling behavior. Due to the fact that the inputs for a PLS model are scaled, the fact the differences between spectra are small in an absolute sense has no impact. These small changes are readily identified as being of key importance.

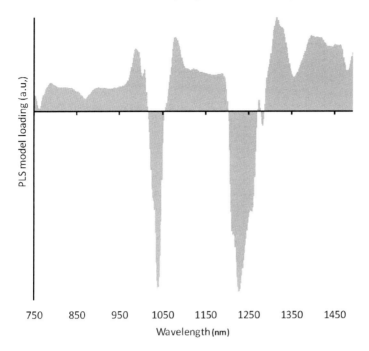

Figure 5.18 Loadings of the PLS model discussed in the text. A negative value indicates a region of the spectrum where, when high peak intensities are observed, lower boiling points can be expected. In contrast, positive value indicate regions in the spectra where high peak intensities correspond to higher boiling samples.

5.5
Other Examples of Parallel Reactor Equipment Applied Beyond Screening – Long-Term Catalyst Performance

The evolution of catalytic activity and selectivity over time is key to the successful implementation of a process at industrial scale. Although not typically associated with parallel reactor technology, the investigation of these topics is ideally suited to the use of such reactor systems. The reason for this is straightforward: access to multiple reactors allows a detailed investigation of multiple catalysts. Conventionally, this type of investigation is applied in the final research stages of a program. Since the typical tests involved are lengthy (up to months on stream) one needs to focus on a single catalyst if only a single reactor is used. The impact of such a catalyst failing in such an extended test is enormous. The research effort would have to be ramped up again, studying the deactivation or selectivity losses in detail. After all this effort the extended performance test would have to be repeated. The setback in time to commercialization could cause significant financial losses. Parallel reactor equipment can address this issue by allowing more than one catalyst to be evaluated in an extended performance test. Using 10–20 catalysts in this stage, rather than just a single catalyst, reduces the risk of late-stage failure tremendously. An additional positive aspect of parallel reactor equipment is the higher consistency of the data obtained for multiple catalysts compared to sequential testing in a single reactor.

Changes in catalyst behavior over time are always of key interest in industry. The longer an amount of catalyst can be utilized the better this is for the economics of the process. In a batch process one would aim for the highest number of times a catalyst can be recycled. In a flow process the objective would be to achieve the longest possible time on stream. In both cases, the performance in terms of activity and selectivity is key. Typically, by making minor changes in the operating conditions the effective lifetime of a catalyst can be extended while maintaining acceptable product specifications. Time-scales and mechanisms of deactivation differ widely between different chemistries and different catalyst types (Figure 5.19) [37].

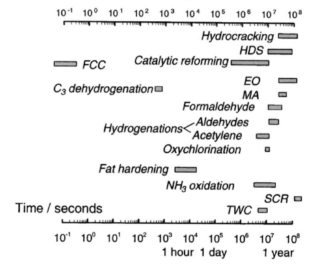

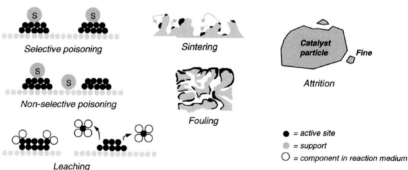

Figure 5.19 Timescale of catalyst deactivation for different processes (a) and possible mechanisms for catalyst deactivation (b) (adapted from [37]).

On the lower extreme of the scale, FCC catalysts show deactivation on a time-scale of seconds. On the other hand, catalysts used in hydrodesulfurization can have an effective lifetime of several years. If deactivation needs to be studied on a time-scale of days to months, parallel reactors are ideally suited. For longer and short time-scales, long-term performance data are typically obtained at larger scale.

Perhaps one of the most well known examples of the importance of stability data is Fischer–Tropsch (FT) synthesis. Here an initial deactivation of the catalyst, by means of filling up the catalyst pores with heavy product deposits, is actually desirable and required to obtain stable catalyst performance. Although historically perceived as a complex reaction to implement in a parallel reactor system, current technology is well capable of handling this reaction [6]. Due to the complex start-up behavior of the reaction, even a simple performance comparison already requires several days of testing time. This alone is already a solid justification for the use of parallel reactor systems. As an example, consider the performance of low-loading Ru catalysts on various supports over a period of about five days (Figure 5.20). Clearly here the initial deactivation has progressed significantly, but still a full steady state has not yet been established. For early screening work this may not be too much of an issue, when comparing the activity of the three supports used (TiO_2, Al_2O_3 and SiO_2) the ranking is clear even in this short test. In later stage research, where differences between individual catalysts are small, the importance of a longer test time will become more apparent. These small differences also demonstrate the other important advantage of parallel reactors: data consistency. When trying to identify small differences in activity or selectivity

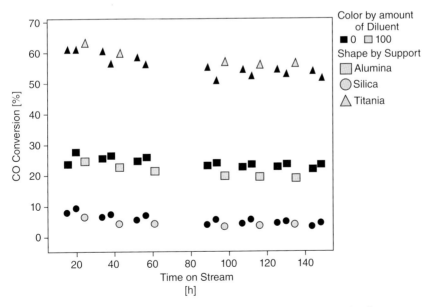

Figure 5.20 Activity of 1 wt% Ru on various oxides with (closed symbols) and without (open symbols) bed dilution using α-alumina. (reproduced from *Catal. Today* 171 (2011) 207–210, with permission).

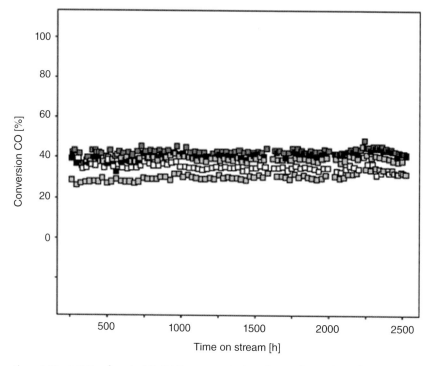

Figure 5.21 Activity of a set of Co/Al$_2$O$_3$ catalysts over a period of three months. Conversion is maintained near constant by adjusting the feed flow during the run to obtain a clear understanding of selectivity trends (reproduced from *Catal. Today* 171 (2011) 207–210, with permission).

between similar catalysts the experimental error when using sequential experimental in a single reactor may well be greater than the difference in performance that is of interest. Longer performance tests at small scale are well within the reach of current technology. As an example, consider the activity of a number of cobalt-based catalysts obtained over a period of three months (Figure 5.21). Here the activity of the catalyst systems is monitored and a constant conversion is maintained during the run by periodically adjusting the flow of syngas to compensate for activity losses. Even when only comparing a few catalysts, parallel reactor equipment here offers a great benefit over a single reactor: three months of test time per catalyst forces the researcher to narrow the scope of the investigation to the bare minimum if a single reactor is used.

Other applications where long test cycles are required are numerous. When catalyst fouling by coking is observed, typically an activity test cycle is followed by a regeneration cycle, and to get a good understanding several cycles are performed back to back. The advantage of parallel reactors here, besides allowing multiple catalysts to be compared with high consistency, is in the use of automated control software allowing the cycles to follow one another with minimal intervention of the operator. The reductive regeneration of catalysts on stream is also easily achieved in a similar manner.

5.6
Concluding Remarks

The previous sections show that there are many applications for parallel reactor equipment that go beyond the more traditional primary screening. In fact, in nearly each stage of catalyst research, there are areas of interest that benefit from not just a single catalyst being studied. With the current trend towards renewable building blocks and products, many new to world processes are under investigation in both industry and academia. This requires the discovery of new catalysts, bio-derived feedstocks are substantially different from petrochemically derived feedstocks. Also, in terms of process development, new inventive engineering solutions are required. Although the biobased field can learn from the achievements of the petrochemical industry over the past decades, there is often too little synergy between existing catalysts and processes to be of direct use. For this reason, it is important to integrate conceptual process design early on in the catalyst discovery effort. By evaluating economics at each stage and always being on the lookout for potential pitfalls, one reduces the risk of late-stage failure.

References

1 Werpy, T., and Petersen, G. (2004) Top Value Added Chemicals from Biomass: Volume I, US-DOE.
2 Holladay, J.E., White, J.F., Bozell, J.J., and Johnson, D. (2007) Top Value-Added Chemicals from Biomass: Volume II, US-DOE.
3 Hendershot, R.J., Snively, C.M., and Lauterbach, J. (2005) *Chem. Eur. J.*, **11**, 806.
4 Turner, H.W., Volpe, A.F., Jr., and Weinberg, W.H. (2009) *Surf. Sci.*, **603**, 1763.
5 Moates, F.C., Somani, M., Annamalai, J., Richardson, J.T., Luss, D., and Wilson, R.C. (1996) *Ind. Eng. Chem. Res.*, **35**, 4801.
6 van der Waal, J.K., Klaus, G., Smit, M., and Lok, C.M. (2011) *Catal. Today*, **171**, 207.
7 Ras, E.J., Maisuls, S., Haesakkers, P., Gruter, G.J., and Rothenberg, G. (2009) *Adv. Synth. Catal.*, **351**, 3175.
8 Van Der Waal, J., Van Putten, R., Ras, E., Lok, M., Gruter, G., Brasz, M., and De Jong, E. (2011) *Cellul. Chem. Technol.*, **45**, 461.
9 Ras, E.J., McKay, B., and Rothenberg, G. (2010) *Top. Catal.*, **53**, 1.
10 Kazi, F.K., Patel, A.D., Serrano-Ruiz, J.C., Dumesic, J.A., and Anex, R.P. (2011) *Chem. Eng. J.*, **169**, 329.
11 Roman-Leshkov, Y., Barrett, C.J., Liu, Z.Y., and Dumesic, J.A. (2007) *Nature*, **447**, 982.
12 Bozell, J.J., Moens, L., Elliott, D.C., Wang, Y., Neuenscwander, G.G., Fitzpatrick, S.W., Bilski, R.J., and Jarnefeld, J.L. (2000) *Resour. Conserv. Recycl.*, **28**, 227.
13 *FreeSlate (former Symyx)* (2013) www.freeslate.com/ (accessed 8 January 2013).
14 *HEL* (2013) www.helgroup.com (accessed 8 January 2013).
15 *Mettler-Toledo* (2013) http://us.mt.com/us/en/home/products/L1_AutochemProducts.html (accessed 8 January 2013).
16 *Premex* (2012) www.premex-reactorag.ch/ (accessed 8 January 2013).
17 *Parr* (2013) www.parrinst.com/ (accessed 8 January 2013).
18 *Flowrence by Avantium* (2013) http://www.avantium.com/chemicals/chemicals-systems/flowrence/ (accessed 8 January 2013).
19 *hte* (2012) www.hte-company.com (accessed 8 January 2013).
20 *Avantium* (2012) www.avantium.com/ (accessed 8 January 2013).
21 Berger, R.J., and Kapteijn, F. (2007) *Ind. Eng. Chem. Res.*, **46**, 3863.

22. Berger, R.J., and Kapteijn, F. (2007) *Ind. Eng. Chem. Res.*, **46**, 3871.
23. Antony, J. (2003) *Design of Experiments for Engineers and Scientists*, Butterworth-Heinemann.
24. Khuri, A.I. (2006) *Response Surface Methodology and Related Topics*, World Scientific Publishing Co. Pte. Ltd.
25. Tamhane, A.C. (2009) *Statistical Analysis of Designed Experiments – Theory and Applications*, Wiley.
26. Ryan, T.P. (2011) *Statistical Methods for Quality Improvement*, Wiley.
27. *Modde* Version 9.1 (2013) http://www.umetrics.com/modde (accessed 8 January 2013).
28. *Design-Expert and Design-Ease* Version 8 (2013) www.statease.com (accessed 8 January 2013).
29. *Miner3D* Version 7.4 (2013) www.miner3d.com/ (accessed 8 January 2013).
30. *Spotfire* Version 5 (2013) spotfire.tibco.com/ (accessed 8 January 2013).
31. *Tableau* Desktop version (2013) www.tableausoftware.com/ (accessed 8 January 2013).
32. *Ggobi* Version 2.1 (2010) www.ggobi.org/ (accessed 8 January 2013).
33. *Mondrian* Version 1.2 (2011) http://rosuda.org/mondrian/ (accessed 8 January 2013).
34. Wold, S., Esbensen, K., and Geladi, P. (1987) *Chemometr. Intell. Lab. Syst.*, **2**, 37.
35. Geladi, P., and Kowalski, B.R. (1986) *Anal. Chim. Acta*, **185**, 1.
36. *SWRI diesel data set* (2005) http://www.eigenvector.com/data/SWRI/index.html (accessed 8 January 2013).
37. Moulijn, J.A., Van Diepen, A.E., and Kapteijn, F. (2001) *Appl. Catal. A*, **212**, 3.

6
Braskem's Ethanol to Polyethylene Process Development

Paulo Luiz de Andrade Coutinho, Augusto Teruo Morita, Luis F. Cassinelli, Antonio Morschbacker, and Roberto Werneck Do Carmo

6.1
Introduction

6.1.1
Overview of Braskem Activities and History

Braskem attracted the attention of the industry in 2010 with the start up of its ethanol-to-ethylene plant in Brazil that allowed the company to produce its "Green PE" and become the largest producer of bioplastics in the world. It is even more surprising that this news was generated by a merely eight-year-old company that grew to become America's largest petrochemicals producer.

Braskem was formed in 2002 by the consolidation of six petrochemical companies, and grew both by the subsequent acquisitions and investment in new plants. Among such acquisitions were five Brazilian companies, followed by an expansion into North America (five plants in the US) and Europe (four plants in Germany). Braskem's 35 industrial units manufacture polymers (polyethylene–PE, polypropylene–PP and poly vinyl chloride–PVC), fuels such as gasoline, and other chemicals such as BTX (benzene, toluene, xylenes), butadiene, isoprene, ethyl tert-butyl ether (ETBE) and caustic soda.

6.1.2
Why Renewable Polymers and Why Green Polyethylene?

Nowadays, there is a strong incentive to produce polymers that have lower impacts upon the environment. Among the candidate technologies are biopolymers, which may prove to have a variety of environmental benefits.

While there is no generally accepted definition of biopolymers, they can be defined either as polymers that biodegrade under certain conditions, or as polymers that are obtained from renewable feedstock. Additionally, some bioplastics are both biodegradable and made from renewable materials [1]. Companies are

Catalytic Process Development for Renewable Materials, First Edition. Edited by Pieter Imhof and Jan Cornelis van der Waal.
© 2013 Wiley-VCH Verlag GmbH & Co. KGaA. Published 2013 by Wiley-VCH Verlag GmbH & Co. KGaA.

currently working on the development of many types of biopolymers, many of which may eventually find relevant commercial applications.

Polyethylene (PE), which is obtained by the polymerization of ethylene, is one of the most important commercially available polymers. Conventional ethylene production is based on naphtha or natural gas, but Braskem selected an innovative renewable route for the production of PE based on the use of ethylene obtained from a biobased resource, ethanol obtained from the fermentation of sugar cane.

The ethylene monomer obtained from ethanol dehydration in the Braskem process (known as "green ethylene") is chemically identical to the monomer produced by the petrochemical route; the only difference is the source of carbon (biobased versus fossil). The polymers obtained from green ethylene are referred to as Green Polyethylene (or "Green PE").

One of the main benefits of Green PE is the utilization of biobased feedstock and the capability to sequester significant quantities of atmospheric CO_2, captured by the sugar cane plant. Another important characteristic of Green PE is that clients may use it as a substitute for fossil-based PE. Moreover, Green PE has the same properties and characteristics as the fossil-based PE [2]. Therefore, Green PE is a drop-in new product that does not require different machines or processing conditions to conventional PE.

6.2
Ethanol and Brazil

In the 1970s, Brazil started a program named Pró-álcool (Brazilian ethanol program) to decrease its dependence on foreign oil. Initially, the program focused on economic considerations, such as incentives for ethanol production from sugar cane [3]. Unfortunately, the program did not address many important social and environmental concerns [3].

As a result, Brazil experienced an impressive decrease in sugar cane ethanol production cost (experience curve), which led to an approximately threefold reduction in production costs and a yearly growth in productivity of 2.6% based on sucrose (which is further transformed to ethanol) [3, 4].

Using land to produce bioplastics raises concerns about the effect on food production and the impact on protected areas. In this regard, it is important to note that the sugar cane areas are distant from the Amazon rain forest (as shown in Figure 6.1) [5], and that Brazil has 22% of the arable land in the world, only 19% of which is cultivated so far [3].

In 2007, the area dedicated for sugar cane was 7.8 million hectares, which is one third of the area used for soybean crops and half that used for corn crops. Approximately half of the sugar cane area is dedicated to biofuels production. Sugar cane plantations for fuel production in Brazil correspond to 5% of cultivated land, 1% of the area of agricultural property, or 0.5% of the area of the country [3].

The cultivation of sugar cane for ethanol generates by-products that contribute significantly to the generation of electricity. Sugar cane mills use bagasse (a residue from sugar extraction) to produce renewable electric energy to power the

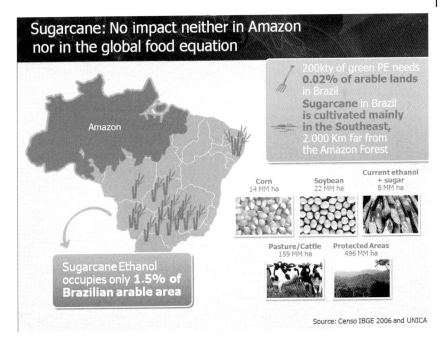

Figure 6.1 Map of Brazil showing the legal Amazon Forest and the location of the sugar cane plantantion. Based on UNICA [5] and Censo IBGE 2006 [6].

mills, and sell excess electricity into the power grid. In fact, this renewable source of electricity is very important as sugar cane bagasse co-generation currently accounts for 6% (or 7.3 million kW) of the installed power generation capacity in Brazil (behind only hydroelectric and natural gas generation) [7].

More recently, in an effort to decrease the use of additives in gasoline and to cut greenhouse gas emission (GHG), the United States has turned to corn-based ethanol. The US has since become the world's largest ethanol producer (Brazil ranks second).

In January, 2012, US federal tax incentives for ethanol producers as well as the Tariff on Brazilian exports into the US expired. Both of these factors will likely reduce the profitability of ethanol. However, the US ethanol industry can survive these transitions since the industry is already established [8, 9]. Aside from the fact that Brazilian ethanol is approximately half the cost of US corn ethanol, transportation costs make Brazilian ethanol less competitive in the US market [10]. Moreover, Brazil imported one-third of US exported ethanol (greater than 1 billion gallons and some 2.5 billion US$) in order to satisfy internal demand [11].

In Western Europe, wheat and sugar beet have been used for ethanol production. It is interesting to note that sugar-beet productivity per hectare is lower than sugar cane, but the grow rate is impressive: between 1961 and 2009 the sugar-cane and sugar-beet productivity per hectare increased 41 and 130%, respectively [12].

When different sources of ethanol are compared, sugar cane has the highest productivity of ethanol per hectare (7500 l per hectare), while beets and corn have productivities of 5500 and 3500 l per hectare respectively [13]. The output/input energy ratio[1]) of sugar cane has a value of 9.3, corn and beet have values of 1.4 and 2, respectively [3], suggesting that ethanol obtained from sugar cane is a preferable choice to produce Braskem's green ethylene and Green PE.

6.3
Commercial Plants for Ethanol Dehydration

Dehydration technology, to produce ethylene from ethanol, is not a new process, and is based upon discoveries dating back to 1783–1797 [14–16]. The first industrial plant was built and operated by Elektrochemische Werke GmbH at Bitterfield in Germany in 1913, where the main use of ethylene was for the production of ethane for refrigeration use. It is interesting to note that during the Second World War ethanol dehydration technology was the only source of ethylene in Germany, Great Britain, and the United States [17].

Since then and until the 1980s companies like Halcon/Scientific Design, ICI, ABB Lummus, Petrobras, Solvay, Petron Scientech and Union Carbide have designed and constructed plants in India, Pakistan, Australia, Peru and Brazil [17, 18]. At least one plant in India has been operating and supplying mono ethylene glycol (MEG) (from India Glycols), which can be used for the production of polyethylene terephthalate (PET). PET is a common material for blow molded plastic bottles that often incorporate a fraction of monomers obtained from renewable feedstock.

6.3.1
Salgema 100 kty Plant

In 1981, Salgema Indústrias Químicas Ltda. began operations in the Brazilian State of Alagoas with their chlor-alkali plant. During preliminary studies for this plant there was a requirement to identify means to dispose of chlorine. The best solution, at that time, was to utilize it for the production of ethylene dichloride (EDC) and vinyl chloride monomer (MVC), which were used for the production of PVC. Using this model, Salgema supplied ethylene and chlorine to another company, Companhia Petroquímica de Camaçari (CPC), to manufacture PVC.

The technology used at Salgema's ethanol dehydration plant was developed by Petrobras, based on an adiabatic process which showed an inside battery limits (ISBL) investment three times lower than the isothermal process that was used in other plants at the time, and a production cost approximately half that of the isothermal process [19].

It is worth noting that, at that time, Salgema received a ten year tax incentive to use the ethanol produced from Alagoas grown sugar cane for ethylene production (by dehydration). Figure 6.2 shows an aerial view of the former dehydration plant.

---

1) Output/input energy ratio is the ratio between the energy in the fuel (output) and the fossil-fuel energy used to make the fuel (input).

Figure 6.2 Aerial view of former chlor-alkali unit located in Maceio, State of Alagoas, Brazil. The arrow marker shows where the ethanol-ethylene dehydration unit was located.

However, the dehydration plant was not able to supply all the ethylene required by the PVC plant. As a result, the ethylene produced from fossil raw materials was brought by sea from the nearby State of Bahia. In that manner, the PVC produced by CPC was obtained from a mix of renewable and petrochemical sources.

The ethanol dehydration plant started its operation with a capacity of 60 kty, which was later increased to 85 and 100 kty in 1984 and 1990, respectively. In the late 1980s the construction of an ethylene pipeline of approximately 600 km from Bahia to Alagoas was completed (partially motivated by the increase in ethanol price due to the price increase and the expiration of tax incentives for ethanol use, as well as the lower price of fossil-derived ethylene). Consequently, the ethanol dehydration plant was shut down in May1992.

In 2002, both Salgema and CPC, then named Trikem, conveyed their assets to other companies, which formed the foundation for Braskem. The knowledge and human capital used in the dehydration plant located in Alagoas ultimately supported the design of a pilot plant and the 200 kty industrial green ethylene plant.

6.3.2
Triunfo 200 kty Plant

In 2010, Braskem started an ethanol dehydration industrial plant, to produce ethylene, with a capacity of 200 kty. The plant, located in Triunfo, Rio Grande do Sul State, Brazil, was built with an investment of 290 million US$. This plant is located within Braskem's Basic Petrochemicals Complex and supplies the existing polymerization facilities with green ethylene with purity and characteristics comparable to fossil or petrochemical ethylene. Figure 6.3 shows a view of Braskem's green ethylene plant.

The decision to invest in this 200 kty green ethylene plant was based on market signals that indicated there was a demand for renewable polymers which could contribute to a reduction in GHG emissions. Moreover, Braskem already

Figure 6.3 200 kty ethanol dehydration to ethylene plant located in Triunfo, RS, Brazil.

possessed the knowledge and experience to conduct the ethanol dehydration process at industrial scale. Despite that, it is important to note that critical improvements were required to increase the ethylene purity and decrease the capital expenditure (CAPEX) and operational expenditure (OPEX) from the technology used in the Salgema 100 kty ethylene plant (for PVC production) to the new Triunfo 200 kty ethylene plant (dedicated to polyethylene polymerization).

6.3.3
MEG Plants

The ethylene produced by ethanol dehydration technology can also be used as raw material for the production of monoethylene glycol (MEG); in this process ethylene is oxidized to ethylene oxide (EO), which is further reacted with water to produce the MEG.

PET is usually produced by the step-growth polymerization process using MEG and purified terephthalic acid as the main raw materials. In the case of the PET Coca-Cola PlantBottle™, the petrochemical MEG is replaced by the renewable MEG. In this process, the ethanol is dehydrated to ethylene, which is further converted to MEG (as mentioned above). The PET Coca-Cola PlantBottle™ currently incorporates up to 30% biobased source by weight to the composition of PET plastic [20][2].

In 2010, two plants in India produced green MEG from ethanol: one from India Glycols Limited at Kashipur, Uttar Pradesh and the other from Reliance Industries

---

2) Until now there has been no industrial production of PTA from renewable sources, but there are several companies working in technologies that could lead to a green PTA, e.g., Gevo, Draths (recently bought by Amyris), Virent, Global Bioenergies, Anellotech, Genomatica, Sabic, Butamax, etc. Also, there is a "PTA-like" R&D effort conduct by Avantium which is focused on the production of 2,5-furandicarboxylic acid.

Ltd at Kurkumbh, Maharashtra, with capacities of approximately 130 and 80 kty, respectively [21]. The green MEG used in PET PlantBottle™ is obtained from sugar cane ethanol and is produced in India by India Glycols [22] where the ethanol is produced from Brazilian sugar cane [23].

The technology to transform ethanol to EO/MEG has been sold by companies like Scientific Design [24, 25] and Petron Scientech [26], where the main technology is based on the adiabatic process. Bruscino mentioned that the integrated process based on an adiabatic reactor does not require further purification of ethylene for EO and MEG production [25]. The EO and MEG sections are very similar to the petrochemical version and are considered a known technology. A joint venture between Toyota Tsusho and Greencol Taiwan started a 100 kty of bio-MEG plant in Kaohsiung, Taiwan, in the second quarter of 2012 [22, 27].

6.3.4
Announced Renewable Polymer Projects

In 2008, Solvay announced a 60 kty dehydration plant to produce PVC and Songyuan Jian Biochemical announced a 300 kty ethanol-based ethylene plant. Nevertheless, no updates on these projects have been found in the literature or news [28–30].

In October 2010, Braskem announced the construction of another production unit utilizing renewable raw materials. This time, the plant will produce green propylene, which is also made from sugar cane ethanol, and will have a minimum production capacity of 30 kty. The start up is projected for late 2013. This will enable production of green polypropylene; which, in its fossil-based version, is the second most used thermoplastic resin in the world [31].

A joint venture between Dow/Mitsui was announced to build an integrated industrial plant in the Brazilian state of Minas Gerais at Usina Santa Vitoria with operations planned for mid-2013. Dow's project dates back to 2007 when it started as a joint venture with Crystalsev at the same location. However, that project was aborted in 2009 [32].

It is worth noting that existing green ethylene plants are located far from the ethanol plants that produce the raw material, and the logistics of ethanol transportation to the plants is reported to contribute contaminants that require special handling at the green ethylene plant [33].

6.4
Legislation and Certification

6.4.1
Ethanol Suppliers Code of Conduct

At the time that this document was written, Braskem's green ethylene plant was supplied with ethanol produced in accordance with a code of conduct. The basic principle of the code is the continuous improvement and compliance with

Brazilian legislation as modeled by the practices described in the São Paulo State Agri-environmental Protocol, the Global Compact, and the National Commitment to Improve Labor Conditions in Sugar cane Plantations. Therefore, Braskem's ethanol suppliers are required to adhere to the measures and good business practices listed below. These measures cover aspects such as [34]:

1) Cane burning,
2) Biodiversity,
3) Good environmental practices,
4) Human and labor rights,
5) Product life cycle analysis.

One year following the introduction of the code of conduct, Braskem S.A. has achieved 93% participation among its ethanol suppliers. Specifically, by the end of 2011, 19 of 21 ethanol plants that supply ethanol to Braskem follow the code of conduct, while the others were in the process of becoming compliant [35].

6.5
Process Description

6.5.1
Reaction

The production of ethylene from ethanol requires the removal of one oxygen and two hydrogen atoms, which is achieved by a dehydration reaction (as mentioned above). However, when ethanol is subjected to thermal or catalytic dehydration, other reactions also occur.

The complexity of these reactions is exemplified by the model proposed by Nguyen and Mao [36] and Phillips and Datta [37], described as the triangular network of reactions shown in Figure 6.4. The desired reaction is reaction 1, whereby ethanol (A) is dehydrated intramolecularly to ethylene (B). However, intermolecular dehydration also occurs (reaction 2), leading to the production of diethyl ether (DEE) (C), which in turn can be dehydrated to form more ethylene (reaction 3). In general, reaction 2 is favored by lower temperatures.

The ethanol intramolecular dehydration reaction is highly endothermic: $\Delta H = +10.82$ kcal mol^{-1}, which corresponds to about 386 Gcal per metric ton of ethylene. This has a direct impact on the choice of the industrial reaction system, since the reactor must be able to carry the reaction at a high selectivity while meeting the heat transfer requirements to operate within the temperature range required for good conversion.

Many other reaction networks have been proposed in the literature. Butt *et al.* proposed a series/parallel network [38] that include the same reactions above and added a fourth reaction whereby DEE (C) decomposes directly into ethanol (A) and ethylene (B). While the specifics may differ for different catalysts, similar networks are typically reported for the catalytic dehydration.

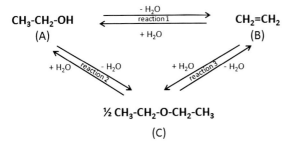

Figure 6.4 Reactions that occur in the dehydration of ethanol. Based on Phillips and Datta [37].

Although the actual mechanisms are still subject to discussion in the literature, it remains that under the right conditions it is possible to achieve high selectivity in the conversion of ethanol to ethylene. DEE may be formed in varied quantities and should not be regarded as a by-product, in the sense that it leads to a loss of reagent, since DEE can be recycled and converted to ethylene. Due to the tight specifications for ethylene composition in the polymer industry, a good understanding of such side reactions helps to avoid some costs associated with the separation of by-products.

6.5.1.1 Catalysts

The dehydration of ethanol may be carried out with a number of different catalysts. Winter and Meng-Teck describe a wide variety of catalysts that have been used in studies: alumina, silica, clay, silica-alumina, various metal oxides, phosphorus oxides, alumina with copper, nickel or chromium oxides as promoters, phosphates, molybdates and sulfuric acid [39]. This list has since been expanded to include many types of zeolites and other catalysts that are discussed below.

Not all of these catalysts have been employed in industrial areas. In the 1950s, supported phosphoric acid was prevalent because of the high purity product, but this was achieved at a high cost because of corrosion and low productivity [17]. Alumina was the preferred catalyst in the 1960s when a number of small plants were built around the world [39]. Most of these plants were shut down when catalytic cracking of naphtha became the choice for ethylene production at large scales. Proprietary catalysts have also been proposed, and Halcon SD Syndol catalyst was used to replace an older catalyst, aluminum oxide and magnesium oxide supported on porous silica, as reported by Kochar et al. [40].

Alumina catalysts are commercially available in pellets having different shapes, ranging from spheres and rings to more elaborate shapes with lower pressure drop. High surface areas and particle equivalent diameters of 3–5 mm are typical. Kagyrmanova et al., for instance, used a commercial catalyst in cylindrical pellets with 5–6 mm height and 3.6–4 mm diameter and a BET surface area of 270 $m^2\ g^{-1}$ [41]. Morávek and Kraus performed a kinetic study using an alumina catalyst with a BET surface area of 206 m² g⁻¹ [42]. Alumina catalysts are active and selective in a temperature range 450–500 °C, with conversions above 99% and ethylene molar selectivity ranging from 97% to 99%. Typical liquid hourly space velocities are around 0.15 to 0.5 h⁻¹, and coke formation is very low if water is present in the feed [17].

Zeolites have also been proposed for ethanol dehydration, requiring lower temperatures than those used with alumina catalysts. Phillips and Datta, for instance, studied different H-ZSM-5 zeolites (Si/Al =25–37.5) that produced ethylene below 300 °C [37]. However, even at these low temperatures coke formation at the Brönsted acid sites – with consequent catalyst deactivation – was observed. They also reported a positive effect of water in the feed and some concern about the effect of binders used in the zeolite [37].

Other catalysts show promise but still have to be proven in the field. NiAPSO-34 and SAPO-34 catalysts, for instance, have been tested with good results. Zhang *et al.* compared these catalysts to gamma-alumina and HZSM-5 (Si/Al = 25), and found that NiAPSO-34 was the best performer with respect to conversion, selectivity and stability [43]. SAPO-11 catalysts were shown by Wu *et al.* to lead, under certain conditions, to high conversions, reaching 99% ethylene in the gas phase product [44].

Catalyst manufacturers offer dehydration catalysts that are useful for ethanol dehydration in their catalogs. During the pilot plant development, Braskem tested about a dozen commercial catalysts with different compositions and shapes. The extensive testing allowed the selection of catalysts that had a superior performance in the adiabatic reaction process. Testing in the pilot plant did not stop with the industrial plant start up, and new catalysts, including custom developments, have shown potential to improve the reaction performance.

6.5.1.2 Side Reactions

The extent of side reactions depends on the catalysts used in the reactors and process conditions. Excessive side reactions result in both loss of ethylene conversion and production of undesirable products, thus having a direct impact on the downstream separation costs. There is no consensus for a model that includes all the reactions involved in ethanol dehydration reactors.

Nevertheless, side reactions can be separated into two classes:

- Class one encompasses reactions that involve either the reactant (ethanol) or a reaction product; these include the dehydrogenation of ethanol that produces acetaldehyde and hydrogen, the dimerization and oligomerization of ethylene that produces C4 and heavier olefins, the hydrogenation of olefins that form corresponding saturated hydrocarbons (ethane from ethylene is a major loss of value in this process), and coke formation on the catalyst.

- Class two encompasses reactions that involve impurities in the feedstock, which include the dehydration of the heavier alcohols present in the ethanol (generally called fusel oils). Propene, for instance, can result from the dehydration of 1-propanol found in fusel oil.

Morschbacker cites a number of by-products that have been observed in ethanol dehydration, such as: acetic acid, ethyl acetate, acetone, methanol, methane, ethane, propane, propylene, n-butane, butylene isomers, hydrocarbons with 5 carbons or more, carbon monoxide, and carbon dioxide [17]. The impact of these side reactions is discussed below as the separation of by-products is described.

6.5.1.3 Fixed Bed, Isothermal Reaction

Isothermal reactors were designed to maintain a nearly constant reaction temperature and heat transfer, even though there is significant temperature variation across the reaction bed (due to the low heat transfer rates in the catalyst).

Fixed-bed reactors are advantageous for heterogeneous catalytic processes because they are typically simpler to operate and require lower capital cost. Early designs used external heating to compensate for the heat required for the reaction. Externally heated fixed-bed reactors were used industrially in the 1960s. Kochar and Marcell discuss small ethylene plants (around 20 kty) in Brazil and Pakistan that used this arrangement [45]. They also mention that catalyst regeneration was a major problem.

Multitubular reactors with a shell-and-tube design have also been used. In the case of multitubular reactors, catalysts are placed in tubes, so that the small bed diameter can reduce radial temperature gradients while heated fluid is circulated around the tubes in the reactor shell.

Kagyrmanova *et al.* present pilot plant longitudinal temperature profiles where a large temperature drop (around 100–110 °C) develops rapidly at the reactor entrance [41]. The temperature then rises slowly and reaches essentially the feed temperature at the reactor exit.

Morais *et al.* also propose multitubular reactors for the dehydration of ethanol and observe the same fast temperature drop and slow recovery but are able to reduce the maximum temperature drop to about 25 °C [46].

6.5.1.4 Fixed Bed, Adiabatic Reaction

Adiabatic reactors were developed in the 1970s as a cost-effective alternative to the isothermal, multitubular reactors, described above. Adiabatic reactors are vertical cylindrical vessels that lack the internals to include a randomly packed bed of catalyst. No heat transfer equipment is provided.

Adiabatic operation requires a compromise between conversion and temperature drop. Different designs utilize different strategies to achieve good performance.

One approach is to use multiple reactors and limit the conversion in the first reactor. By converting only part of the reactant, a lower temperature drop takes place. For example, the process of Kochar *et al.* [40], describes vaporized ethanol sent to a reaction area composed of four reactors operated in series . The reactor effluents are sent to a furnace where they are reheated before entering the next reactor. They also show four reactor beds with a single furnace with different heating zones [40].

Another approach was proposed by Petrobras in the 1970s. Although it also includes an arrangement of reactors in series with intermediate heating, the reactors operate to near complete conversion. To reduce the temperature drop, a heat-carrying fluid (e.g., steam) is added to the feed of the first reactor. In downstream reactors, the effluent from the previous reactor acts as the heat-carrying fluid. The Petrobras technology has been used in the Salgema plant described above.

A similar approach is described by Bruscino [25]. In a process designed for chemical grade ethylene, the heat-carrying fluid is steam in a closed-loop recycling system [25]. Although there is no need for more than one reactor, the total steam flow is greater and leads to significantly higher operating costs.

Adiabatic operation results in a temperature profile along the catalyst bed. Typically, the reactor feed must be heated so that the exit temperature is high enough to avoid low or incomplete conversion. The temperature drop in the catalyst bed can reach 100 °C.

The use of steam as a heat-carrying fluid balances the obvious disadvantages (feed dilution and equilibrium shift to the left since water is a reaction product) with a positive effect on selectivity. Baratelli has shown that higher steam flow to the reactors results in lower formation of some by-products, especially those that lead to carbon loss (e.g., butene isomers) [19].

6.5.1.5 Fluidized Bed Reaction

A major problem with adiabatic reactors is that they inherently require the reaction to occur within a wide temperature range. With a complex reaction network, such as the one found in ethanol dehydration, the temperature variation can have an impact on the overall reaction selectivity. Fluidized bed reactors provide a tempting alternative since they provide a way to complete the reaction in a nearly isothermic environment that might, in theory, lead to an optimized reaction temperature.

Oil companies use fluidized beds extensively for gas oil cracking in the FCC process. Thus, it is not surprising that the fluidized bed route was proposed by oil companies and technology licensors in the 1970s. Nevertheless, the fluidized bed route has a number of drawbacks: the fluidized bed reactors are more difficult to operate compared to fixed-bed reactors; and the capital expense is higher.

Fluidized beds are an excellent choice when the endothermic reaction is carried out in conditions where a significant amount of coking occurs, as is the case with FCC. Burning the coke in a regenerator provides the heat required to maintain the constant temperature in the reactor. Tsao and Zasloff describe a fluidized bed system where the regenerator uses heated air to remove carbon and tars from the catalyst [47].

The high price of bioethanol imposes a high selectivity requirement – which provides motivation to limit coke generation. The use of existing, high-selectivity catalysts, would require additional heat input, increasing the capital cost of the fluidized bed system.

Recent patent applications, such as US 7,867,378 B2 [48], propose co-processing of ethanol and hydrocarbons from oil refining using a fluidized bed with separate reaction zones.

To date, there is no industrial ethanol dehydration in operation that uses fluid bed reactors.

6.5.2
Removal of Impurities

6.5.2.1 Unreacted Ethanol and Oxygenates

Reactor effluents contain a significant amount of water. In addition to water formed in the reaction, water is contained in the hydrous ethanol feed as well as any added water contained in the heat-carrying fluid in adiabatic reactors. The water is separated by cooling the reactor effluent to around 40 °C and sending the reactor effluent to a quench tower.

Most of the unreacted ethanol and the water-soluble oxygenates exit the quench tower at the bottom with the water. They can be easily removed by re-heating and distillation. Unreacted ethanol and DEE can be recovered and sent to the reactor feed stream. Acetaldehyde, which leads to undesirable side reactions (as described above), is typically sent to the fuel gas stream to burn in the furnace. However, DEE is not very soluble in water, so the removal in the quench tower is only partial.

The gas separated at the top of the quench tower is a raw ethylene stream containing ethylene (from 90 to 99.5% depending on the reactor design and catalyst), hydrocarbons, light gases (e.g., hydrogen, CO, CO_2), and small amounts of oxygenates. In some cases, a second tower is used to wash the raw ethylene stream with cool water to achieve further separation of oxygenates.

6.5.2.2 CO_2 and Acids

The raw ethylene contains small amounts (from 200 to 1000 ppmv) of carbon dioxide, which is easily removed by caustic washing. Any acid compounds such as acetic acid are also removed at this stage.

6.5.3
Ethylene Purification

Purification of the raw ethylene depends on the application. In integrated ethylene/MEG plants, as described by Bruscino [25], the raw ethylene is sent directly to the next plant. If the ethylene is required to be of polymer grade, further purification is a requirement.

Cryogenic distillation has been described as an effective separation technology to remove impurities from the ethylene [17, 49]. However, a major concern is the removal of carbon monoxide, a known inhibitor of polymerization catalysts.

In a typical separation, two distillation columns are used. Ethylene is obtained as a side stream from the first column, where heavier compounds (e.g., ethane, propene, butenes, and DEE) are removed at the bottom, and lighter compounds (e.g., hydrogen, carbon monoxide, and methane) are separated at the top. The second column is used to remove the remaining carbon monoxide.

Other separation technologies such as membranes can be used, but have not been found in industrial applications.

6.6
Polymerization

Apart from the biocontent, the ethylene specification required for the ethanol dehydration plant meets the same specifications imposed by existing polymerization facilities designed for petrochemical/fossil-based ethylene. The first 200 kty ethanol dehydration unit was installed by Braskem at its Triunfo site, where the polymerization capacity was already available. This permitted concentration of investments on the ethylene plant. Ethylene pipelines transport the green ethylene to the polymerization unit to process both Green PE and petrochemical/fossil-based ethylene.

As previously mentioned, the polymerization process that produces Green PE occurs in the same way as conventional ethylene polymerization, and uses essentially the same operational conditions. New green ethylene plants can use the same approach (using existing polymerization units) or may be implemented as integrated projects that use available technology for the polymer plant. When both green and petrochemical ethylene are processed at the same plant, special attention must be given to monomer transitions in order to ensure that the client will receive a product with the required biocontent.

6.7
Conclusion

Braskem's recent advances in Green PE were made possible by the conjunction of several factors, such as previous experience with ethanol dehydration (e.g., production of ethylene for a PVC production complex), increased demand arising from end user awareness of sustainability issues, Braskem growth into a global player with close customer relationships, and the existence of infrastructure such as operating polymerization units.

After the launch of the first industrial unit for green ethylene and with plans to invest in new biobased plants, Braskem is now recognized as a leader in the biopolymers area. Further, biobased developments are supported by Braskem's 2020 vision to be the world leader in sustainable chemistry, to innovate to serve people better.

The Green PE bioplastic is also proof that catalytic processes can be used to complement biotechnology in the production of biobased products with efficient energy use and significant reduction in GHG emissions.

Acknowledgments

The authors are very grateful to Jon Scott Larson for support in the preparation and revision of the manuscript.

References

1. European Bioplastics Bioplastics. http://en.european-bioplastics.org/bioplastics/ (accessed 25 February 2012).
2. Morschbacker, A., and Thielen, M. (2010) Basics of bio-polyethylene. *Bioplastics Mag.*, Sept 2010. pp. 52–55.
3. BNDES, and CGEE (2008) Sugarcane-based ethanol – Energy for Sustainable Development. *Bioetanol*, p. 189, http://www.bioetanoldecana.org/en/download/bioetanol.pdf (accessed 27 February 2012).
4. van den Wall Bake, J.D., et al. (2009) Explaining the experience curve: cost reductions of Brazilian ethanol from sugarcane. *Biomass Bioenergy*, 33 (4), 644–648.
5. UNICA The Industry – Production Map. *UNICA – SugarCane Industry Association*, http://english.unica.com.br/content/show.asp?cntCode=%7BD6C39D36-69BA-458D-A95C-815C87E4404D%7D (accessed 28 February 2012).
6. IBGE (2006) Censo Agropecuário 2006. http://www.ibge.gov.br/home/estatistica/economia/agropecuaria/censoagro/2006/default.shtm (accessed 12 April 2012).
7. ANEEL (2012) Banco de Informações de Geração. *ANEEL – Agência Nacional de Energia Elétrica*, http://www.aneel.gov.br/aplicacoes/capacidadebrasil/OperacaoCapacidadeBrasil.asp (accessed 28 February 2012).
8. New York Times (2012) After Three Decades, Tax Credit for Ethanol Expires. *New York Times*. http://www.nytimes.com/2012/01/02/business/energy-environment/after-three-decades-federal-tax-credit-for-ethanol-expires.html (accessed 27 February 2012).
9. The truth about cars (2011) U.S. Congress Stops Ethanol Subsidies & Tariff on Brazilian Imports. *The truth about cars*. http://www.thetruthaboutcars.com/2011/12/u-s-congress-stops-ethanol-subsidies-tariff-on-brazilian-imports/ (accessed 27 February 2012).
10. Crago, C.L., et al. (2010) Competitiveness of Brazilian Sugarcane Ethanol Compared to US Corn Ethanol. *Agricultural and Applied Economics Association – University of Minnesota Department of Applied Economics*, http://ageconsearch.umn.edu/bitstream/60895/2/Crago_CostofCornandSugarcaneEthanol_AAEA.pdf (accessed 27 February 2012).
11. Western Farm Press (2012) Brazil's Tax on Imported Ethanol Sparks Investigation Cry, Western Farm Press, http://westernfarmpress.com/government/brazil-s-tax-imported-ethanol-sparks-investigation-cry (accessed 27 February 2012).
12. Kuhlmann, T. (2011) Converting European Sugar Beets into High High-Value Products. *German-Brazilian Workshop on Value Creation from Bio-resources*. São Paulo: s.n., 03 16, 2011.
13. Ethanol Summit 2011 (2011) Ethanol Summit 2011. *UNICA*, http://english.unica.com.br/multimedia/ (accessed 28 February 2012).
14. Fourcroy (1797) *Ann. Chim.* January 1797, pp. 48–71.
15. Deiman, J.R., et al. (1795) *Crell's Chem. Ann.*, 195 (2), 195–205, 310–316, 430–440.
16. Priestley, J., and Banks, J. (1783) Experiments relating to phlogiston, and the seeming conversion of water into air. *Phil. Trans. R. Soc. London*, 73, 398–434.
17. Morschbacker, A. (2009) Bio-ethanol based ethylene. *J. Macromol. Sci. Part C.*, 49, 79–84.
18. Marcos, C.A.N. *Chemicals from Ethanol*, SRI Consulting, Menlo Park, 2007.
19. Baratelli, F., Jr. (1980) Projeto Eteno de Álcool. *B. Téc Petrobras*, 23 (2), 91–100.
20. The Coca-Cola Company (2012) PlantBottle® Basics. http://www.thecoca-colacompany.com/citizenship/plantbottle_basics.html (accessed 21 February 2012).
21. Chinn, H., and Kumamoto, T. (2010) *CEH Marketing Research Report*. s.l.: SRI Consulting, 652.4000 A.
22. Coca-Cola to announce PlantBottle partners (2011) *ICIS Green Chemicals*. [Online] 12 14, 2011, http://www.icis.com/blogs/green-chemicals/2011/12/coca-cola-to-announce-plantbot.html (accessed 21 February 2012).

23 The Coca-Cola Company (2012) PlantBottle Frequently Asked Question. http://www.thecoca-colacompany.com/citizenship/plantbottle_faq.html (accessed 21 February 2012).

24 Scientific Design Company, Inc. (2012) [Online] http://www.scidesign.com/ (accessed 21 February 2012).

25 Bruscino, M. (2010) Scientific design's ethanol to monoethylene glycol technology. *Hydrocarbon World*, **5** (2), 12–16.

26 Petron Scientech Inc. (2012) [Online] http://www.petronscientech.com/technologies.htm (accessed 21 February 2012).

27 Guzmán, D. (2012) http://greenchemicalsblog.com/2012/10/26/toyota-tsushos-bio-pet-rolls-out/ (accessed 4 January 2013).

28 Plastics Today (2008) Solvay Indupa Invests in Sugar-cane-derived Ethylene for PVC. http://www.plasticstoday.com/articles/solvay-indupa-invests-sugar-cane-derived-ethylene-pvc (accessed 21 February 2012).

29 Tan, F. (2005) Songyuan Ji'an Biochem Eyes Ethanol-to-ethylene Unit. *ICIS*. [Online] 10 31, 2005, http://www.icis.com/Articles/2005/10/31/1017275/songyuan-jian-biochem-eyes-ethanol-to-ethylene-unit.html (accessed 21 February 2012).

30 Lundgren, A., and Hjertberg, T. (2010) Ethylene from renewable resources, in *Surfactants from Renewable Resources*, (eds M. Kjellin and I. Johansson), John Wiley & Sons, Ltd, Chichester, UK, p. 6.

31 Braskem S.A. (2011) From Brazil to the World. *Braskem – I'm green*, http://www.braskem.com.br/plasticoverde/eng/unidade-industrial.html (accessed 21 February 2012).

32 Biofuels Digest (2011) The Sugar Rush: Dow, Mitsui Revive Major Renewables Project in Brazil. *Biofuels Digest*. [Online] 07 20, 2011, http://www.biofuelsdigest.com/bdigest/2011/07/20/the-sugar-rush-dow-mitsui-revive-major-renewables-project-in-brazil/ (accessed 21 February 2012).

33 Carmo, R.W., Belloli, R., and Morschbacker, A. (2012) Green polyethylene. *Bol. Inf. CETEA*, 24, **1**, 1–5.

34 Braskem S.A. (2011) Braskem – Code of Conduct for Suppliers of Ethanol. [Online] http://www.braskem.com.br/plasticoverde/doc/bras2011-codigo-de-conduta.pdf (accessed 12 February 2012).

35 UDOP (2011) Código de conduta do etanol na Braskem completa um ano com adesão de 93%. União *dos produtores de bioenergia*. [Online] 11 19, 2011, http://www.udop.com.br/index.php?item=noticias&cod=1078879 (accessed 21 February 2012).

36 Nguyen, T.M., and Mao, R.L.V. (1990) Conversion of ethanol in aqueous solution over ZSM-5 zeolites. *Appl. Catal.*, **58**, 119–129.

37 Phillips, C.B., and Datta, R. (1997) Production of ethylene from hydrous ethanol on H-ZSM-5 under mild conditions. *Ind. Eng. Chem. Res.*, **36**, 4466–4475.

38 Butt, J.B., Bliss, H., and Walker, C.A. (1969) Reaction rates in a recycling system – dehydration of ethanol and diethyl ether over alumina. *AIChE J.*, **8** (1), 42–47.

39 Winter, O., and Meng-Teck, E. (1976) Make ethylene from ethanol. *Hydrocarbon Process.*, **Nov**, 125–133.

40 Kochar, N.K., Merims, R., and Padia, A.S. (1981) Ethylene from ethanol. *Chem. Eng. Progress*, **77** (6), 66–70.

41 Kagyrmanova, A.P., Chumachenko, V.A., Korotkikh, V.N., Kashkin, V.N., and Noskov, A.S. (2011) Catalytic dehydration of bioethanol to ethylene: pilot-scale studies and process. *Chem. Eng. J.*, **176**, 188–194.

42 Morávek, V., and Kraus, M. (1986) Kinetics of individual steps in reaction network ethanol-ether-ethylene-water on alumina. *Collect. Czech. Chern. Commun.*, **51**, 763–773.

43 Zhang, X., Wang, R., Yang, X., and Zhang, F. (2008) Comparison of four catalysts in the catalytic dehydration of ethanol to ethylene. *Microporous Mesoporous Mater.*, **116**, 210–215.

44 Wu, L., Shi, X., Cui, Q., Wang, H., and Huang, H. (2011) Effects of the SAPO-11 synthetic process on dehydration of ethanol to ethylene. *Front. Chem. Sci. Eng.*, **5** (1), 60–66.

45 Kochar, N.K., and Marcell, R.L. (1980) Ethylene from ethanol: the economics are improved. *Chem. Eng.*, **87** (2), 80–81.

46 Morais, E.R., Lunelli, B.H., Jaimes, R.R., Victorino, I.R.S., Maciel, M.R.W., and

Maciel Filho, R. (2011) Development of an industrial multitubular fixed bed catalytic reactor as CAPE-OPEN unit operation model applied to ethene production by ethanol dehydration process. *Chem. Eng. Trans.*, **24**, 403–408.

47 Tsao, U., and Zasloff, H.B. (1979) Production of ethylene from ethanol. US Patent 4,134,926.

48 Pinho, A.R., Cabral, J.A.R., and Leite, L.F. Process for converting ethanol and hydrocarbons in a fluidized catalytic cracking unit, US patent 7,867,378, 2011.

49 *Ethylene from ethanol*. [Online] http://www.chematur.se/sok/download/Ethylene_rev_0904.pdf (accessed 21 February 2012).

7
Fats and Oils as Raw Material for the Chemical Industry
Aalbert (Bart) Zwijnenburg

7.1
Introduction – Setting the Scene, Definitions

Fats and oils are an important energy source in human nutrition (the highest of the three basic foods that consist of fats and oils, proteins, and carbohydrates). Besides the high energy content, many fats and oils contain fatty acids that are essential to health and cannot be synthesized by the human body [1].

At the start of a chapter on fats and oils as renewable raw materials, it may be wise to start to set the scene using definitions. The term fats and oils has been used with different meanings depending on origin, time and legal requirements (to name a few) so it is necessary to determine the scope of this chapter. According to Markley's definition [2] fats and oils refer to the glyceride esters of the fatty acids, which are obtained from plants or animals by physical processes such as cooking or steaming, pressing, and solvent extraction. Humans have used fats and oils for food and other applications since prehistoric times, because these renewable resources can be easily isolated from their source. As the fats and oils are insoluble in water, animal fats can be easily obtained from fatty tissue upon boiling in water. Vegetable oils are obtained by pressing of oilseeds.

Fats and oils are used interchangeably, the material solid at ambient temperature being denoted as fat, the liquid state as oil [1].

Oils and fats contribute significantly to the energy intake of the human diet. This chapter will focus on their use for chemicals, but it will be shown that the food applications will have an effect on the process design, especially when it comes to raw material selection.

Markley's definition may need to be revised in the future. In the 1980s, fats and oils mainly consisted of triglycerides of fatty acids but, especially in the last decade, new biotechnological processes have become available.

7.2
Why Fats and Oils Need Catalytic Transformation

Chemically, fats and oils are defined as consisting of triglycerides, or esters of fatty acids formed by the reaction of glycerol and three molecules of fatty acids. Fatty acids are designated as such because they were originally found to be a component of vegetable or animal fats [2].

For chemicals production, several strategies are being employed. Most of the performance characteristics of the products are dependent on the active group.

7.2.1
Carboxylic Acids

Fatty acids are derived from the glycerides by the so-called fat splitting process. These materials are used as such or converted into other oleochemicals such as alcohols and esters. A byproduct of the fatty acid production is glycerol. The main characteristic of the fatty acid is the carboxylic acid group, which in combination with the hydrophobic nature of the rest of the molecule makes it ideal for soap manufacture [3]. Fatty acids can also be polymerized by making use of the unsaturated carbon–carbon bonds and, ideally, retaining the carboxylic acid function. Major products are isostearic acid and dimer fatty acids [3–5].

7.2.2
Alcohols

Fatty acids can be converted into fatty alcohols that similarly find use in the detergent industry. Their characteristics will be discussed further in Section 7.4.

7.2.3
Amines and Amides

Fatty acids can be further converted into fatty amines and amides, by reactions involving nitrogen. In this way, the starting acidic materials are converted into neutral or basic products [3].

7.2.4
Esters

The main product by the ester route is the methyl ester (the reaction product of glyceride or fatty acid with methanol) for the biodiesel industry. Interestingly, this route is aimed at reducing the effect of the carboxylic acid group and making the overall molecule highly apolar [6]. It is also possible to make other molecules soluble in fatty matter by linking them to fatty acids. Besides biodiesel, many other esters exist and find use as solvents or other high performance products. Esterification is carried out both by acid–base and enzymatic catalysts, but these will not be the focus of this chapter.

Most of these above products are based on the chemistry involving the carboxylic acid/ester group in the glyceride raw material. However, most of the raw materials will be partly unsaturated in the fatty acid chain. One of the main characteristics of oils and fats is the so-called degree of unsaturation that can be present in the fatty acid part. Historically, this unsaturation has been expressed by the so-called iodine value (IV) (based on the reaction of the double bond with iodine). Typically, the higher the unsaturation, the lower the melting point. For example, a triglyceride based on stearic (C18) acid will have a melting point of 72 °C, whereas glycerol oleate (one double bond for each fatty acid) will have a melting point of 1.3 °C, and the corresponding linoleate (two double bonds per fatty acid) and linolenate (three double bonds per fatty acid) melt at −3.1 and −5.8 °C, respectively [7]. The degree of unsaturation also has an effect on the oxidative stability, the higher the unsaturation, the higher the chance for reaction with oxygen.

The melting point (and related degree of unsaturation) is one of the main characteristics that determines what fatty acid composition is formed during biosynthesis (lipogenesis). Although unsaturated fatty acids have a lower energy density, their lower melting point allows organisms to grow at lower temperatures. This effect is illustrated in Figure 7.1.

This is the main reason why tropical oils (such as palm oil) have a lower IV than oils from temperate regions, such as sunflower and soybean. The highly unsaturated oils (more than one double bond per fatty acid moiety) are therefore found in temperate regions, whereas the tropical regions yield more saturated materials.

The invention of catalytic hydrogenation of liquid oils at the start of the twentieth century [12, 13] enabled the use of highly unsaturated oils, such as whale oil and cottonseed oil, in both food and chemical applications. Since the 1950s, the main

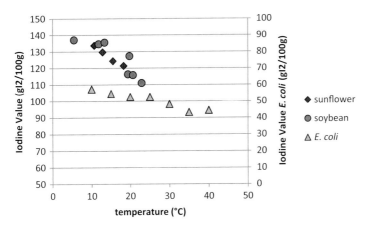

Figure 7.1 Degree of unsaturation (expressed as iodine value [8]) as a function of growth temperature. Soybean oil data were taken from [9], sunflower oil data from [10]. *E. coli* data (plotted on right axis) were taken from [11].

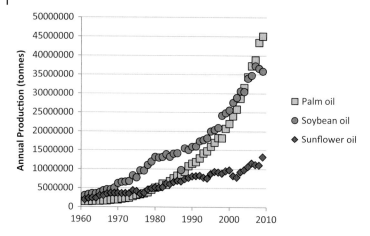

Figure 7.2 Annual production from 1960 to 2010 of some vegetable oils [14].

sources of oils and fats are soybean oil (the oil being a byproduct from the soybean meal production for the feed industry) and palm oil. Since 2003, palm oil has overtaken soybean as the main oil source, as can be seen in Figure 7.2.

Although there has been a shift from unsaturated oils (such as soybean, typical IV 130) in favor of the tropical oils (such as palm, typical IV 60), there is still a need for saturation, both to increase the melting points of the materials, so that they are solid at ambient temperatures, and to increase the oxidative stability.

For saturation of the naturally derived materials, after its invention more than a century ago, catalytic hydrogenation is still the main technique in use today, and is expected to be for the foreseeable future. At the end of the twentieth century, the trend has been towards higher saturated materials as can be seen in Figure 7.2. As many researchers nowadays focus on new types of oils that are derived from areas with marginal food production (and to accommodate the need for water and other nutrients) the types of oil may change in nature. Algae-derived triglyceride oils have been named as possible sources [15]. Clearly, the exact composition of the fatty acid chains may be very different from currently used vegetable oils, most notably in the degree of unsaturation [6]. Hydrogenation to arrive at the desired end product properties may therefore be needed. Another influence may be the higher water content of algae- or biology-derived oils, as discussed in detail below.

A rough indication of the hydrogenation catalysts needed for these materials is given in Figure 7.3. The selection criteria are based on the acidity of the medium and the water content. At high acidities, precious metal catalysts (such as Pd/C) must be used to withstand the process conditions, but at lower acidities nickel catalysts come into play. In view of their lower metal costs, these are favored in industry. More detailed selection criteria will be discussed in Section 7.3.3.

In the following sections, a systematic approach to process design based on oils and fats as renewable materials will be discussed.

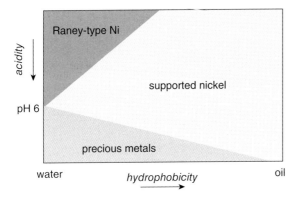

Figure 7.3 Hydrogenation catalyst selection chart based on the acidity and water content of the material to be hydrogenated.

7.3
Catalytic Process Development – Conceptual

Any new process starts by generating ideas and an initial assessment is needed to decide whether or not these ideas are worth pursuing. For fats and oils there are some specifics that can be included. These are roughly based on the selection of the right feedstock, using biological or chemical conversion processes (or a combination), continuous or batch processes, and initial catalyst selection.

As 50% of the current fats and oils production is used for food consumption [1], any new process that potentially uses food-grade sources will need to be carefully evaluated. Given a renewed interest in biotechnological processes, low-grade sugar sources have also become available for production of fatty compounds.

7.3.1
Biology or Chemical Routes?

Conceptually, one can consider a Van Krevelen [16] diagram that plots hydrogen and oxygen contents of different materials. Two plots for renewable feedstocks and oleochemical products are given in Figure 7.4.

Traditional Van Krevelen diagrams plot hydrogen and oxygen contents. In this modified diagram, for the *y*-axis, the hydrogen content is replaced by the oxidation state of the carbon, calculated from the hydrogen and oxygen contents of the molecule. In this diagram, going from right to left, oxygen content is decreasing. In general, lower oxygen contents correlate to a lower water solubility. By converting carbohydrates into fatty compounds, the carbon becomes more reduced, partly by reducing the oxygen content of the molecule.

Interestingly, using biological processes (lipogenesis), it is possible to decrease the oxygen content, but the minimum carbon oxidation state (maximum reduction potential) seems to be around −1.65. Even for generating the unsaturated fatty acids (dehydrogenation), reducing power is needed for lipogenesis processes [17].

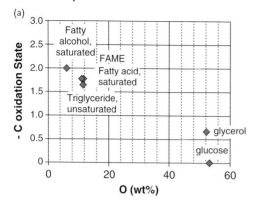
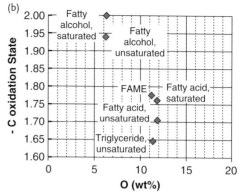

Figure 7.4 (a) Modified Van Krevelen diagram for renewables. C oxidation state and O content of fatty compounds have been calculated for a 50:50 mixture of C16 and C18 compounds. Unsaturated compounds are a 50:50 mixture of mono-unsaturated and fully saturated C16 and C18 compounds, thereby resembling palm oil-based materials. (b) Modified Van Krevelen diagram for oleochemicals, as (a) but zoomed in on the oleochemicals.

New biotechnology processes have recently become available that partly seem to overcome this barrier and these will be discussed in Section 7.5. However, for higher reduced carbon values it will still be necessary to consider chemical processes such as hydrogenation and hydrogenolysis.

7.3.2
How to Select between Slurry and Fixed-Bed Operations?

Should the above considerations have led to the choice of a chemical process, the initial focus will be on continuous operations, preferably in fixed-bed catalytic reactors. This ensures an easy catalyst separation and involves lower labor costs per unit product.

Typical feedstocks for products based on fats and oils can originate from low-quality materials (especially when feedstocks should not compete with the food chain). For example, waste oils from rendering plants and restaurants, as well as oils derived from biotechnological processes can be used. Typically the latter originate from aqueous suspensions and as such can contain high levels of water and salt impurities. However, in other feedstocks the levels of contaminants such as sulfur, phosphorus, chlorine, nitrogen, free fatty acids (FFA), water, soaps, and so on can be high.

The catalytic performance in the hydrogenation reaction is directly related to the purity of the feedstock. Purification of oils and fats is becoming increasingly costly due to the high energy prices. Distillation of feedstocks with high levels of contaminants has become an expensive operation, and less efficient separation leads to an increased level of heavies/lights, sterols/tocopherols, and water in the oil. Similar reasoning applies to bleaching and deodorization. There are, however, several ways to apply hydrogenation catalysts for poison-rich feedstocks in such a way that the most effective performance can be achieved, so that the optimal cost balance is achieved. This is illustrated in the following application example [18].

The most common method of hydrogenation in slurry-phase reactors is through a single dosage of catalyst for each batch of oil. In this case, the amount of catalyst poisoned by the impurities is limited. Typical values for a nickel catalyst for rapeseed oil hydrogenation are given in Figure 7.5.

For feedstocks with low levels of contaminants this is indeed generally the optimal method, but for feedstocks with high levels of contaminants catalyst poisoning could bring the need for an alternative sequential dosing method, as shown in Figure 7.6. This example applies to the hydrogenation of a tallow-derived fatty acid, which contains a high level of sulfur (38 ppm).

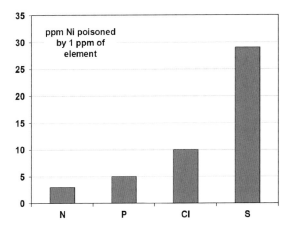

Figure 7.5 Poisoning of nickel catalyst by impurities present in rapeseed oil, taken from [19].

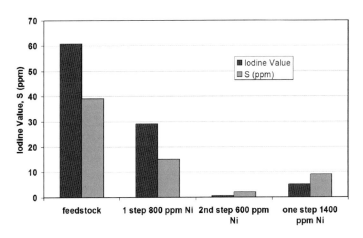

Figure 7.6 Hydrogenation of tallow fatty acid with high sulfur content.

Single dosage of 1400 ppm nickel will give a hydrogenated product with an IV around 5 g per 100 g and a sulfur content which at 9 ppm is still high. In the alternative sequential dosing approach 800 ppm nickel is added to the system in the first step. After a given time a second charge of 600 ppm fresh nickel is added. As can be seen from the graph the end IV (<1 g/100 g) and sulfur (<2 ppm) values are much lower than with the single dosage, although overall the same catalyst amount has been used.

For single dosage the sulfur will deposit about equally onto all the nickel catalyst particles that are present, thereby decreasing its activity. In the sequential dosage approach most sulfur will deposit onto the first charge of catalyst. Although the activity will be affected more than for the single dose approach, the catalyst has a second function to act as a sink for the poison. For the second charge the catalyst will be much less affected by the poisons left in the feedstock and, therefore, will maintain its activity for a longer time.

Flexibility like this is a key advantage of slurry-based operations and, therefore, they should also be considered in the conceptual design of processes.

7.3.3
How to Choose between Nickel and Palladium?

During development of catalytic processes, catalyst selection is important for the overall cost-effectiveness and economics. In the feasibility stage, usually activity and selectivity are determined in pilot systems. However, at this stage information on the processing cost can also be obtained. This will be illustrated in the following example.

Hydrogenation of unsaturated compounds can be carried out over many metal catalysts, most often nickel and palladium are used. Early on in the development process a selection can be made between these two metals, based on the existing information. A case study on a slurry catalytic process is laid out in Table 7.1.

In Table 7.1, the metal dosage (wt% of metal in the slurry system) is given in the first line. This value can be taken from laboratory experiments and reflects the amount of catalyst required to achieve the desired conversion and product yield

Table 7.1 Case study on catalytic hydrogenation – nickel versus palladium.

		Ni	Pd
1	Metal dosage (%)	0.05	0.0005
2	Metal price ($/kg)	17.6	20962
3	Metal cost ($/kg product)	0.009	0.10
4	Typical product loss (%)	5	1.0
5	Metal cost incl. loss ($/kg product)	0.009	0.11
6	Metal reclaim (%)	50	95
7	Incl. metal reclaim value ($/kg product)	0.0046	0.0053

in a certain cycle time. In general, palladium catalysts are more active than nickel, which is reflected in the lower metal dosage.

The second line reflects the metal costs, these can be found from readily available sources [20, 21]. Combination of the two gives the cost of metal (nickel or palladium) to achieve the desired product.

For slurry processes, the catalyst needs to be separated from the product. When more catalyst is needed (e.g., for nickel), the filter cake obtained in a slurry process can retain more product and hence increase the costs. This is reflected in lines 4 and 5 of Table 7.1. Assuming that both catalysts can be used only once, the filter cake is sent for metal reclaim. Line 6 indicates the typical metal value that will be compensated by a metal reclaimer. The overall outcome will reflect the cost of metal per product unit. In this specific case study, palladium proves to be slightly more expensive than nickel.

The metal prices as given in Table 7.1 were accessed on 11 October 2012. Metal prices can vary considerably and it is highly recommended to repeat the analysis for a few high/low values (the referenced websites also give historical data) for assessing the process economics. However, these calculations are based on readily accessible data and will provide a good guidance for the way forward.

In this section it was shown how to conceptually design a process using fats and oils as feedstock. This will be applied in the following section on fatty alcohols.

7.4
Fatty Alcohols: Then and Now, a Case Study

Fatty alcohols are, for the scope of this chapter, defined as even-numbered primary alcohols with 6 to 22 carbon atoms, derived from fats and oils. Competing routes for similar alcohols are based on hydroformylation (the so-called Oxo-synthesis [22]) or ethylene oligomerization and will not be discussed in this section. The advantage of the renewable feedstock is the absence of branched material, whereas the Oxo process yields some branched products [3]. The other synthetic alcohol processes, originally developed by Conoco and Ethyl Corporation, do yield linear fatty alcohols, albeit with a lower carbon number than the natural sourced processes [23]. The current demand for fatty alcohols is around 3 million tonnes annually, roughly two thirds coming from renewable sources [24].

Fatty alcohols mainly find use in the detergent industry. In the 1920s sulfonated fatty alcohols were shown to react neutral and to be insensitive to water hardness – making them ideal for washing powder [25]. This sparked the interest in finding viable manufacturing routes for fatty alcohols. In the past, fatty alcohols were recovered from whale oil (more specifically sperm oil, from the head cavity of a sperm whale), by saponification of the fatty alcohol–fatty acid esters (or wax esters). One of the earliest accounts is by Chevreul [26]. Another source is jojoba oil, but again for this feedstock supply is limited [27]. Clearly, this led to the search for alternative feedstocks.

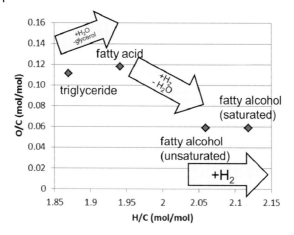

Figure 7.7 Van Krevelen diagram showing the reduction of feedstock to fatty alcohols.

Again, a Van Krevelen diagram (Figure 7.7) is shown for the conversion of triglycerides to fatty alcohols. Part of the reduction is carried out by removing glycerol (by fat splitting through hydrogenolysis). The next step that is required is the reduction of the fatty acids to the alcohols.

The reduction of fatty acids to fatty alcohols was developed in the early 1930s [25]. The non-catalytic route using sodium metal (Bouveault-Blanc reduction) was the first commercial route [28]. The main advantages were the low pressures and high yields, but the stoichiometric yield of sodium-alkoxylates and improvements in the catalytic processes have favored the latter route [29].

7.4.1
Catalyst Selection

The starting material for the reduction process is usually fatty acids. It is possible to use triglycerides directly as the starting material for a reduction process, but this involves a co-production of glycerol (and glycerol conversion products such as 1,2-propane diol) [30]. Most current processes are therefore based on fatty acids.

For conversion of fatty acids into fatty alcohols by catalytic reduction, the starting material is reduced by adding hydrogen and removing water – hydrogenolysis. The hydrogenolysis reaction needs to be carried out at relatively high temperature (>150 °C) and hydrogen pressure (>50 bar). Under these conditions other reactions, such as the further reduction of fatty alcohols to alkanes, can occur. The main metals used for hydrogenolysis catalysis are ruthenium [31] and copper [29]. Compared to copper, other base metals such as nickel and cobalt exhibit the tendency to convert the desired fatty alcohol product to alkanes by deoxygenation and decarboxylation reactions [29]. For the manufacture of biofuels, this alkane manufacture has received more interest in the past decade and is currently commercialized by Neste Oil [32]. Similar to the hydrogenation catalysts, depicted in Figure

7.3, the choice between ruthenium and copper depends on the acidity of the feedstock. Fatty acids as such will be too corrosive for copper catalysts and will require the use of ruthenium. In fact, the sensitivity of copper to fatty acids and corresponding soap formation have excluded the use of copper for the hydrogenation of edible oils [33, 34]. Clearly, ruthenium catalysts will be more expensive than copper catalysts and hence much of the process design has historically been aimed at using copper catalysts with fatty acid feedstocks. Fatty acids can be readily converted to esters, which are more benign to copper-based catalysts. Hence, most processes are actually based on the reduction of fatty acid esters to fatty alcohols.

7.4.2
Slurry versus Fixed-Bed Processes

The first catalytic reduction processes were based on slurry operation, for the reasons mentioned in Section 7.3.2. The improvements in fatty acid processing have made fixed-bed operations possible by decreasing the amount of variability and catalyst poisons, such as sulfur.

However, slurry processes can still be applied for fatty alcohol synthesis, for example, the process originally developed by Lurgi (now Air Liquide) [35]. The advantage of the Lurgi slurry phase process is the option to use fatty acids directly. The fatty acid feedstock is mixed with a 250-fold excess of a copper catalyst suspension of fatty alcohol product [35, 36]. The fatty alcohol rapidly converts the fatty acid to so-called wax esters, subsequent hydrogenolysis yields fatty alcohol. The main disadvantage of the slurry process is the need for catalyst recycle via separation processes such as filtration.

Fixed-bed processes using copper catalysts need to operate using fatty ester feedstock in order to ensure a sufficient catalyst lifetime. Basically two variants are still in use today: the vapor process by Davy Process Technology [37] and trickle-phase processes by Lurgi/Air Liquide. Compared to 5 kg for the slurry process [35], the catalyst consumption is less than 1 kg per tonne of fatty alcohol product for fixed-bed processes [38]. The different fixed-bed processes have different pros and cons. The Davy Process Technology process has the advantage of a low pressure (40 bar, [36]) and makes use of a methyl ester feedstock. The Lurgi/Air Liquide process can run with methyl esters or wax esters [38]. The wax ester process can operate without the need for recycling methanol, thereby decreasing the investment and operating costs.

By using these high temperature and hydrogen pressure processes, any carbon–carbon double bond in the fatty acid-based feedstock will be fully saturated, which is often desired. Minor byproducts are the aforementioned alkanes and aldehydes. Alkanes need to be removed by distillation, whereas aldehyde traces can be removed by hydrogenation under relatively mild conditions [18, 38].

Clearly, these processes make use of high pressure hydrogen, which involves considerable investment and operation cost. Also other feedstocks than triglycerides/fatty acid have recently received increased interest in order to prevent competition with the food chain. Recent developments will be discussed in the next section.

7.5
Conclusion and Outlook: Development Challenges for the Future

Chemical transformations using catalysts are likely to be required for production of renewable chemicals and similar design considerations can be used. At the end of the twentieth century there has been a shift from unsaturated oils (like soybean) to more saturated fats (like palm). This trend may be partly reversed as an increased demand for fats and oils will need to be served from the temperate regions as well, and may not come from only the tropical areas. Also, new biotechnological processes that operate at lower temperatures will have an energetic advantage. It was shown in the earlier sections that these lower temperatures will yield more unsaturated products.

In the introduction, algae-based processes have been mentioned. Besides these new processes that are designed to yield traditional feedstocks (triglycerides and fatty acids), several new companies are involved in biosynthetic pathways for making fatty materials from (low-grade) sugar sources. Some of these new products (fatty alcohols, fatty acid methyl esters and farnesene) are shown in Figure 7.8.

The farnesene process, patented by Amyris [39], has the advantage of yielding a product without any oxygen. The hydrophobic nature renders separation of this compound from the bio-reactor (aqueous phase) relatively easy. However, when fatty alcohol-type products are needed, it may be useful to produce oxygenates, such as fatty acid methyl ester (FAME) by OPX [40] or fatty alcohols by LS9 and Genomatica [41, 42]. These new processes produce higher reduced carbon (above the 1.65 threshold for traditional oleochemicals). However, the patent literature

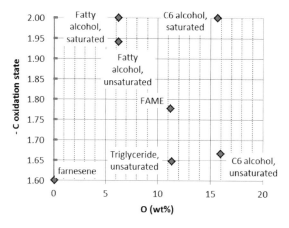

Figure 7.8 Modified Van Krevelen diagram showing new fatty compounds (fatty alcohols, fatty acid methyl esters and farnesene) obtainable by bio-synthetic pathways.

on the fatty alcohols does seem to indicate high yields involve shorter carbon chains and are at least partly unsaturated. In Figure 7.8, as a reference, data on hexanol and hexenol (C6-alcohol) are given, showing the lower reduction degree of the carbon in these materials. Also, aldehydes can be a byproduct using these biotechnology routes.

Similar design characteristics will play a part for these new feedstocks based on biosynthetic pathways. The chemical processes are an ideal addition to new biotechnological processes as they increase the carbon reduction and can be used to decrease water solubility, enabling easier separation. Initially, slurry catalytic operations will be used, but for high volume chemicals there will be a drive towards continuous processes. In general this will mean development in upstream purification as well as new catalysts with increased stability. Precious metal-based catalysts exhibit high activity and stability in aqueous enviroments, but in view of the cost and availability base metal catalysts will also be the subject of further research.

In this respect, the fatty alcohols case may also be an example for future products. Initially fatty alcohols were recovered from natural resources with little chemical modification. The increased demand for the fatty alcohol products led to new sources that were more abundantly available but needed more chemical transformation. The first chemical processes were slurry-based but increased upstream purification and catalyst stability enabled fixed-bed continuous processes to be used. New products from (new) fats and oils may follow the same development path.

References

1 O'Brien, R.D. (2004) *Fats and Oils. Formulating and Processing for Applications*, 2nd edn, CRC Press, Boca Raton, FL.
2 Markley, K.S. (ed.) (1983) *Fatty Acids. Their Chemistry, Properties, Production, and Uses*, 2nd edn, Fats and Oils. A series of Monographs on the Chemistry and Technology of Fats, Oils, and Related Products, eds H.A. Boekenoogen, *et al.*, Robert E. Krieger Publishing Company, Malabar, FL.
3 Johnson, R.W., and Fritz, E. (eds) (1989) *Fatty Acids in Industry*, Marcel Dekker, New York and Basel.
4 Koster, R.M. (2003) *Solid Acid Catalysed Conversions of Oleochemicals*, University of Amsterdam, Amsterdam.
5 Den Otter, I.M.J.A.M. (1970) The dimerization of oleic acid with a montmorillonite catalyst I: important process parameters; some main reactions. *Fette Seifen Anstrichmittel*, **72** (8), 667–673.
6 Hoekman, S.K., *et al.* (2012) Review of biodiesel composition, properties, and specifications. *Renew. Sust. Energ. Rev.*, **16** (1), 143–169.
7 Meara, M.L. (1978) *Physical Properties of Oils and Fats*, Scientific and technical surveys, vol. 110, Leatherhead Food Research Association, London.
8 AOCS (2009) Recommended Practice Cd 1c-85 Calculated Iodine Value, in *Official Methods and Recommended Practices of the AOCS*, AOCS, Urbana, IL.
9 Tsukamoto, C., *et al.* (1995) Factors affecting isoflavone content in soybean seeds: changes in isoflavones, saponins, and composition of fatty acids at different temperatures during seed development. *J. Agric. Food Chem.*, **43** (5), 1184–1192.
10 Lajara, J., Diaz, U., and Quidiello, R. (1990) Definite influence of location and

climatic conditions on the fatty acid composition of sunflower seed oil. *J. Am. Oil Chem. Soc.*, **67** (10), 618–623.

11 Marr, A.G., and Ingraham, J.L. (1962) Effect of temperature on the composition of fatty acids in Escherichia Coli. *J. Bacteriol.*, **84** (6), 1260–1267.

12 Koetsier, W.T., and Zwijnenburg, A. (2011) C. Martin Lok's contribution to the industrial development of heterogeneous hydrogenation catalysts. *Catal. Today*, **163** (1), 10–12.

13 Wilson, C. (1954) *The History of Unilever: A Study in Economic Growth and Social Change*, Cassell & Company Ltd, London.

14 Food and Agriculture Organization of the United Nations (2011) Food and Agricultural Commodities Production, http://faostat.fao.org/ (accessed 13 October 2011).

15 Norsker, N.-H., et al. (2011) Microalgal production – a close look at the economics. *Biotechnol. Adv.*, **29** (1), 24–27.

16 Van Krevelen, D.W. (1993) *Coal: Typology, Chemistry, Physics, Constitution.* 3rd edn, Elsevier, Amsterdam.

17 Brown, H.A., and Murphy, R.C. (2009) Working towards an exegesis for lipids in biology. *Nat. Chem. Biol.*, **5** (9), 602–606.

18 Dupain, X., and Zwijnenburg, A. (2010) Novel catalytic solutions in oleochemicals processes. *Oils and Fats International (OFI) Magazine*, (1), 32–34.

19 Klauenberg, G. (1984) Hydrierung von Rüböl – Einfluß von Prozeßbedingungen und Produktqualitäten auf Verlauf und Ergebnis der Hydrierung. *Fette Seifen Anstrichmittel*, **86** (S1), 513–520.

20 Johnson Matthey Plc (2012) http://www.platinum.matthey.com/ (accessed 11 October 2012).

21 London Metal Exchange (2012) http://www.lme.com/nickel.asp (accessed 11 October 2012).

22 Bohnen, H.-W., and Cornils, B. (2002) *Hydroformylation of Alkenes: An Industrial View of the Status and Importance,* Advances in Catalysis, Academic Press, pp. 1–64.

23 Washecheck, P.H. (1981) *Monohydric Alcohols Manufacture, Applications, and Chemistry*, ACS Symposium Series, ed. E.J. Wickson. American Chemical Society, Washington, DC.

24 Noweck, K., and Grafahrend, W. (2000) Fatty Alcohols, in *Ullmann's Encyclopedia of Industrial Chemistry*, Wiley-VCH Verlag GmbH & Co. KGaA.

25 Schrauth, W. (1928) Verfahren zur Darstellung höher molekularer Sulfonsäuren. DE 542048, assigned to Deutsche Hydrierwerke AG (DEHYDAG).

26 Chevreul, M.E. (1817) De la cétine (spermaceti). *Ann. Chim. Phys.*, **7**, 155–181.

27 Wisniak, J. (1994) Potential uses of jojoba oil and meal – a review. *Ind. Crops Prod.*, **3** (1–2), 43–68.

28 Wang, Z. (2010) Bouveault-Blanc Reduction, in *Comprehensive Organic Name Reactions and Reagents*, John Wiley & Sons, Inc.

29 Rittmeister, W. (1956) Fettalkohole, in *Ullmanns Encyklopädie der technischen Chemie* (ed. H. Buchholz-Meisenheimer), Urban & Schwarzenberg, München-Berlin, pp. 437–453.

30 Demmering, G., Heck, S., and Friesenhagen, L. (1994) Process for the production of fatty alcohols. US 5,364,986, assigned to Henkel AG.

31 Mendes, M.J., et al. (2001) Hydrogenation of oleic acid over ruthenium catalysts. *Appl. Catal. A Gen.*, **217** (1–2), 253–262.

32 Jakkula, J., et al. (2004) Process for producing a hydrocarbon component of biological origin. US 7,232,935 assigned to Neste Oil.

33 Okkerse, C., et al. (1967) Selective hydrogenation of soybean oil in the presence of copper catalysts. *J. Am. Oil Chem. Soc.*, **44** (2), 152–156.

34 Patterson, H.B.W. (1994) *Hydrogenation of Fats and Oils: Theory and Practice*, AOCS Press, Champaign IL.

35 Voeste, T., and Buchold, H. (1984) Production of fatty alcohols from fatty acids. *J. Am. Oil Chem. Soc.*, **61** (2), 350–352.

36 Van de Scheur, F.T. (1994) *New Developments in the Hydrogenolysis of Esters on Copper-Based Catalysts*, University of Amsterdam, Amsterdam.

37 DP Technology (2012) Natural Detergent Alcohols, http://www.davyprotech.com/ (accessed 21.10.2012).
38 Lurgi (2012) Fatty Alcohol Technology, http://lurgi.info/ (accessed 21 October 2012).
39 Renninger, N.S., *et al.* (2010) Production of isoprenoids. US 7,659,097, assigned to Amyris Biotechnologies Inc.
40 Lynch, M.D. (2011) Production of an organic acid and/or related chemicals. WO 2011063363, assigned to OPX Biotechnologies Inc.
41 Hu, Z., and Valle, F. (2012) Enhanced production of fatty acid derivatives. US 8,283,143, assigned to LS9 Inc.
42 Sun, J., Pharkya, P., and Burgard, A.P. (2011) Primary alcohol producing organisms. US 7,977,084, assigned to Genomatica.

8
Production of Aromatic Chemicals from Biobased Feedstock

David Dodds and Bob Humphreys

8.1
Introduction

An estimated 170 billion tonnes of biomass are generated globally each year [1], and the Billion-Ton study conducted by the US Department of Energy and US Department of Agriculture [2] concluded that about 1.3 billion tons of biomass are available annually in the US alone. Given the current politics and economics surrounding the continued input of fossil carbon into the atmosphere, the global oil supply, and the desire of all governments for local economic development, biomass is now considered a very attractive starting material for commodity chemical production. Further, production of commodity chemicals is generally a better use of biomass relative to the production of biofuels [3].

Aromatic chemicals are uniquely important in the pantheon of commodity chemicals. Although not produced in volumes even close to that of the non-aromatic molecules used for fuel purposes, the annual global production of the three largest aromatic chemicals is currently estimated at 103 million metric tonnes total, with benzene at 44 million metric tonnes, toluene at 22 million metric tonnes, and xylenes at 37 million tonnes [4]. All of these are found in the benzene–toluene–xylene (BTX) process stream in petrochemical refineries. Benzene is most commonly used to make phenol and styrene. Much of the toluene is disproportionated to give benzene and xylene isomers, and essentially all of the xylene stream is isomerized to *para*-xylene (PX) for the production of terephthalic acid (TA). Terephthalic acid is polymerized with ethylene glycol to make polyethylene terephthalate (PET), which is currently the most popular target for the biobased chemicals industry. The current global consumption of PET is estimated at approximately 51 million metric tonnes, which requires approximately 28 million metric tonnes of PX [5].

The attraction of BTX produced from biomass is driven largely by the use of BTX to produce PX, which in turn is used to produce TA and PET. As an historical note, the molecule now called terephthalic acid was first synthesized in 1847 via

Catalytic Process Development for Renewable Materials, First Edition. Edited by Pieter Imhof and Jan Cornelis van der Waal.
© 2013 Wiley-VCH Verlag GmbH & Co. KGaA. Published 2013 by Wiley-VCH Verlag GmbH & Co. KGaA.

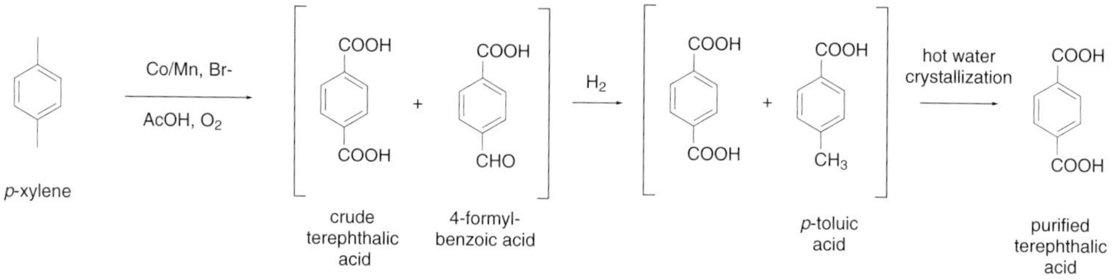

Figure 8.1 PX conversion to PTA.

the nitric acid oxidation of turpentine, or "terebenthine" in French [6]. The author of the paper, A. Cailiot, wrote:

> "Je désignerai le premier de ces acides, celui qui est insoluble, sous le nom d'*acide téréphthalique*."

The current process for the oxidation of PX to produce purified terephthalic acid represents several decades of optimization of the original Mid-Century process, also known as the Amoco process [7], summarized in Figure 8.1. In this process, PX is oxidized to TA using oxygen in acetic acid over a Co/Mn catalyst in the presence of a source of bromide ion to give crude terephthalic acid. During the process, the oxidation of the first methyl group on PX renders the second methyl group much less reactive, and partial oxidation of the PX can occur, giving 4-formyl benzoic acid as the major impurity. This is catalytically reduced to toluic acid which is removed in the high-temperature hot water re-crystallization of the resulting mixture to leave purified terephthalic acid (PTA). The process is very efficient, with an overall yield of 95%, and is of sufficient commercial value that improvements continue to be published over half a century after the Mid-Century patent was granted [8].

Various efforts to render biomass to commodity chemicals, and especially aromatic chemicals, have been made and can be divided into biological and non-biological approaches. These approaches can be subdivided further: those producing BTX streams from biomass for insertion into current chemical processes running from petroleum-derived BTX, and those directed at specific chemicals, and especially TA. The non-biological approaches to the transformation of biomass into BTX will be considered first, followed by the biological methods.

8.2
Chemical Routes to Aromatic Chemicals from Biomass

This section will focus on conversion of biomass to BTX using a bio-refinery concept that is analogous to the production of BTX at a petroleum refinery. Either biomass itself or biomass-derived raw materials are processed directly to produce fuels and chemicals without the intervention of any biological step. Biomass refining processes, like petroleum refining, involve chemical conversions at elevated temperatures and pressures, and rely heavily on heterogeneous catalyst technology.

Attractions of this approach for converting biomass to chemicals include:

1) There is no need to invest in R&D and capital to develop fermentation.

2) A fully integrated, continuous refining process is possible, as is partial energy self-sufficiency through burning of biomass (e.g., lignin) or co-products (e.g., hydrocarbon gases, coke).

3) The potential exists to develop a high-margin chemicals business analogous to petrochemicals, with much of the fixed cost shouldered by a much larger fuels business, with the ability to make full use of existing downstream BTX processes and equipment.

At the same time, there are some challenges:

1) A bio-based fuels and chemicals business should utilize non-food biomass as the primary feedstock to avoid impacting or appearing to impact global food supplies.

2) Biomass has higher oxygen content than petroleum, which means that conversion of biomass to more highly reduced hydrocarbons will require additional hydrogen compared to petroleum; the hydrogen may come from external sources or sacrifice of biomass.

3) Biomass refining will require robust catalyst technology that can cope with the composition of biomass. On the other hand, biomass refiners may benefit greatly from the huge base of catalyst technology and experience that has been developed in petroleum refining over many decades by both refiners and catalyst suppliers.

Many types of biomass are mentioned as potential sources of raw material for a bio-refinery employing chemical process technology, including, but not limited to: agricultural products and agricultural waste; forest industry products and waste; and municipal waste, such as discarded wood. Irrespective of the source, it is assumed here that for successful commercialization of biomass-to-BTX technology, economic control of biomass feedstock type and quality will be important and has yet to be demonstrated. Assuming that only non-food sources of biomass will be considered, the biomass will be predominantly structural (i.e. cellulose/hemicelluloses/lignin).

An overview of three technologies for transforming biomass via chemical process technology to fuels and chemicals with emphasis on BTX will be presented below: technology that utilizes aqueous sugar feedstock and hydroprocessing; technology that employs high temperature processing of chopped biomass as feedstock; and, technology that uses pyrolysis bio-oil produced by high temperature processing of biomass. Effort will be made to discuss chemistry and give examples of reduction to practice. A fourth technology that can convert fats into aromatics also will be touched on briefly.

8.2.1
Process Chemistry

The process chemistry applied to chemical refining of biomass to fuels and chemicals including aromatic chemicals can include pyrolysis, hydrogenation and hydrogenolysis, reforming, condensation, and cyclization. Process chemistries depend heavily on catalysts, usually heterogeneous, to control products. This section will give a short overview of process chemistries that are important in conversion of biomass to aromatics chemicals, including BTX in reactions that are employed in direct, thermal processing. Each step will produce a mixture of products.

8.2.1.1 Pyrolysis

Direct pyrolysis of biomass to produce bio-oil (also known as pyrolysis oil) has been explored in great detail [9]. The biomass must be dried thoroughly and ground to an appropriate particle size before thermal treatment. Pyrolysis is conducted in the absence of oxygen. The pyrolysis temperature typically is in the range of 450–600 °C. Fast pyrolysis (very rapid ramping of temperature) followed by rapid quenching of the pyrolysis gases is important for good yields of bio-oil [9a]. Pyrolysis of pure cellulose produces levoglucosan as the major product, and a mechanism for levoglucosan formation has been proposed [9a, m]. In contrast, pyrolysis of biomass takes place via many competing reactions. Trace (ppm) quantities of metal salts can affect the products formed [10].

The resulting, complex mixture of products can include low molecular weight alcohols and diols, aldehydes, ketones, carboxylic acids, furans, saccharides, phenolics, substituted phenolics and polyhydroxylated aromatics. Raw bio-oil is acidic and corrosive, a result of the high content of low molecular weight carboxylic acids. It contains a substantial amount of water and can be separated into water-soluble and water-insoluble fractions. Figure 8.2 [11] gives some indication of the chemical classes expected from a typical biomass pyrolysis.

8.2.1.2 Hydrogenation and Hydrogenolysis

Catalytic hydrogenation using molecular hydrogen and a catalyst results in the addition of hydrogen across the π-bond of double bonds such as those found in olefins, aldehydes and ketones. In biomass conversion, the latter two reactions are important, converting the aldehyde and ketone functions to alcohols, yielding sugar alcohols [12]. For example, glucose, an aldohexose, is converted to sorbitol while fructose, a ketohexose, is converted to a mixture of sorbitol and mannitol [12b]. These reactions are shown in the upper part of Figure 8.3. Catalytic hydrogenation of actual biomass-derived sugar mixtures will result in a more complex product mixture. Catalytic hydrogenation can be carried out under relatively mild conditions using supported noble metal catalysts [12c]. Reduction of aldehyde and

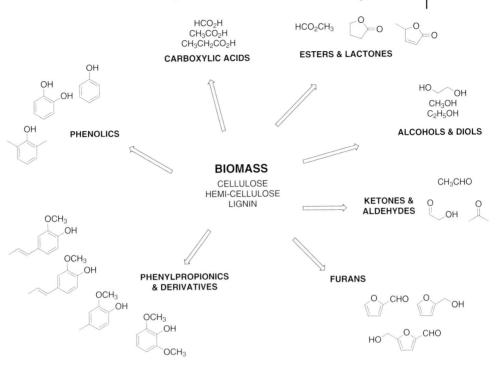

Figure 8.2 Typical products from fast pyrolysis of biomass.

ketone functionality (and presumably other, thermally sensitive groups) by catalytic hydrogenation may have an impact on yield in some processes by limiting fragmentation and condensation via retro-aldol and related reactions.

Catalytic hydrogenolysis of carbohydrates results in addition of molecular hydrogen across C–O and C–C bonds. Hydrogenolysis of carbohydrates has been reviewed recently [12d]. A wide array of supported base metal and noble metal catalysts has been used for hydrodeoxygenation [12e] along with molecular hydrogen. Processing conditions are typically more severe than those used for hydrogenation [12f]. Hydrogenolysis at a C–O bond results in replacement of C–O bond by a C–H bond with concomitant generation of water while reaction at C–C bonds results in fragmentation of the carbohydrate. The lower reaction in Figure 8.3 gives an example hydrogenolysis of a sugar alcohol. The process can result in conversion of carbohydrate-derived molecules in bio-oil fractions to mixtures of mono-oxygenated compounds (e.g., alcohols, furan derivatives), hydrocarbons and olefins. Hydrogenolysis increases the ratio of hydrogen to carbon in the processed biomass, a step that is critical to upgrading biomass for fuel applications [12g].

The range of processes and conditions used in industrial hydrogenation and hydrogenolysis is very wide. The catalysts employed can be proprietary, so specific details about catalysts (structure, preparation, modifiers, catalyst activation) may not be available in the open literature.

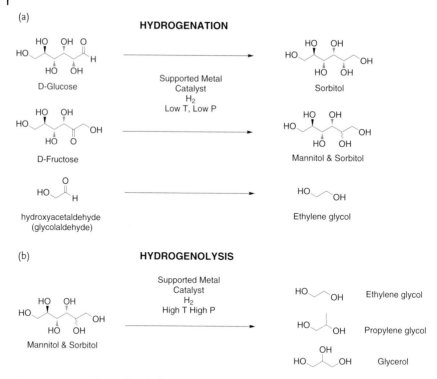

Figure 8.3 Exemplary catalytic hydrogenation and catalytic hydrogenolysis in bio-oil conversion.

8.2.1.3 Catalytic Reforming

Catalytic reforming was developed by the petroleum industry to convert naphthas (mixtures of hydrocarbons produced during petroleum cracking) into a mixture of more branched hydrocarbons and aromatic molecules with higher octane number [13]. The process also generates molecular hydrogen, a result of dehydration during formation of small aromatic molecules (e.g., benzene, toluene) and olefins. The ratio of hydrocarbon to aromatic content can be adjusted to respond to market needs. Catalytic reforming can be applied in biomass upgrading, as will be discussed in more detail below.

Also, steam reforming has been applied to biomass conversion to give synthesis gas (a mixture of $CO + H_2$) [14]. This hydrogen can be used to reduce reliance on externally sourced hydrogen in biomass upgrading.

8.2.1.4 Zeolite Treatment

Zeolites are crystalline, microporous, aluminosilicate materials. Micropore size and the chemical environment within the micropores depend on zeolite structure and exact composition. Because micropore size is well defined and comparable to molecular dimensions, zeolites can be size and shape selective in their adsorption properties [15]. Size-shape selectivity coupled with a predictable chemical environment within the micropores make zeolites useful for selective chemical catalysis.

Zeolites are widely applied as catalysts and adsorbents in chemical and petroleum processing.

Application of zeolites to catalyze the formation of aromatics (e.g., benzene, toluene, xylene, or collectively, BTX) from naphthas is well known [13]. Zeolite technology also has been applied to upgrading of biomass to catalyze the formation of aromatic molecules from biomass pyrolysis gases or as part of the reforming process [16].

8.2.2
Technology Examples

8.2.2.1 Conversion of Biomass-Derived Sugars to Aromatics Including BTX

Technology has been reported for conversion of biomass-derived sugars to hydrocarbons that are useful as fuels and industrial chemicals, including aromatics, with technology initially established by Cortright and Dumesic [17].

The technology employs an aqueous sugar feed-stream derived from conventional sugar sources, grain crops, or a broad range of non-food biomass. Conversion of the sugar feed-stream to products, including aromatics, occurs in several steps in a continuous process [17a, j]:

1) The aqueous sugar stream can be subjected to catalytic hydrogenation and/or hydrogenolysis. The typical reaction temperature is in the range of 100–150 °C. This hydrotreating step converts the mostly 5- and 6-carbon sugars to sugar alcohols or lower molecular weight oxygenated compounds, as exemplified in Figure 8.3. The process relies on heterogeneous catalysts, with patent examples including a supported ruthenium-based catalyst [18].

2) The product stream from the hydrotreating step then enters a process known as aqueous phase reforming (APR) to reduce the oxygen content of the feed. Process temperature in the range of 150–265 °C and a PT/Re containing catalysts are reported [19]. Products from the process include "mono-oxygenates" (mono-oxygenated compounds such as alcohols, tetrahydrofurans, ketones, and aldehydes), hydrocarbons, and molecular hydrogen. Dehydrogenation occurs during the process providing hydrogen in-situ for the hydrodeoxygenation process. The process can produce excess hydrogen, which can be recycled or used in step 1 above to reduce the need for outsourced hydrogen [17j]. Product distribution (paraffins vs. oxygenates, type and molecular weight of oxygenate) can be controlled by reaction conditions and catalyst.

3) The APR product stream then passes through a process employing acid catalysts, basic catalysts or mixed acid/base catalysts. Product distributions depend on feed composition, catalyst and reaction conditions. Conditions and catalyst [20] are reported that convert the feed to predominantly aromatics, including aromatics in the commercially significant C6–C9 range at a typical reaction temperature of 375 °C. Examples for production of aromatics include modified ZSM-5 zeolite catalysts. It is believed that acidic catalysis promotes reactions such as dehydration of oxygenates to form olefins, oligomerization and cyclization of the olefins to produce aromatics, and hydrogen transfer to form paraffins [21].

Table 8.1 Conversion of sucrose/xylose (93/7) and sorbitol to paraffins and aromatics.

Feed	Product yield (wt% of feed carbon)			
	Total C1–C3	C5+ Paraffins	C5+ Olefins	Aromatics
Sucrose/xylose (93/7)	22.3	20	0.8	25.0
Sorbitol	18.1	11.3	7.8	22.3

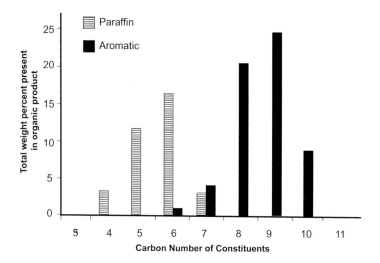

Figure 8.4 Paraffin and aromatic product distribution from Table 8.1 data [17a].

The overall process generates mixtures of hydrocarbons including aromatics in the C6–C9 range, the composition depending on process conditions and catalysts.

Referenced patents have been chosen specifically because they show production of hydrocarbons including aromatics. Data cited in Table 8.1 were chosen to illustrate the hydrocarbon product from two different feeds, demonstrating that significant amounts of aromatics are produced in the C6–C9 range in the product stream. Examples of aromatic product yields and distributions are shown in Table 8.1 and Figure 8.4 [22]. Data for the sucrose/xylose and sorbitol feeds are provided here. Catalysts and conditions for each step of the process (hydrogenation for sucrose/xylose feed only, APR/deoxygenation and condensation for sucrose/xylose and sorbitol feeds) vary with feed type. Condensation catalysts are based on modified ZSM-5.

It is apparent from the data in Table 8.1 and Figure 8.4 that significant yields of aromatic compounds in the C6–C9 range can be obtained directly from the process described. The aromatics have been processed further to produce crystalline p-xylene (PX) [23].

8.2.2.2 Pyrolysis of Solid Biomass to Aromatic Chemicals

Direct production of aromatic chemicals including BTX has been demonstrated by fast pyrolysis of carbohydrates and biomass in the presence of a suitable catalyst

[24]. The process involves rapid heating of biomass under conditions where the gases generated immediately encounter catalyst. Catalysts for formation of aromatic chemicals include examples based on zeolites such as ZSM-5. Conversion of solid biomass to aromatic chemicals occurs in the following steps [24a]:

1) Preparation of the biomass source by drying and grinding to the desired particle size range.

2) Particulate biomass is fed into a reactor containing catalyst (zeolites such as ZSM-5) and is pyrolized. Very rapid heating converts the biomass to pyrolysis gases, which rapidly enter the catalyst bed. Catalyst is chosen to favor aromatic products. The reported pyrolysis temperature is 600 °C.

Some examples of product distributions from pyrolysis of various solid biomass sources are shown in Figures 8.5 and 8.6. Data show that >20% of the biomass carbon can be converted to aromatics (Figure 8.5), with BTX a significant portion of the total aromatics (Figure 8.6) [25].

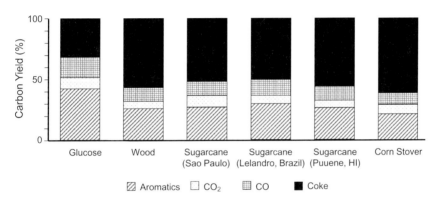

Figure 8.5 Product distribution from biomass sources.

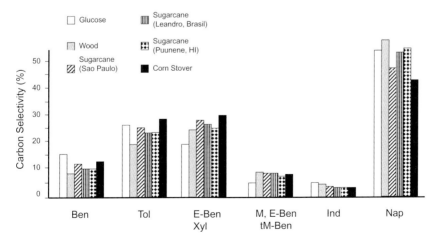

Figure 8.6 Yields of aromatic chemicals from pyrolysis of various biomass sources.

Pyrolysis of wood in a fluidized-bed reactor at 600 °C over a ZSM-5 catalyst gave a mixture of aromatic and olefinic products, including benzene, toluene, xylenes, naphthalene, ethylene and propylene (% carbon yields based on biomass consumed: 3.4, 5.6, 2.4, 0.6, 5.5, 2.8) [26].

8.2.2.3 Upgrading Bio-Oils to Aromatics

Technology for converting bio-oils to mixtures of aromatic chemicals and olefins has been reported; an overview of the technology has been published [27]. Bio-oil (BO) can be converted directly or separated into a water-soluble phase (water-soluble bio-oil or WSBO) and an organic phase before conversion. The composition of the water-soluble portion of oak wood bio-oil has been analyzed [28]. Table 8.2 shows some product data reported for conversion of bio-oil to hydrocarbons. Only selections of data for BO and WSBO are shown. The first two data sets were generated by subjecting the BO or WSBO to a low temperature hydrogenation followed by passing the feed over a zeolite ZSM-5 catalyst. For the second data set (WSBO only), two hydrogenation steps (low and high temperature) were followed by zeolite treatment. Hydrogenations were conducted with a Ru/C catalyst at 125 °C and 5.2 Mpa, and a Pt/C catalyst at 250 °C and 10 MPa. Zeolite treatment was conducted at 600 °C. Significant yields of olefins (ethylene, propylene, butylene) and aromatics were obtained. Data in Table 8.2 show that BTX comprises the major portion of aromatic chemicals produced [29].

8.2.2.4 Aromatic Chemicals from Other Renewable Raw Materials

Several patent applications have been published that cover conversion of fats and oils to chemicals by steam cracking. In one process [31], the fats and oils are refined, fractionated by crystallization, and the solid, mostly saturated fats are hydrodeoxygenated to a "bio-naphtha" stream consisting predominantly of paraf-

Table 8.2 Conversion of bio-oil to alkanes, olefins and aromatics [30].

Feed	Process	C1–C6 Alkanes (% C yield)	Olefins (selectivity, %)			Olefins (% C yield)	Aromatics (selectivity, %)				Aromatics (% C yield)
			C2	C3	C4		B	T	X	EB	
BO	Ru/H$_2$ Zeolite	2	52.2	35.9	11.4	18.2	16.9	37.2	38.5	3.4	14.4
WSBO	Ru/H$_2$ Zeolite	7.5	31.8	55.4	12.8	30.2	17.6	45.5	31.3	2.6	21.6
WSBO	Ru/H$_2$ Pt/H$_2$ Zeolite	15.0	32.0	53.8	14.2	43.0	27.0	49.3	19.1	2.3	18.3

BO = DOE bio-oil, WSBO = water-soluble portion of bio-oil.
C2 = ethylene, C3 = propylene, C4 = butane.
B = benzene, T = toluene, X = xylene, EB = ethylbenzene.

Table 8.3 Yields of C6–C9 products from steam cracking of palm fatty acids.

C6–C9 yield, wt% / Cracking temp	760 °C	820 °C
Benzene	4.56	7.40
C6 Nonaromatic	2.27	0.93
Toluene	1.94	2.40
C7 Nonaromatic	0.83	0.17
Ethylbenzene/xylenes	0.63	0.56
Styrene	0.70	1.19
C8 Nonaromatic	0.35	0.10
C9 Aromatic	0.90	1.19
C9 Nonaromatic	0.19	0.02

fins. The hydrodeoxygenation process employs supported catalysts (Ni/W/Mo-type oxides, possibly sulfided). Alternatively, the fats can be saponified and the fatty acids can be decarboxylated catalytically or the salts resulting from neutralization can be decarboxylated thermally to give bio-naphtha. The bio-naphtha is then steam cracked at >800 °C to give ethylene, propylene, other olefins, aromatics (including BTX), and "bio-gasoline". Aromatic chemical yields are in the 7–8% range and the overall yield of high-value hydrocarbons from the process, including aromatics, can be >70% based on the bio-naphtha feed. Aromatics and other products can be separated by existing fractionation processes used in the petrochemical industry.

Other patent applications [32] describe a process that uses a hydrogenated fatty acid feed stream, which involves prior saponification of the fats and oils, followed by isolation of the fatty acid stream by distillation and hydrogenation to remove all but 5% of the unsaturation (i.e. 5 mol% of the fatty acids contain at least one double bond). The steam cracking process is conducted at 750–900 °C and does not employ catalysts. Yields of C6–C9 products, including aromatic and non-aromatic compounds, from steam cracking of palm oil fatty acids are shown in Table 8.3 [33]. Reported data show formation of aromatic chemicals in the valuable C6–C9 range.

8.2.3
Summary

Technology has been reported for conversion of biomass and biomass-derived raw materials directly to aromatics, including BTX, as well as to fuels and other chemicals using processes that are analogous to those employed for conversion of petroleum to fuels and chemicals. Heterogeneous catalyst technology is critical for most steps of biomass-to-BTX transformation. Biomass itself is also a potential source of excess hydrogen required to address the high oxygen of biomass feedstock, providing a possible route to renewable BTX that avoids intervening biological steps such as fermentation.

8.3
Biological Routes to Specific Aromatic Chemicals

8.3.1
PTA via PX

The non-biological routes from biomass to BTX and especially PX have been discussed in the previous section; biological routes to aromatic chemicals will now be considered. As with the non-biological routes to PX, the ultimate goal is in fact terephthalic acid and PET. As no anabolic biological routes directly expressing PX have yet been reported, the most comparable routes are those which are based on deliberately engineered metabolic pathways to produce materials that can then be chemically transformed to PX.

8.3.1.1 PX via Isobutanol and Isobutylene
The production of PX via the dimerization of isobutylene has been known for some decades [34] and patent applications on this transformation continue to be filed in recent times [35]. Theoretically, two molecules of isobutylene can be condensed to give a single molecule of iso-octene, which cyclizes to 1,4-dimethyl cyclohexane and subsequently can be dehydrogenated to give PX. In the process, iso-octane (which will not cyclize) is also produced and this decreases the overall yield of PX to approximately 30 to 40%; more recent efforts are directed at improving the yield of PX in the dehydrocyclization reaction [36].

Two approaches to isobutylene production via engineered pathways are currently being pursued and both proceed from pyruvate available by glycolysis. One route continues along the metabolic pathway to the branched amino acid valine, while the other proceeds via polyhydroxyalkanoate biosynthesis.

8.3.1.2 Valine Pathway to Isobutylene
The production of PX from biomass in order to make use of the known chemistry from isobutene begins with the biological production of isobutanol. This is achieved via interception of the pathway to the branched amino acid valine and is illustrated in Figure 8.7 [37, 38].

Four enzymes are required by this pathway; acetolactate synthase (ALS, EC 2.2.1.6), ketol-acid reductoisomerase (KARI, EC 1.1.1.86), dihydroxyacid dehydratase (DHAD, EC 4.2.1.9), 2-ketoacid decarboxylase (KIVD, EC 1.2.1.25), isovaleraldehyde dehydrogenase (IDH, EC 1.2.1.n). The most unusual enzyme is the KARI which catalyzes two reactions, a reduction and an isomerization, reportedly at the same active site [39]. The theoretical stoichiometry is:

$$1 \text{ glucose} \rightarrow 1 \text{ iBuOH} + 2 \text{ CO}_2 + \text{H}_2\text{O}$$

thus giving a 100% theoretical molar yield, although two moles of CO_2 are lost in this pathway giving a 41% theoretical weight/weight yield of isobutanol from glucose. A practical isobutanol yield of approximately 60% of theoretical is reported [40].

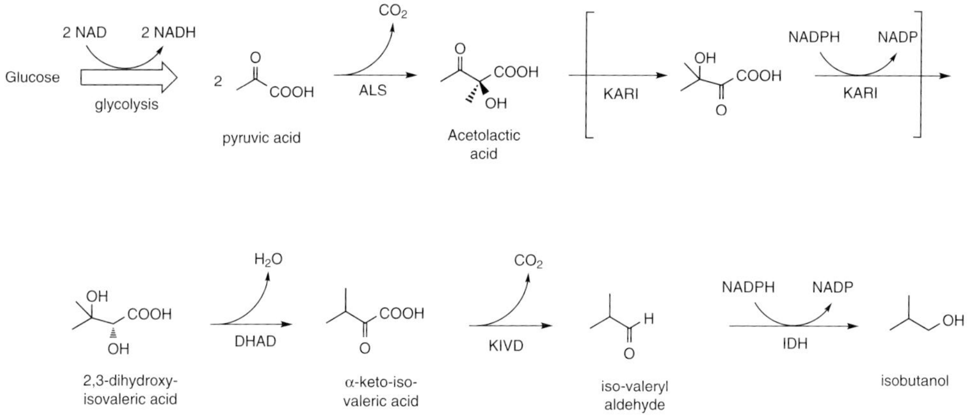

Figure 8.7 Isobutanol from glucose, via branched amino acid metabolism.

As with all efforts at producing commodity chemicals from biomass, the optimization of the metabolic process is critical. As some yeasts generally have higher glycolytic flux than *E. coli*, especially *S. cerevisiae* used for ethanol production, a more productive process may be possible if the pathway illustrated above is engineered into such a yeast host [41, 42]. Under anaerobic conditions, yeasts used for the production of ethanol will take pyruvate through lactate to give ethanol, and carbon flux to the isobutanol pathway will be lost. Blocking the metabolic route to ethanol is thus highly desirable if a yeast such as *S. cerevisiae* or *K. lactis* is to be used as the production organism, and disabling the pyruvate decarboxylase enzyme in the metabolic pathway to ethanol has been taught in published patent applications [43]. Blocking the metabolism of pyruvate to lactate or to formate plus acetyl-CoA, and of the metabolism of acetyl-CoA to ethanol has also been revealed [44]. Other improvements, such as the balancing of the NADH/NADPH pools of reducing equivalents in order to increase flux to the desired isobutanol product have been disclosed, either by using trans-reduction of NADH and NADPH [41, 45], or altering the actual generation of reducing equivalents by employing the Entner–Doudoroff glycolysis pathway rather than the Embden–Meyerhof [46], or engineering other pathways [47] for production of reducing equivalents. Improvements of individual enzymes in the pathway to isobutanol itself include altering the cofactor requirement of the KARI enzyme from NADPH to NADH [46] and the use of transcriptional activator genes for enhanced expression of the DHAD enzyme [48].

Isobutanol can have a deleterious effect, and continuous removal of it from the fermentation is required to achieve useful yields. Dehydration of isobutanol is then achieved using a variety of commercial zeolite catalysts and this can be tuned to give all different isomers of butene, both isobutene and double-bond isomers of *n*-butene [49]. Dehydration is followed by dimerization of the isobutene, although this step can be combined with the dehydration [50].

An example of the fermentative production of isobutanol and its conversion to PX is given in the patent literature [50, 51] and illustrated below. The dehydration is performed on isobutanol vapor that contains water-vapor and proceeds in 95% yield using BASF AL-3996 dehydration catalyst at 290 °C. The vapor stream is then oligomerized in the presence of HZSM-5 zeolite, and this is fed to the dehydrocyclization reaction which is catalyzed by a chromium oxide catalyst, BASF-1145E at 550 °C, giving a cyclization yield to PX of about 38% in this last step. The PX may then be oxidized to PTA using conventional processes. The entire set of process yields is reported as follows [50]; 100 kg glucose produces 18.7 kg of PX; this is a mass yield of 18.7%. However, the process is reported to also produce 19.1 kg iso-octene plus 26.3 kg isobutylene. If the isobutylene and iso-octene streams were continuously recycled, presumably the output of the process would exceed 60% mass yield.

8.3.1.3 Direct Biological Isobutylene Production

The conversion of biomass to provide a four carbon moiety for elaboration to PX has been approached differently through the technology of Marliere et al. While the downstream chemistry for the oligomerization and dehydrocyclization of isobutene to PX remains unchanged, the biological part of the process makes isobutylene directly rather than first making isobutanol. This can theoretically be accomplished by appending a biological dehydration step to an isobutanol pathway, and the patent literature reveals the use of oleate hydratase (EC 4.2.1.53) [52], or prenyl isoflavinoid hydratases EC 4.2.1.n for dehydration [53].

An alternative route is possible via the metabolic intermediate 3-hydroxyisovaleric acid (3HIVA). Transformation of the 3HIVA metabolite directly to isobutylene is possible via an enzyme of the mevalonic diphosphate decarboxylase (MDH) family. This catalyzes a combined dehydration/decarboxylation as revealed in the patent literature; this is shown in Figure 8.8. The enzymatic reaction requires ATP, and as the 3HIVA substrate molecule lacks the diphosphate group of the natural substrate mevalonic acid diphosphate, addition of a catalytic amount of alkyldiphosphate cofactor is reported as required [54].

The 3HIVA intermediate is theoretically available via manipulation of the pathways normally used to generate polyhydroxyalkanoates (PHAs) as storage polymers in some micro-organisms [55], however, no example of an organism with this complete pathway is given in any published patent applications. The patent literature [56] does reveal a preferred route that uses the first part of the pathway for the biological production of acetone and this is shown in Figure 8.9. This route requires the use of micro-organisms that are homoacetogens, such as *Clostridia*

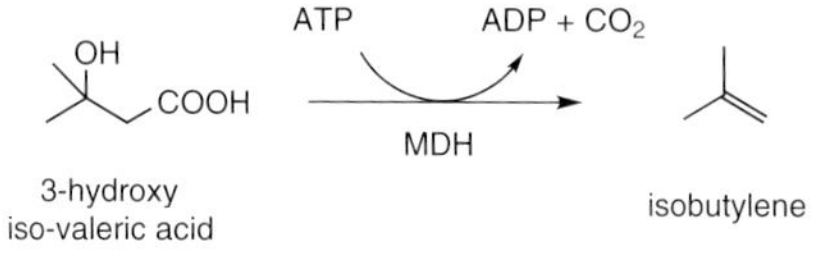

Figure 8.8 Reaction catalyzed by mevalonic diphosphate decarboxylase.

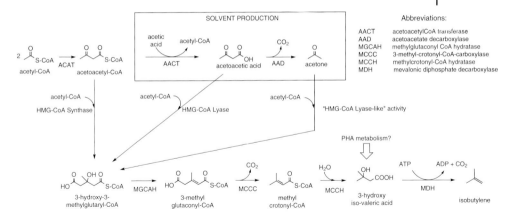

Figure 8.9 Routes to isobutylene from acetyl-CoA.

sp., that are capable of producing 3 mol of acetyl-CoA from 1 mol of glucose. From the acetyl-CoA pool, condensation of 2 mol of acetyl-CoA via acetyl-CoA-acetyl transferase (ACAT, EC 2.3.1.9) produces 1 mol of acetoacetyl-CoA. Under normal solvent-producing metabolism, this is displaced by an available carboxylic acid (such as acetic or butyric), catalyzed by acetoacetyl-CoA transferase (AACT, EC 2.8.3.8) to give free acetoacetic acid. Decarboxylation of the acetoacetic acid via acetoacetate decarboxylase (AAD, EC 4.1.1.4) completes the production of acetone. Interception of either the acetoacetyl-CoA (via HMG-CoA synthase, EC 2.3.3.10) or the free acetoacetic acid (via HMG-CoA lyase, EC 4.1.3.4) and the condensation of an additional acetyl-CoA in either case produces the required HMG-CoA intermediate. Three enzymes are required following the production of HMG-CoA; methylglutaconyl-CoA hydratase (MGCAH) catalyzes the dehydration of HMG to 3-methyl-glutaconyl-CoA, 3-methyl-crotonyl-CoA-carboxylase (MCCC) catalyzes the subsequent decarboxylation step to produce methylcrotonyl-CoA, and methylcrotonyl-CoA hydratase (MCCH) adds water across the olefin to yield the required 3HIVA. This route is also reviewed in the open literature [57]. A further variation is the use of an enzyme with activity similar to HMG-CoA-lyase that is capable of condensing acetyl-CoA directly with acetone to produce 3HIVA [58], all of these routes are summarized in Figure 8.9.

The same patent application [58] teaches that the methyl crotonyl intermediate can be produced in analogous manner by a series of enzymes from the polyketide synthesis (PKS) system, the difference being that the various thioester intermediates are attached to the acyl carrier proteins of the PKS system, rather than coenzyme A.

Not surprisingly, the theoretical stoichiometry for all of these routes is the same as for the metabolic route to isobutylene via isobutanol:

$$1 \text{ glucose} \rightarrow 1 \text{ isobutylene} + 2\ CO_2 + 2\ H_2O$$

The direct production of isobutene in a fermentor provides an advantage as it is a gas with very low water solubility and can be collected directly from the headspace

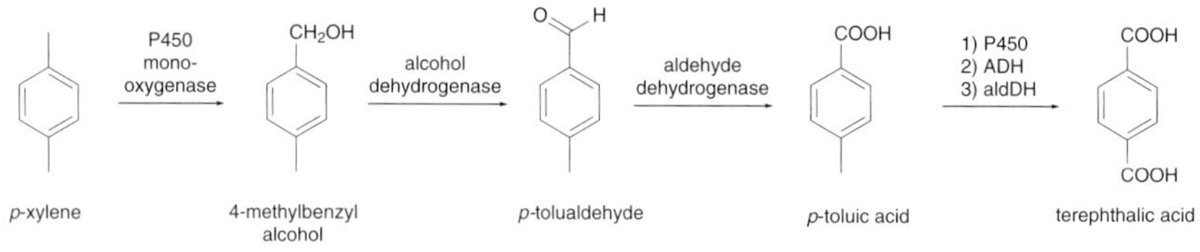

Figure 8.10 Biological oxidations of PX giving terephthalic acid.

of the fermentor. The open literature reports that a carbon balance of 0.66 is expected from glucose to isobutylene [59].

8.3.1.4 Biological Oxidation of PX to PTA

As efficient as the current Amoco chemical process [7] for the oxidation of PX to PTA is, the loss of the acetic acid by oxidation to CO_2 is considered the largest single industrial consumption of acetic acid. Further, the high temperatures and corrosive nature of the process for the oxidation require special materials for construction of the reactors.

Biological approaches to the oxidation of PX to PTA have been explored, and several patents have been granted [60]. The oxidation is performed by bacteria of the Proteobacteria group and proceeds through a multi-step process beginning with oxidation via a P450 monooxygenase (Figure 8.10). This first step oxidizes PX to methylbenzyl alcohol. The benzylic alcohol is then oxidized via a standard alcohol dehydrogenase (ADH) to *p*-tolualdehyde and a second dehydrogenase (aldDH) completes the oxidation of the original methyl group by taking the *p*-tolualdehyde to *p*-toluic acid. The three-step oxidation is then repeated on the remaining methyl group.

It is not reported in the patent literature if the oxidation of one methyl group must proceed all the way to the carboxylic acid before oxidation of the second methyl group can occur [61], but the use of biological oxidation of PX to TA has been in the open literature for at least a decade [62]. While the use of the published organisms would allow a completely biological pathway from PX produced by biological methods through to terephthalic acid, no large scale example is given in the literature and P450 enzymes have generally low turnover numbers. Conversely, by using acetic acid produced from biological sources, such as woody biomass [63], the Amoco oxidation process could be made "green", at least to the degree that all carbon atoms in the starting material and the reaction solvent would be from renewable biomass.

8.3.2
Aromatics via HMF Production

It is possible to produce substances other than BTX and PX from biomass via non-biological routes. Under conditions considerably less energetic than those used for gasification or pyrolysis, carbohydrate biomass will undergo not only hydrolysis to monosaccharides in the presence of heat and acid, but also a series

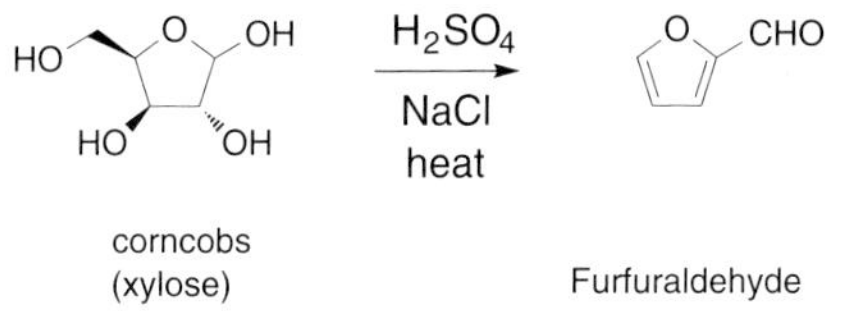

Figure 8.11 Synthesis of furfuraldehyde from xylose.

of dehydration reactions to give furan derivatives. Furans are aromatic molecules and warrant consideration here.

Corn cobs and oat hulls contain a very high ratio of hemicellulose to cellulose, and have been long recognized as practically useful sources of xylose. The direct treatment of such raw biomass with strong acids and heat will produce large amounts of furfuraldehyde; this was known in the nineteenth century [64] and the Quaker Oats company filed patents for furfuraldehyde production in the 1920s and 1930s (Figure 8.11) [65]. As Roger Adams noted [66],

> "Carbohydrate materials such as corn cobs, wood, the hulls of oats, rice, peanuts, etc., when heated with steam under pressure or distilled with dilute hydrochloric or sulfuric acids, yield appreciable quantities of furfural. Practically all the furfural now prepared technically is from oat hulls. The literature on the preparation of furfural is too extensive for citation here. This literature has been reviewed in "Furfural and Its Derivatives," Bulletin 2 (1928)."

Extension of this chemistry to 6-carbon sugars has also been known since that time and has enjoyed considerable publication and patenting [67]. The treatment of 6-carbon sugars, such as glucose, with heat and acid produces hydroxymethylfurfural (HMF), and yields in the 50–75% range are reported in the earlier literature and patents. A wide variety of supported acid catalysts can be used, and this general process can also be adapted to produce levulinic acid as well as HMF [68]. Sucrose and other disaccharides can also be used with hydrolysis to individual sugars (e.g., fructose and glucose from sucrose) presumably occurring as the first step [69] (Figure 8.12).

Oxidation of both the hydroxy group and the aldehyde of HMF to carboxylic acids produces furanedicarboxylic acid (FDCA). This aromatic molecule is in fact the only aromatic compound identified in the 2004 report by the US Department of Energy *Top Value Added Chemicals from Biomass; Volume I* [70]. Chemical oxidation of HMF has been known for many decades [71], and perhaps not surprisingly the catalysts appear to be very similar to those used for the oxidation of PX to PTA; a combination of Co/Mn plus Br [72], and the use of gold catalysts has also been recently published [73]. Biological oxidative methods have appeared more recently, including the use of chloroperoxidase [74] and an oxidoreductase [75].

Work by Dumesic *et al.* [76] shows that HMF can be full reduced to give 2,5-dimethyl furan, also an aromatic compound although with utility as a fuel, rather than a chemical feedstock or monomer. In this process, carbohydrates

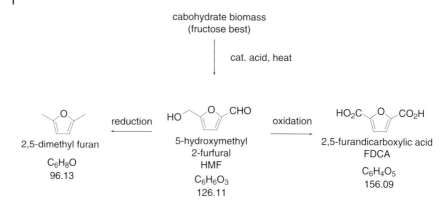

Figure 8.12 Furans produced from carbohydrates.

are converted to HMF in a biphasic extractive reaction system using NaCl as a dehydrating agent, and several reactor conditions are reported to give conversions of fructose to HMF in the 80% range. Fructose is clearly the best carbohydrate starting material for conversion to HMF in this system, and several conditions showing yields of HMF from fructose of approximately 84% are reported [77]. Hydrogenolysis of the HMF over a Cu/Ru catalyst is reported to give yields of 2,5-dimethylfuran at approximately 70%. To add a biological aspect to the chemical process, a combination of enzymatic isomerization of either glucose or mannose to fructose, followed by mild chemical conversion has been revealed in the patent literature, and arranged as a continuous process for the production of HMF [78].

8.3.2.1 Preparation of PTA via HMF

FDCA can be used as the diene in a Diels–Alder reaction with ethylene to give the [2.2.1] oxa-bicylic compound as the immediate product. At the reaction temperature of approximately 200 °C, this intermediate undergoes dexoygenation and aromatization (formally, a single dehydration) to give PTA. (Figure 8.13) A catalyst for the Diels–Alder reaction is not required, according to the patent examples, but the usual types of catalysts (e.g., Lewis acids) may be used. The Diels–Alder reaction can proceed with either FDCA to give PTA, or the dimethyl ester of FDCA to give DMT [79]. This reaction avoids the issues of p-xylene oxidation and the attendant impurities, but the low yields quoted in the patent examples of less than 1% of the desired PTA or dimethyl terephthalate (DMT) product make this an impractical process. The low yield is presumably a function of the reported reversibility of the Diels–Alder addition of ethylene to FDCA.

Conversely, HMF can be reduced to 2,5-dimethylfuran first and then the Diels–Alder reaction with ethylene performed. The subsequent spontaneous dehydration reaction gives PX rather than PTA, but in considerably higher yields that is, 30–65% [80]. The PX product can then be taken on to PTA using the current commercial oxidation and high-temperature hot water purification process.

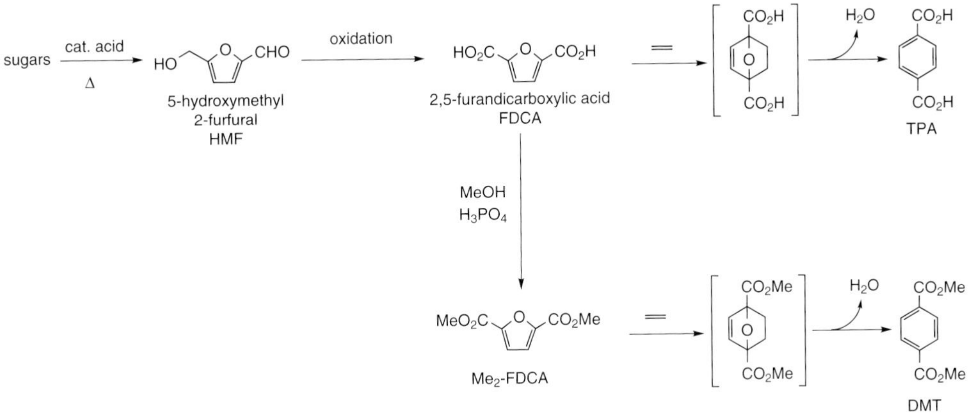

Figure 8.13 Routes to terephthalic acid and dimethyl terephthalate from HMF.

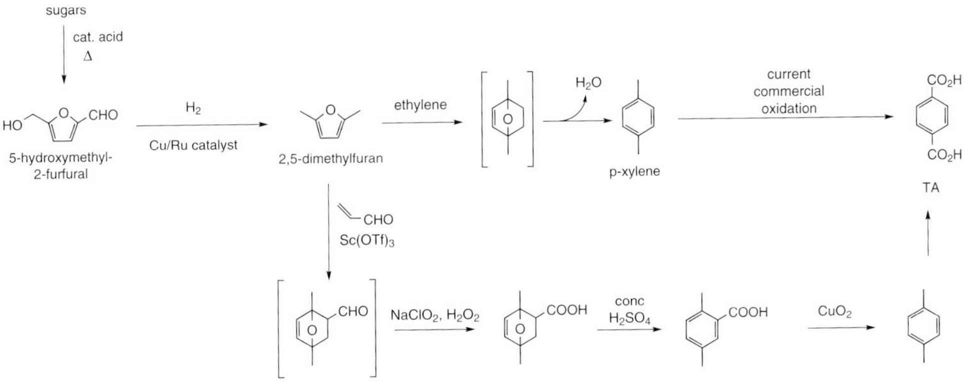

Figure 8.14 Routes to terephthalic acid via PX starting with HMF.

An academic variation has been published, in which acrolein is used instead of ethylene as the dienophile in the Diels–Alder reaction. Separate oxidation and aromatization steps are then performed to take the Diels–Alder product to 2,5-dimethyl benzoic acid, after which decarboxylation and dehydration give the desired PX product [81]. Both routes are shown in Figure 8.14.

A different route has been taken by Avantium to address the market for a bio-based packaging material. Rather than produce PTA for condensation with ethylene glycol to make PET, Avantium is pursuing the polyester of ethylene glycol condensed directly with 2,5-furandicarboxylic acid itself, with the polymer being termed PEF [82]. This polymer has been described previously by Celanese [83], but not prepared from biomass sources. The PEF polymer is reported to have better barrier properties for oxygen, carbon dioxide and water than PET [84].

As part of the process, the conversion of carbohydrates to alkyl ethers of HMF has been revealed [85]; by using alcohols during the dehydration of the biomass,

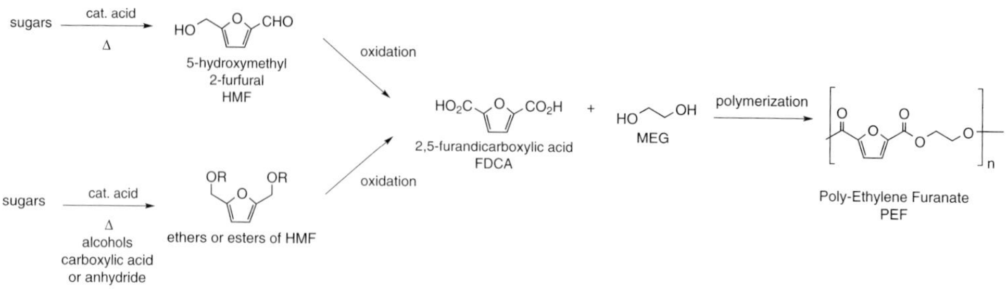

Figure 8.15 Overall route from sugars to PEF.

which would normally give HMF, the alkyl ethers are formed and these are reported to be more stable than HMF itself. An improved oxidation process using alkyl ethers of HMF is described and claimed to be novel [86]; the ethers can be formed when alcohols are present during the dehydration process. Esters of HMF, and specially the acetate ester of HMF formed by using acetic acid or acetic anhydride during dehydration, have been described and are also claimed to give an improved oxidation process [85]. An improved method of polymer formation reported to reduce color bodies and other impurities completes the overall process description of transforming glucose to the final PEF polymer [87] (Figure 8.15).

8.3.2.2 Yield Summary of HMF Routes to PTA

Overall mass yields in the patent literature just referenced, from carbohydrate to HMF and on to FDCA appear to be approximately 80+% for the production of HMF (the highest yields coming from fructose), and approximately 70% molar yields for the oxidation of HMF to FDCA with yields just over 80% when using ethers of HMF. Correcting for the difference in molecular weights of HMF ($C_6H_6O_3$, MW = 126.11) and FDCA ($C_6H_4O_5$, MW = 156.09) this gives a mass yield of approximately 93% from HMF to FDCA, for an overall mass yield of about 75% for the production of FDCA from biomass.

The yields of PX from biomass vary widely. The theoretical molar yield of isobutanol or isobutylene from glucose is 100%, with the practical molar yield of isobutanol reported in patent examples to be approximately 60%, and the practical molar yield of isobutylene is projected to be higher at around 66%. Adjusting for the molecular weights of glucose and isobutanol/isobutylene, this gives mass yields of around 25%. The currently reported practical mass yield for the conversion of these four-carbon molecules to PX is approximately 40%. The molar yield for the commercial process for the oxidation of PX to PTA is in the 94–95% range; a mass yield of 147%. Thus current technologies are reported to give approximately a 14% to 15% mass yield of glucose to PTA. Assuming that the biological process could reach 85% of theoretical, and the cyclization process improved to give a 60% yield of PX, the isobutanol/isobutylene processes could give an overall mass yield of biomass to PTA of slightly over 40% via biobased PX.

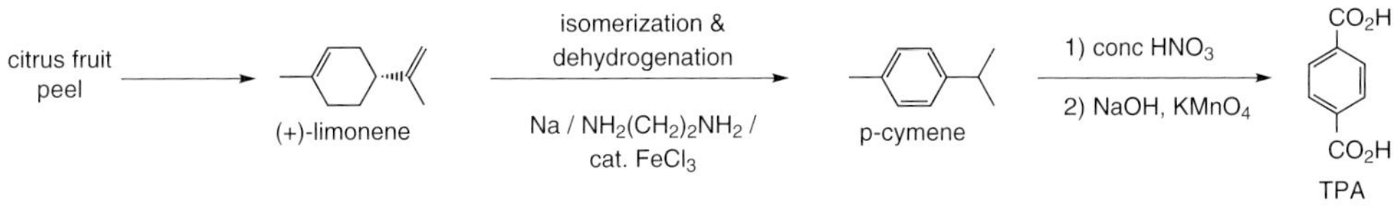

8.3 Biological Routes to Specific Aromatic Chemicals

Figure 8.16 Limonene to terephthalic acid.

8.3.3 Limonene to PTA

An interesting route unrelated to any of the previous routes has been described by Berti *et al.* [88] and is shown in Figure 8.16. The starting biomass is (+) limonene, currently extracted on a large-scale from citrus peel and used in a variety of products from cleaning solvents to flavors and fragrances. Limonene is first isomerized and aromatized using sodium and ethylenediamine to give p-cymene. This is followed by a two-step chemical oxidation, nitric acid followed by basic permanganate, to yield PTA. The isomerization and combined oxidation yields are reported as 99% and 85%, respectively. The chemistry followed for the isomerization and aromatization of limonene to p-cymene has been described in earlier literature [89], as has the nitric acid oxidation of p-cymene to p-toluic acid. At the time of writing, limonene had a higher market price than PTA so it seems unlikely that this route will be practical, regardless of yield.

8.3.4 The Common Aromatic Pathway

8.3.4.1 Background

Aromatic compounds are ubiquitous in Nature, and occur as the products of a variety of metabolic pathways, for example, the A ring in the estrogen steroids (ultimately from the isoprene pathway) and tetracycline (via a polyaromatic intermediate, ultimately from the polyketide pathway). However, the major pathway for production of aromatic compounds in Nature is the one that ultimately ends with lignin and provides the aromatic amino acids and other essential aromatic metabolites along the way. This pathway is known by several names depending on which section is of interest: the shikimic acid pathway, chorismate pathway, aromatic amino acid pathway or phenylpropanoid pathway; most prosaically it is called the common aromatic pathway. The entire pathway is known and available in open databases [90] and reviewed in the open literature [91] and is summarized briefly in Figure 8.17. The pathway is considered "common" up to chorismic acid, after which the pathway branches.

The pathway begins with the condensation of phosphoenyl pyruvate (PEP) produced from glycolysis, and erythrose-4-phospate (E4P) produced by the pentose phosphate pathway, to give 3-deoxy-D-*arabino*-heptulosonic acid 7-phosphate (DAHP) as the first metabolite that commits carbon to the pathway. Pyruvate can also be condensed with E4P to give DAHP, and this allows carbon from PEP

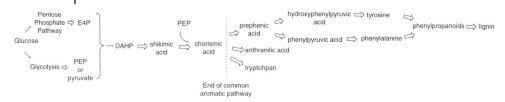

Figure 8.17 Overview of common aromatic pathway.

metabolism to be recovered and input into the pathway. The pathway proceeds through shikimic and chorismic acids, produces the three aromatic amino acids tryptophan, tyrosine and phenylalanine, and ultimately ends with sinapyl, coumaryl and coniferyl alcohols and the related acids: cinnamic, coumaric, caffeic, ferulic and sinapic, which polymerize under various conditions to produce lignin.

The upper part of the pathway, from DAHP to shikimic acid, has enjoyed considerable investigation and manipulation, largely by Draths and Frost who have published extensively in both the open [92] and the patent literature, and their work has been reviewed numerous times as well [3, 93]. The entire common aromatic pathway, and the branches to aromatic compounds using either biological or chemical reactions is summarized in Figure 8.18.

The condensation of PEP with E4P is performed by DAHP synthase (AroF) and commits the carbon input from the pentose phosphate and glycolytic pathways to the common aromatic pathway. Not surprisingly, it is the first potential bottleneck to carbon flux through the entire pathway, and correct expression of AroF is required [94]. AroF occurs in Nature as several isozymes which are sensitive to feedback inhibition by products of the lower parts of the pathway, that is, tyrosine, phenylalanine and tryptophan, and a feedback-insensitive AroF activity is obviously advantageous. Balancing and optimization of carbon flux into and within the pentose phosphate pathway to achieve the required ratio of E4P to PEP or pyruvate also requires consideration [95]. Rearrangement of DAHP by the next enzyme in the pathway, AroB (3-dehydroquinate synthase) with elimination of the phosphate group gives 3-dehydroquinic acid (DHQ), which forms the 6-membered ring that will eventually be the phenyl ring in all downstream aromatic compounds.

DAHP can also be formed from the condensation of E4P with pyruvate via mutants of the KDPGal aldolase enzyme (DgoA) instead of AroF [96]. This arrangement leaves the PEP from glycolysis available for other metabolic uses, for example glucose transport via the PTS system [97], while recovering the carbon that would otherwise be lost through the conversion of PEP to pyruvate. The ideal carbon flux through from glucose to DAHP is illustrated in Figure 8.19 [93a, 97–100]. This diagram addresses only the carbon flux from glucose to the DAHP intermediate at the beginning of the common aromatic pathway, and does not include the generation or consumption of ATP; an act essential for any cell consuming a carbon source and performing active metabolism.

Figure 8.19 shows that under ideal conditions, 7 molecules of glucose (each containing 6 carbon atoms) will produce 6 molecules of DAHP (each containing

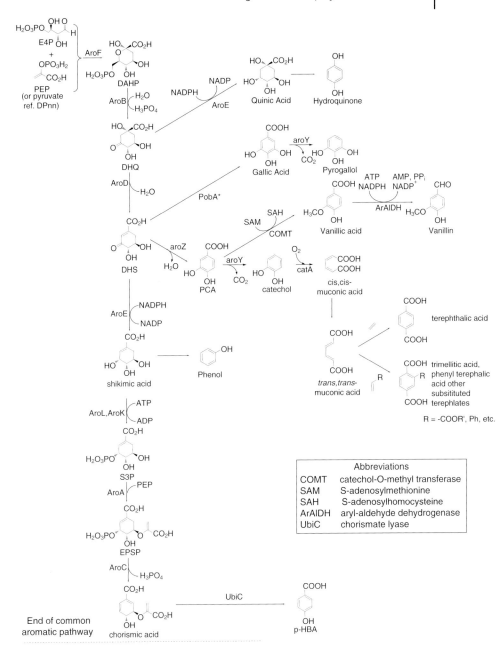

Figure 8.18 Details of engineering of the common aromatic pathway.

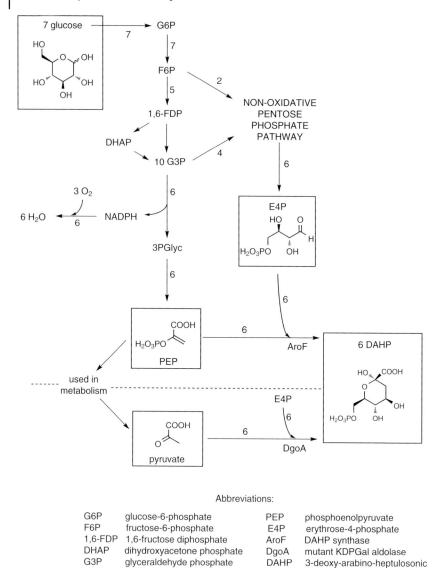

Figure 8.19 Carbon flux from glucose to DAHP.

7 carbon atoms). Although no carbon is lost, the *molar* ratio of DAHP to glucose is thus 6/7 (i.e., 85.7%) and the theoretical stoichiometry is:

$$7 \text{ glucose} + 3 \text{ O}_2 \rightarrow 6 \text{ DAHP} + 6 \text{ H}_2\text{O}$$

Muconic Acid Route to PTA The aromatic molecule of principle interest is terephthalic acid, which does not occur in the common aromatic pathway. Draths and Frost have shown that this molecule can be reached via a combination of metabolic engineering to divert carbon flow from the common aromatic pathway to the non-

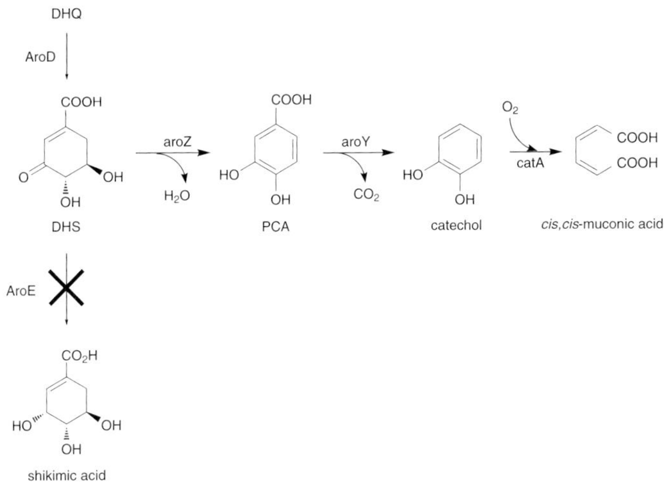

Figure 8.20 Details of enzymatic steps, DHS to *cis,cis*-muconic acid.

aromatic diacid muconic acid [92e, f, h]. A Diels–Alder condensation of the muconic acid with ethylene or other dienophiles leads to a variety of aromatic compounds not otherwise accessible from the common aromatic pathway.

A considerable amount of patent literature exists covering this particular route [101–103]. To produce muconic acid via fermentation of glucose in an engineered micro-organism, the common aromatic pathway is blocked immediately before the shikimic acid intermediate by deleting or disabling the AroE enzyme. In the absence of any other manipulation, this would cause the build up of dehydroshikimic acid (DHS). However, by inserting three heterologous genes, AroZ and AroY, both from *Klebsiella pneumoniae*, and the dioxygenase CatA from *Acenitobacter calcoaceticus*, the carbon flux is allowed to proceed from DHS to *cis, cis*-muconic acid. In these three enzymatic steps (illustrated in Figure 8.20), DHS is dehydrated to give protocatechuic acid (PCA) which is then decarboxylated to catechol; this decarboxylation forces the loss of one of the seven carbon atoms from the initial DAHP molecule as carbon dioxide. The CatA dioxygenase adds a molecule of oxygen between the vicinal hydroxy groups, thus cleaving the phenyl ring of catechol. Catechol is toxic, so the CatA activity must be expressed at a level that prevents the build-up of catechol and drives the carbon flux to the relatively non-toxic *cis,cis*-muconic acid as fast as the catechol is formed. This changes the stoichiometry from that given for DAHP previously, to the following:

$$7 \text{ glucose} + 9 \text{ O}_2 \rightarrow 6 \text{ muconic acid} + 6 \text{ CO}_2 + 12 \text{ H}_2\text{O}$$

The *cis,cis*-muconic acid is isolated from the biological production medium (i.e., the fermentation broth) and must be isomerized to the *trans,trans*- isomer in order to be used in the subsequent Diels–Alder condensation. Using ethylene as the dienophile gives 1,4-cyclohexene dicarboxylic acid. Aromatization via loss of two molecules of hydrogen produces terephthalic acid. By using other dienophiles, substituted terephthalic acids may be made; acrylate will give trimellitate, and styrene will yield phenyl terephthalic acid. This chemistry is summarized in Figure 8.21.

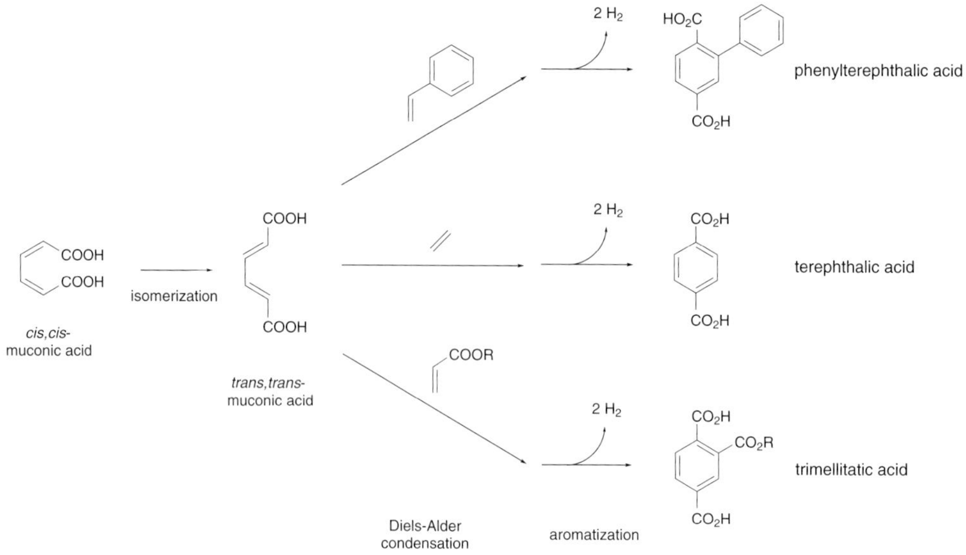

Figure 8.21 Chemical reactions of cis,cis-muconic acid to terephthalates.

Alternate Routes to Terephthalic Acid An imaginative proposal by Osterhout utilizes the common aromatic pathway to produce p-toluic acid instead of pHBA by inserting a different starting material at the very beginning of the pathway in place of DAHP [104]. In the proposed pathway, pyruvate is first condensed with glyceraldhyde-3-phosphate (G3P) to give 1-deoxy-D-xylulose-5-phosphate (DXP). This condensation is actually the first step in the non-mevalonic acid pathway of isoprenoid biosynthesis and is catalyzed by deoxyxylulose-5-phospate synthase (DXS, EC 2.2.1.7). The second step, reduction and isomerization of the 1-deoxyxylulose-5-phosphate to 2-methyl-erythritol-4-phosphate is the second step in that pathway, catalyzed by DXP reductoisomerase (DXR, EC 1.1.1.267). A diol dehydratase is required for the third step, and Osterhout provides several candidates but presents no data; presumably such an enzyme would need to leave the methyl group with the same relative stereochemistry as the hydroxy group in the 2-position of E4P. These three steps result in the compound 2-hydroxy-3-methyl-4-oxobutoxy phosphonate which is proposed to take the place of E4P in the condensation with PEP catalyzed by an appropriately engineered version of the enzyme AroF. Instead of producing DAHP, this condensation produces a DAHP analog with a methyl group instead of a hydroxy group at the 4-position. The proposed route shows this methyl group persisting through the entire series of enzymatic reactions in the common aromatic pathway, at which point it remains as the 4-methyl group in p-toluic acid. The proposed route is summarized in Figure 8.22 and ends with the same multi-step biological oxidation proposed above for the oxidation of PX to give terephthalic acid [60].

8.3 Biological Routes to Specific Aromatic Chemicals

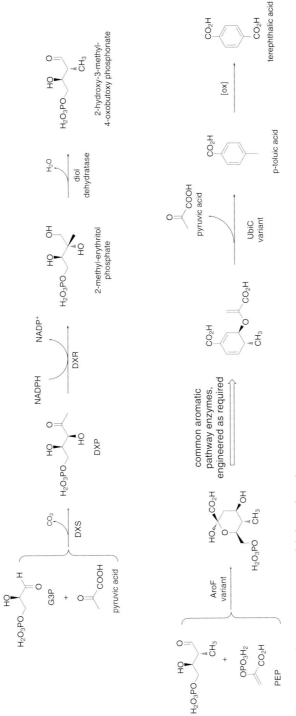

Figure 8.22 Alternate route to terephthalic acid via chorismate.

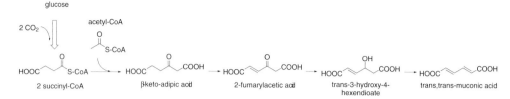

Figure 8.23 Biological route to *trans,trans*-muconic acid.

A patent application by Burk and Osterhout [105] proposes biological routes to take glucose to muconic acid completely outside the common aromatic pathway; the muconic acid is then available for the Diels–Alder chemistry previously described. While a number of routes are proposed, the simplest begins with the formation of succinyl-CoA, the pathways to which are well known and have previously been exploited [106]. The formation of succinyl-CoA has the very significant advantage of incorporating CO_2 into the succinyl-CoA intermediate, giving a theoretical carbon stoichiometry greater than 100%. In the proposed route, succinyl-CoA is condensed with acetyl-CoA to give β-keto-adipoyl-CoA. The Wood–Lungdahl pathway is proposed for avoiding the loss of CO_2 by decarboxylation of pyruvate in standard glycolysis. The resulting β-keto-adipic acid is taken to 2-fumarylacetate via a reductase operating in the oxidizing direction. This is reduced to *trans*-3-hydroxy-4-hexendioate via a dehydrogenase followed by a dehydratase give *trans,trans*-muconic acid. General examples of enzymes catalyzing the various required reactions are given, but no data for a completed pathway is disclosed. The pathway is summarized in Figure 8.23, and the patent application states that such a pathway should allow the following stoichiometry:

$$11 \text{ glucose} + 6 \text{ } CO_2 \rightarrow 12 \text{ muconic acid} + 30 \text{ } H_2O$$

8.3.4.2 Other Aromatic Compounds from the Common Aromatic Pathway

As Draths and Frost have published, a large number of aromatic chemicals can be produced from branches intermediate in the common aromatic pathway. After DAHP is formed, it is dephosphorylated to 3-dehydroquinic acid (DHQ). DHQ itself is a useful anti-oxidant [107] and may be chemically oxidized to yield quinic acid, which can be further oxidized to hydroquinone.

Quinic Acid and Hydroquinone By blocking the pathway before the formation of DHS, that is, by deleting or otherwise disabling the AroD activity that takes DHQ to DHS, the DHQ intermediate can be forced to accumulate. The same enzyme that takes DHS to shikimic acid, AroE (shikimate dehydrogenase) will also reduce DHS to quinic acid, and metabolic pathway constructs lacking AroD activity which will ferment glucose to quinic acid have been disclosed. Chemical oxidation of the quinic acid by acidified NaOCl, followed by heating [108] yields hydroquinone [109]. Production of hydroquinone in this manner avoids the use of benzene (Figure 8.24).

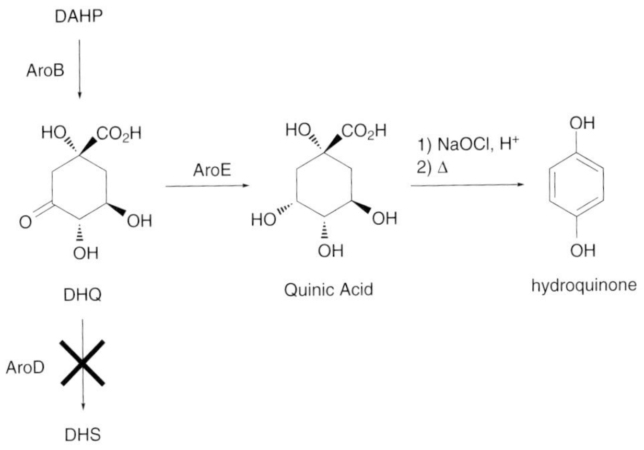

Figure 8.24 DHQ to hydroquinone.

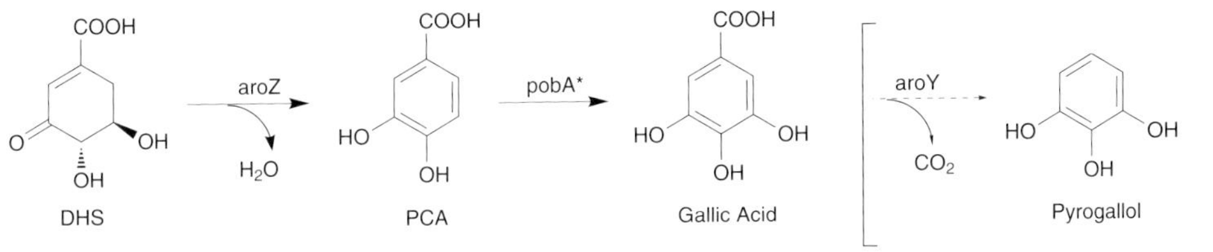

Figure 8.25 DHS to gallic acid and pyrogallol.

Gallic Acid and Pyrogallol A slightly different rearrangement of the enzymes in the pathway leading to muconic acid will allow the biological production of gallic acid, and its decarboxylated derivative pyrogallol; this is shown in Figure 8.25. This is accomplished by using the pathway with the AroE activity disabled, as done for the production of muconic acid. However, by replacing the AroY activity (which would otherwise take the PCA intermediate on to catechol) with a mutant of p-hydroxybenzoate hydroxylase, PobA*, the PCA is oxidized to form gallic acid [110]. The same literature indicates that it is possible to biologically produce pyrogallol by utilizing the same AroY decarboxylase activity that takes PCA to catechol. However, this oxidation must be done in a separate biological reaction to avoid the formation of catechol from PCA by the AroY activity [111].

Vanillic Acid and Vanillin Figure 8.26 shows the same pathway as in the gallic acid construct, that is, AroE blocked to prevent formation of shikimic acid, and this again forces the accumulation of DHS and its conversion to PCA via AroZ. By replacing the AroY and catA activities (used to take PCA to muconic acid) with catechol-O-methyl transferase (COMT) and arylaldehyde dehydrogenase (ArAlDH), vanillin may be produced. The COMT activity requires S-adenosylmethionine as the methyl donor, and S-adenosylhomocysteine is the by-product. The arylaldehyde dehydrogenase requires both ATP and NADPH to reduce the carboxylic acid to an aldehyde [112].

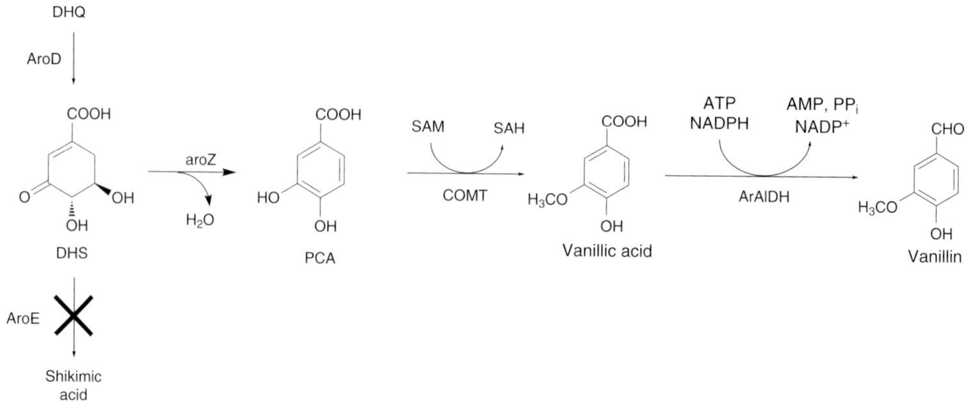

Figure 8.26 Metabolic route from DHS to vanillin.

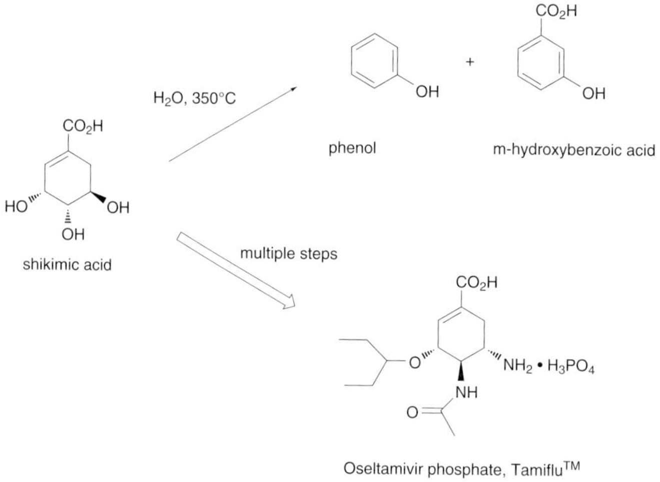

Figure 8.27 Commercial chemicals form shikimic acid.

Shikimic Acid and Phenol The earliest commercial target from the common aromatic pathway was shikimic acid, and blocking the pathway at this point by disabling the shikimate kinase enzymes AroK and AroL, as well as the activity that allows transport of shikimic acid back into the cell, ShiA [113]. Shikimic acid has been used as the starting material for commercial production of the neuraminidase inhibitor marketed as Tamiflu® [93a, 114].

By treating the shikimic acid produced by the AroK/AroL blocked pathway with near-critical water at 350 °C, phenol as well as *meta*-hydroxybenzoic acid is produced in yields of 53% and 18% respectively. By appending this single nonbiological step to the engineered common aromatic pathway, phenol may be produced without the use of benzene [115] (Figure 8.27).

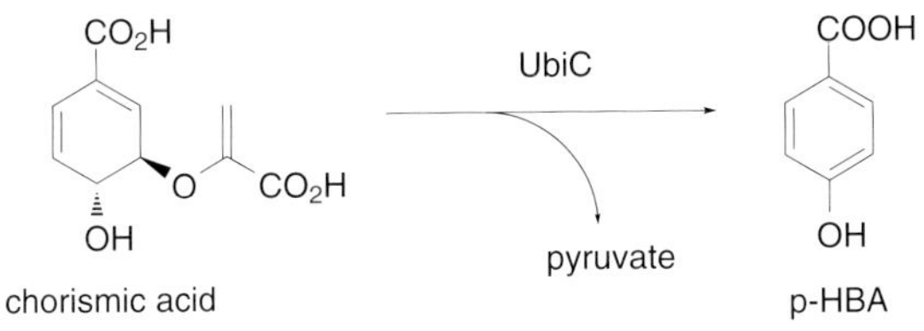

Figure 8.28 Enzymatic reaction to *p*-hydroxybenzoic acid.

p-Hydroxybenzoic Acid   *p*-Hydroxybenzoic acid (pHBA) is a key monomer in the production of liquid crystalline polymers (LCPs) that have high-value applications in the thermoplastics market, and it is possible to produce this compound via the common aromatic pathway. The final compound in the pathway is chorismic acid; after this, the pathway splits into several branches leading to the aromatic amino acids and other metabolites. It is possible to produce pHBA from chorismic acid via the enzyme chorismate lyase, also called chorismate pyruvate lyase (UbiC). Initial work by Amaratunga [116] achieved this by first selecting for an *E. coli* strain that was an over-producer of chorismic acid, and then introducing the UbiC gene. Later work by Barker and Frost [117] used a host strain of *E. coli* in which the three enzymes allowing chorismic acid to proceed into aromatic amino acid biosynthesis had been deleted; chorismate mutase, prephenate dehydratase, and anthranilate synthase (TrpE). Barker and Frost noted that formation of post-chorismic acid metabolic products occurred even with the multiple deletions, presumably because chorismic acid can rearrange to prephenic acid without enzymatic catalysis, and dehydration and decarboxylation of the resulting prephenic acid is driven by the stabilization achieved by reaching the phenylpyruvate product [100] (Figure 8.28).

Mandelic Acid   Yang *et al.* disabled the routes from chorismic acid to tryptophan and hydroxyphenylpyruvate, leaving only the path to phenylpyruvate open [118]. By blocking the next step, which would take phenylpyruvate to phenylalanine, and inserting the 4-hydroxymandelate synthase gene from *A. orientalis*, the production of S-mandelic acid was accomplished. 4-Hydroxymadelate synthase (HmaS, EC 1.13.11.43) catalyzes both the oxidative decarboxylation of phenylpyruvic acid as well as the oxidation of the benzylic carbon (Figure 8.29).

Styrene/Hydroxystyrene   Proceeding slightly farther along the pathways, through phenylalanine and tyrosine, the metabolites cinnamic acid and *p*-hydroxycinnamic acid (coumaric acid) may be produced by deamination of these two amino acids with an ammonia lyase, rather than a transaminase or amino acid oxidase. Phenylalanine/tyrosine ammonia lyases (PAL, TAL) will deaminate phenylalanine and tyrosine to give cinnamic and p-hydroxycinnamic acids respectively. Sariaslani and coworkers at Dupont have exploited these intermediates to reach 4-hydroxystyrene by both chemo-biological and completely biological routes [119]. *p*-Hydroxycinnamic acid will decarboxylate under basic conditions via a quinone

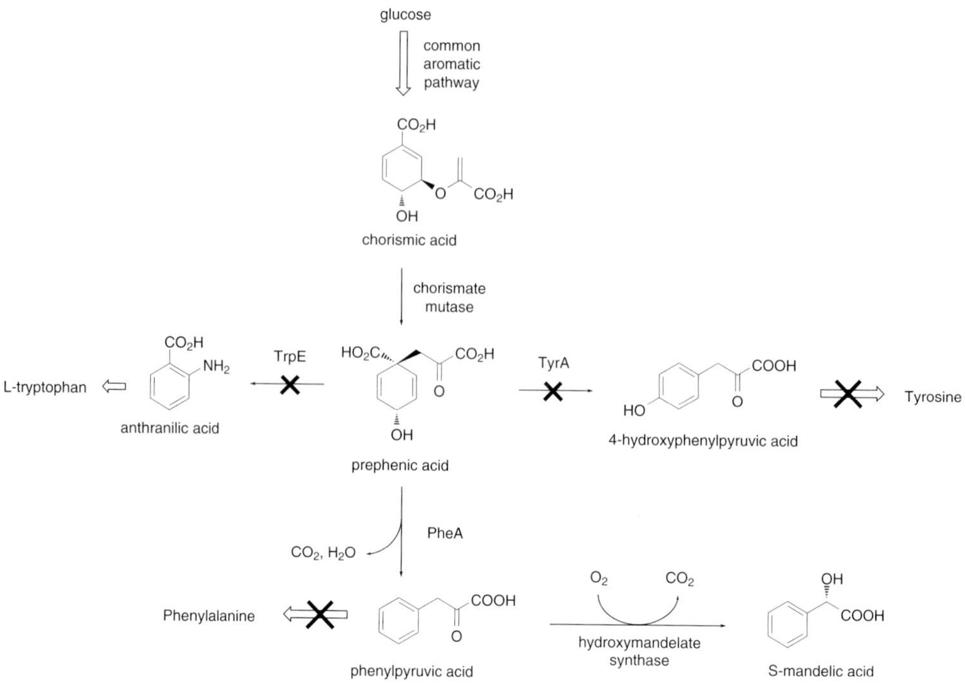

Figure 8.29 Biological route to mandelic acid.

intermediate, however, the cloning of a *p*-hydroxycinnamic acid decarboxylase (PDC) allows a completely biological route [120]. Further work by Ben-Bassat, Haynie, and coworkers at Dupont [121], and by Verhoef *et al.* at TNO have lead to improved processes for producing 4-hydroxystyrene from *p*-hydroxycinnamic acid [122].

In mid-2011, McKenna and Neilsen extended the use of a decarboxylase to cinnamic acid itself (*trans*-cinnamic acid decarboxylase, CADC), and thus the production of styrene. These authors noted that to the best of their knowledge, this was the first report of styrene produced by a microbial system [123]. Figure 8.30 summarizes the routes to styrene and 4-hydroxystyrene.

Anthranilic Acid and Aniline There is a plethora of aromatic molecules produced by the various metabolic pathways that diverge from the end of the common aromatic pathway. Further, there is now considerable knowledge of many metabolic pathways of a number of organisms; these have been mapped and are publicly available [90a, b]. This general situation makes the use of *in silico* design of novel pathways attractive, and several patent applications have been published which directly address compounds that can be derived via manipulation of the common aromatic pathway.

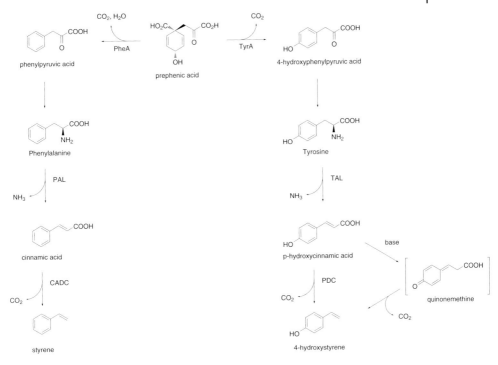

Figure 8.30 Biological routes to styrenes.

Pharkya proposes a pathway utilizing the addition of an amine group from an appropriate amine donor, such as glutamine, to form several well-known industrial aromatic compounds: anthranilic acid, p-aminobenzoic acid, and aniline [124]. This is summarized in Figure 8.31.

8.3.5
Other Routes to Aromatic Compounds

8.3.5.1 Tetrahydroxybenzene and Pyrogallol

In addition to the approach to pyrogallol via the common aromatic pathway, Draths and Frost demonstrated a different route to such polyhydroxylated benzenes (Figure 8.32). In this work, the enzyme *myo*-inositol-phosphate synthase takes glucose-6-phosphate to *myo*-inositol. Dehydrogenase activity oxidizes this to *myo*-2-inosose. Treatment of the *myo*-inosose with hot aqueous acid leads to dehydration and aromatization to give 1,2,3,4-tetrahydroxybenzene [125]. Proceeding further with non-biological chemistry, catalytic hydrogenation of the 1,2,3,4-tetrahydroxybenzene, followed by acid-catalyzed aromatization affords pyrogallol [126, 127]. The approach to aromatics via inositol has a higher theoretical yield than via the common aromatic pathway as the six carbon atoms of the initial glucose molecule are preserved throughout the process.

8 Production of Aromatic Chemicals from Biobased Feedstock

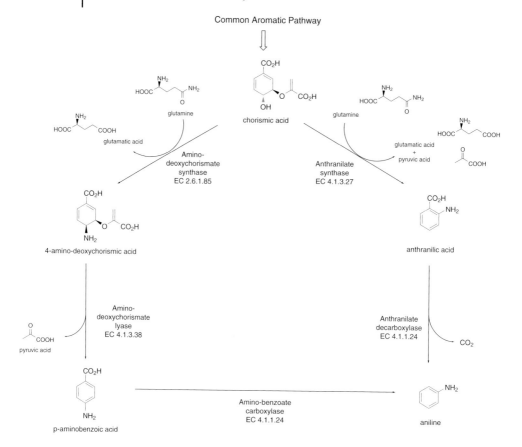

Figure 8.31 Biological routes to p-aminobenzoic acid and aniline.

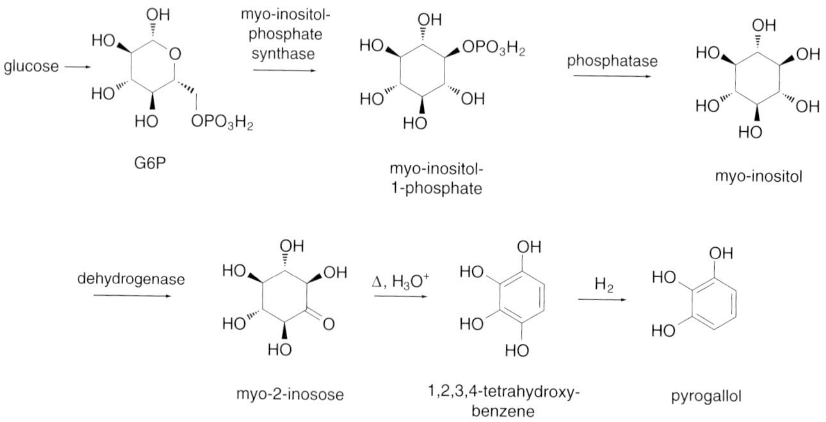

Figure 8.32 Biological production of polyhydroxybenzenes.

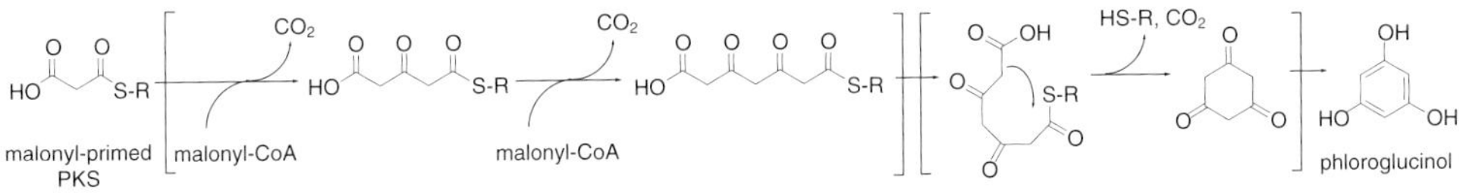

Figure 8.33 Biological production of phloroglucinol.

8.3.5.2 Phloroglucinol

The polyketide synthesis machinery may also be used to produce aromatic molecules [128]. However, the use of the malonyl-CoA pool as the final carbon source, and the necessary decarboxylation that occurs with each condensation makes the carbon-efficiency of this route generally low. Frost et al. describe the use of phloroglucinol synthase (PhlD) which condenses three molecules of malonyl-CoA to give phloroglucinol plus three CO_2 [129] (Figure 8.33).

8.3.5.3 Chalcones, Stilbenes, Vanillin and Lignans

It is worth noting that there are non-commodity, higher-value aromatic molecules that also come from the end of the common aromatic pathway. The commercial opportunities for these could support development and commercialization of the less structurally interesting but much larger volume commodity aromatics. The chemicals that will be briefly considered are the chalcones, stilbenes, and lignans, plus the single compound vanillin. These are shown in Figure 8.34.

Pursuing the common aromatic pathway through either phenylalanine or tyrosine leads to p-coumaloryl-CoA, and this can provide an initial acyl group for condensation with three malonyl-COA moieties, essentially linking the aromatic pathway with polyketide synthesis. Condensation with three malonyl-CoA moieties by the stilbene synthase complex with the extrusion of four molecules of CO_2 gives the stilbene skeleton, while condensation by the closely related chalcone synthase with the extrusion of three molecules of CO_2 leads to the chalcone skeleton. The currently most noted compound in these two groups of chemicals is resveratrol [130], a stilbene cited with a variety of therapeutic effects generally attributed to its anti-oxidant properties [131]. Generally, the "phyto-phenolics" that is, the phenylpropanoids beyond the end of the common aromatic pathway are considered to have wide therapeutic potential [132].

Vanillin was addressed previously above, via an engineered pathway from protocatechuic acid. In naturally occurring pathways it is formed from ferulic acid and, at least in theory, ferulic acid should be isolable from at least some of the current biomass pretreatment processes, especially those using corn fiber as their input biomass as this is a source rich in ferulate [133]. Enzymatic decarboxylations that work on ferulic acid are known and patented [134], but this specific enzymatic technology does not appear to be used commercially. The fermentation of ferulic acid via two microorganisms [135] is reported by Rhodia to produce vanillin (Figure 8.35).

Earlier in the twentieth century, vanillin was produced at large scale from lignin sulfonates recovered from the sulfite liquor of pulp mills. The process [136] treats

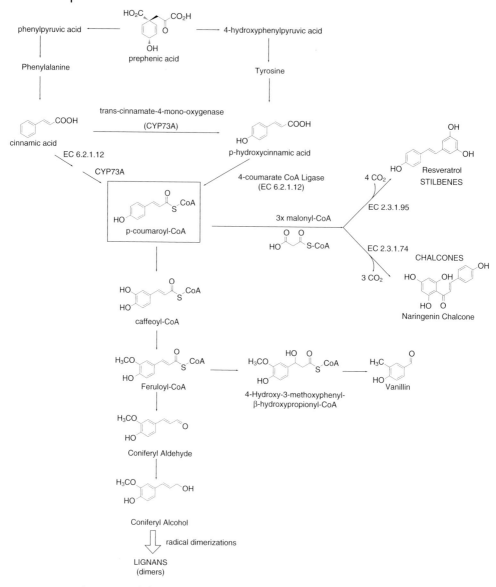

Figure 8.34 Chalcones, stilbenes, vanillin and lignans following the end of the common aromatic pathway.

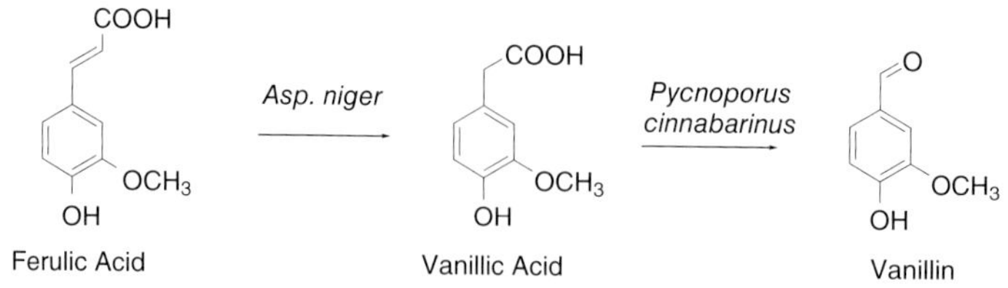

Figure 8.35 Double fermentation of ferulic acid to vanillin.

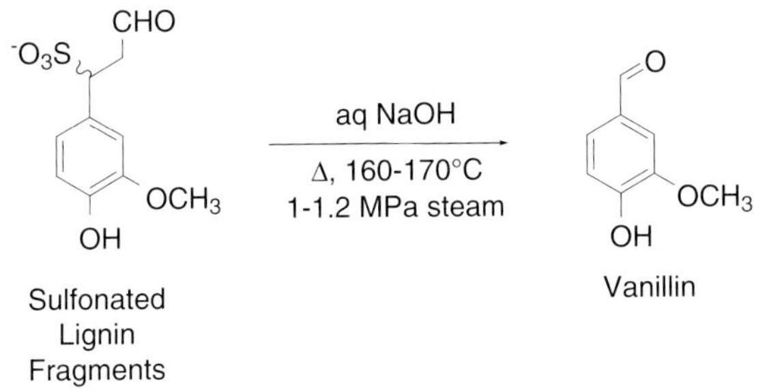

Figure 8.36 Vanillin from sulfite liquor.

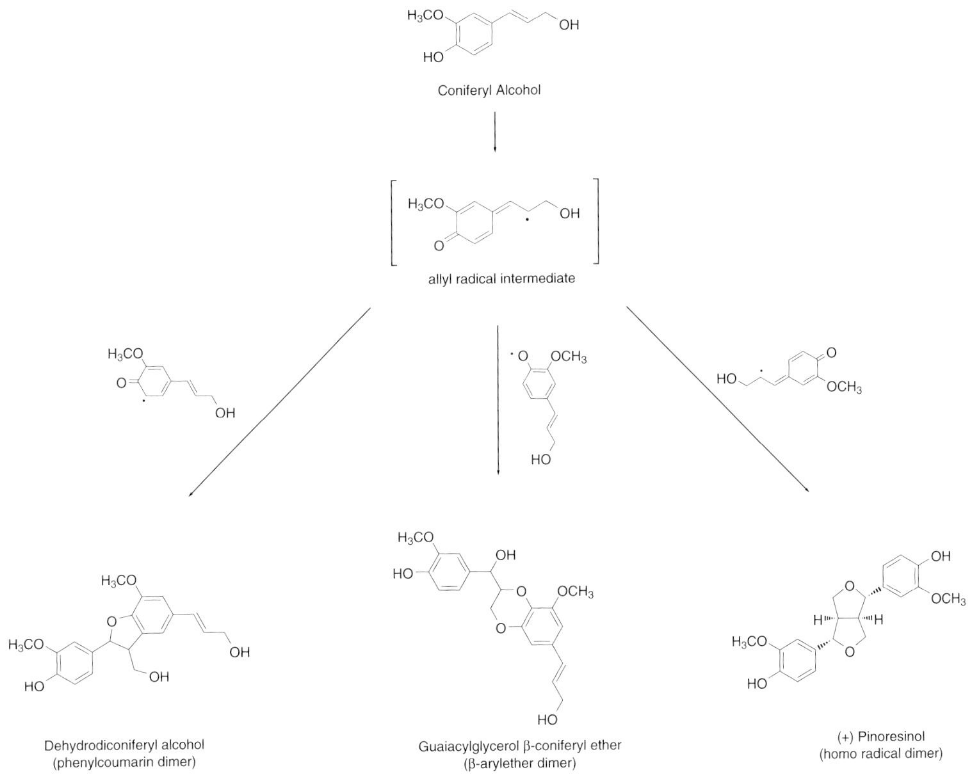

Figure 8.37 Initial lignan biosynthesis from coniferyl alcohol.

the liquor with strong aqueous base and uses aerial oxidation to reach the final vanillin product. This process was used to produce 60% of the world's vanillin supply in the 1980s at a single pulp mill in Canada [137] (Figure 8.36).

The lignans are another group of phenylpropanoids that are formed via the dimerization of coniferyl alcohol, presumably via the initial common allyl radical intermediate. Condensation of the initial radical with its resonance isomers leads to the three general lignan skeletons shown in Figure 8.37, and further metabolism

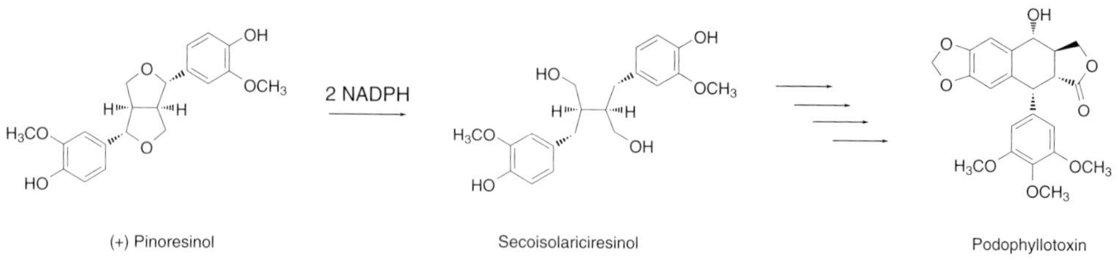

Figure 8.38 Pinoresinol metabolism to podophyllotoxin.

(both oxidation and reduction) produces the wide range of lignan natural products with anti-cancer and other therapeutic benefits [138]. One of the initial lignan structures, (+) pinoresinol, is metabolized to podophyllotoxin which enjoys considerable study as an anti-cancer agent (Figure 8.38).

Vanillin has been produced from lignin in sufficient commercial volume (the Canadian plant reportedly produced 3400 metric tonnes in 1981 [137]) at least to be considered a near-commodity aromatic chemical. However, the stilbenes, chalcones and lignans remain natural products present in very low amounts in biomass, and are not currently isolated in sufficient volumes to be considered commodity chemicals. Although their metabolic pathways are largely known, and directed efforts to generate podophyllotoxin by cell culture have been published [139], the value of these natural products in at least the near future is most likely in pharmaceutical and nutraceutical markets. But as noted earlier in the case of ferulic acid, there may be adventitious process streams in routes to large-volume, commodity biobased aromatic chemicals in which these natural products are present in quantities high enough to provide useful revenue while development of the large-scale operations are underway.

8.4
Lignin – The Last Frontier

Pursuing the metabolic opportunities past the end of the common aromatic pathway leads to compounds generally labeled "hydroxycinnamates", or more generally, "phenylpropionates" [140]. Cinnamic acid, p-coumaric acid, caffeic acid, ferulic and sinapic acids all present themselves as various levels of oxidation and methylation occur on the route that ultimately ends with lignin. These metabolic intermediates, shown in Figure 8.39, could be produced as specialty chemicals in their own right via the type of metabolic engineering reviewed above, or isolated from existing biomass by conventional methods, isolated from pulping liquor streams, or from the effluent streams of various biomass pretreatment methods used to render cellulose more amenable to hydrolysis [141–144].

The ultimate aromatic compound formed by Nature, and at the farthest end of the common aromatic pathway is lignin. Lignin, from the Latin for "wood", is the second largest biological polymer (after cellulose) and the only one composed

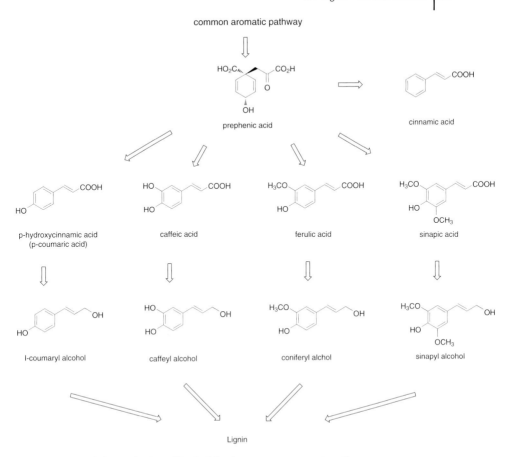

Figure 8.39 Metabolic production of lignin following common aromatic pathway.

entirely of aromatic subunits. At an average presence of 20% in biomass, and an estimated 170 billion tonnes of biomass generated globally each year [1], there are approximately 34 billion metric tonnes of lignin generated each year globally.

The carbon efficiency of lignin production from "sugar" (i.e., sucrose, the sugar most generally used to transport carbon in the vasculature of plants) is important if lignin is to be considered as a useful feedstock. The efficiency depends on which of several related metabolic routes are considered. The estimate of the carbon efficiency of lignin production varies from a low of 60.7% for lignin production via sinapyl alcohol (itself produced from phenylalanine metabolism), to a high of 76.9% for lignin production via p-coumaryl alcohol produced (itself produced from tyrosine metabolism) [145]. As lignin is composed of a combination of sinapyl and p-coumaryl monomers, produced from the tyrosine and phenylalanine pathways, the carbon efficiency of lignin production from sucrose will be close to the average of these two boundary values, or approximately 68 to 70%. Thus, the common aromatic pathway followed from input carbohydrate to the ultimate metabolic product is a remarkably efficient process, comprising a minimum of 24 steps.

As previously noted by the combined US DOE and USDA report [2], about 1.3 billion tons of biomass could reasonably be collected each year within the US. Assuming the same 20% average lignin content as above, this means about 260 million metric tonnes of lignin are available from biomass each year in the US alone. The 2007 report of chemicals available from biomass further concluded that even using then-existing lignin conversion technologies (as of 2007, and assuming a 10% efficiency across all conversion processes), only about 46 million metric tonnes of lignin would be needed to supply the US domestic demand for BTX, phenol, and terephthalic acid [146]. There is thus over five times the necessary amount of lignin available in the US to satisfy the domestic US consumption of these major aromatic commodity chemicals – if lignin could be transmuted from the polymer that it is, into its individual monomeric units.

With the combined annual global production of the three largest aromatic chemicals estimated at 103 million metric tonnes total [4], the global amount of lignin is more than sufficient to supply global aromatic chemicals, even with collection and conversion efficiencies both at 10%; that is, only 1% of the total global lignin collected and converted.

In 2004, world lignin production (exclusively from the pulp and paper industry) was only about 1 million tonnes as lignosulfates and 100 000 tonnes as Kraft lignin [147], and this represented only about 2% of the lignin available in the pulp and paper industry.

8.5
Considerations for Scale-Up and Commercialization

The existing biotech industry is historically based in the production of human therapeutics by what was truly revolutionary technology. The biotech industry that now exists is the result of several decades of technical, commercial and policy development that could accept long-term programs for the commercial development of therapeutics that could not be created any other way and which promised benefits that were otherwise unachievable. There was no incumbent industry that was a target for replacement and efficiencies of scale and manufacturing, cost of starting materials and competition in an established product marketplace were not considerations beyond the general effort providing sufficient return on investment. This is completely different from the biobased chemicals industry that is now developing. The incumbent petrochemicals industry is extremely large and efficient, and the products targeted for introduction by the biobased chemicals industry are generally not new compounds, but existing commodity chemicals with well established markets, supply chains, and prices. In this situation, the effects of global macro-economics and government policies established in response to local and global political forces will have very large and nearly immediate effects on both the incumbent petrochemical industry as well as the nascent biobased chemicals industry. It is beyond the scope of this chapter to discuss these forces, but there are basic principles that must be considered for any commercialization and scale-up effort of any biobased chemical process.

Yield: the efficiency of the conversion of input biomass to output product. Chemists and biologists generally consider this in terms of molar efficiency as this is the best way to understand the efficiency of the reaction pathway involved in whatever is being made. Engineers will consider yield in terms of mass (tonnes in vs. tonnes out) as this allows immediate application to design and costing. Both considerations are essential – one does not supercede the other. For commodity chemicals the largest cost-contributor will be the price of the biomass used. The yield (efficiency) of any proposed process must meet a calculated threshold for commercial viability before it can be considered for further development.

Concentration: this term for chemical processes is synonymous with the word titer, used for fermentation processes. If a process requires a large volume to produce a given unit of product (e.g., one tonne) then the cost of equipment and plant, that is the capital cost, will exceed the upper limit of commercial viability. Generally, chemical processes target concentrations of around 20% (by weight) although this varies widely. For fermentation processes, the current threshold is generally to have titers greater that 100 g l^{-1}, and the downstream process steps immediately following a fermentation should seek to increase this concentration. Current ethanol fermentations produce ethanol at approximately 150 g l^{-1}, or 15%. This is a useful benchmark when starting process development, but should not be considered as more than a guide. Of specific note regarding fermentations that are operated as fed-batch processes, into which a stream of glucose or other fermentable carbon is added: the concentrations of the glucose or other sugar/fermentable carbon must be sufficiently high to avoid increasing the final fermentation volume excessively. Concentrations of sugars of at least 50% by weight are required for fed-batch fermentations.

Productivity: the efficiency of the process in terms of both equipment and time. A process may have a very high yield (that is, the efficiency of the conversion of input biomass to output product may exceed the threshold calculated for commercial viability), but if the process requires very large-volume equipment or a long period of time to run, it will likely not be economic. The cost of purchasing the equipment and building the plant will be very high, as just noted. Further, the cost of operating it to achieve production of a given unit of output product (e.g., 1 tonne) will also be high, leading to high operating costs per unit of output product, and high operating costs may overwhelm otherwise reasonable capital costs, especially if expensive utilities (e.g., heating or cooling) are required. For fermentation processes, the productivity is generally expressed in grams per liter per hour, and a useful benchmark is around 2.5 g l^{-1} h^{-1}.

Carbon efficiency and redox balance: This concept deserves careful consideration by the biobased chemicals industry. The carbon found in the feedstock for the petrochemical industry (i.e., either oil or gas) is generally highly reduced, methane (CH_4) being the most reduced form of carbon, that is, a redox value of −4. Feedstock for the biobased chemicals industry is most generally carbohydrate (either as sugars or cellulose) and this is in the middle of the range of redox potential available to carbon. Considering formaldehyde as the simplest carbohydrate (CH_2O), the carbon atom has a redox value of zero. Carbon dioxide is the most oxidized form of carbon with a redox value of +4. To reach the redox level of hydrocarbons, biological processes must transform at least some of the input carbohydrate to more oxidized forms (generally CO_2) in order to provide sufficient reducing equivalents to transform the remainder of the carbohydrate feedstock to the desired, more reduced product. Thus a net deoxygenation of input carbohydrate biomass is required to reach the redox level needed to produce hydrocarbon products, regardless of the process used (either totally biological or totally non-biological), and the combination of hydrogen balance and carbon balance is a necessary consideration. As lignin is more reduced than carbohydrates, it would provide an advantage as a starting material in this respect.

8.6
Conclusion

The potential of using biomass to produce chemicals has been recognized for decades [148–150] and the advantages of producing chemicals rather than fuels from biomass has been noted earlier [3]. It is clear that there remains enormous and as yet untapped potential for the production of aromatic commodity chemicals from both manipulation of the common aromatic pathway and the deconstruction of the pathway's ultimate aromatic product, lignin.

References

1 Klass, D.L. (1980) *Biomass for Renewable Energy, Fuels, and Chemicals*, Academic Press.
2 US Department of Energy, US Department of Agriculture (2005) Biomass as Feedstock for a Bioenergy and Bioproducts Industry: The Technical Feasibility of a Billion-Ton Annual Supply.
3 Dodds, D.R., and Gross, R.A. (2007) Chemicals from biomass. *Science*, **318**, 1250–1251.
4 Nexant Chem Systems (2012) *Global Chemicals Demand Estimate*, White Plains, New York.
5 Camara Greiner, E.O. (October 2009) CEH Marketing Research Report: PET Polymer, report number 580.1150K, SRI Consulting.
6 Cailliot, A. (1847) Etudes sur L'Essence de Terebenthine. *Ann. Chim. Phys.*, **3** (21), 27–40.
7 Saffer, A., and Barker, R.S. (1958) Preparation of aromatic polycarboxylic acids, US Patent 2833816, May 16, 1958.
8 (a) Zuo, X., Niu, F., Snavely, K., Subramaniam, B., and Busch, D.H. (2010) Liquid phase oxidation of p-xylene to terephthalic acid at medium-high temperatures: multiple benefits of CO2-expanded liquids. *Green Chem.*, **12**, 260–267; (b) Chavan, S.A., Srinivas, D., and Ratnasamy, P. (2001) Selective oxidation of para-xylene to terephthalic acid by oxo-bridged Co/Mn cluster complexes encapsulated in zeolite–Y. *J. Catal.*, **204**, 409–419.

9 (a) Sims, B. (2011) The Bio-Oil Barons, *Biorefining Magazine*, http://biomassmagazine.com/articles/6757/the-bio-oil-barons (accessed 11 January 2013); (b) Vispute, T. (2011) *Pyrolysis Oils: Characterization, Stability Analysis, and Catalytic Upgrading to Fuels and Chemicals*, Open Access Dissertations. Paper 349. Chapter 3, p. 22 http://scholarworks.umass.edu/open_access_dissertations/349 (accessed 11 January 2013); (c) Ringer, M., Putsche, V., and Scahill, J. (2006) *Large-scale pyrolysis oil production: a technology assessment and economic analysis*, National Renewable Energy Laboratory Technical Report NREL/TP-510-37779; (d) Venderbosch, R.H., and Prins, W. (2011) *Fast Pyrolysis*, Chapter 5 in *Thermochemical Processing of Biomass: Conversion into Fuels, Chemicals, and Power*, Robert C. Brown, Ed., John Wiley & Sons, UK; (e) Ortiz-Toral, P.J. (2011) *Steam reforming of water-soluble fast pyrolysis bio-oil: Studies on bio-oil composition effect, carbon deposition, and catalyst modifications*, Theses and Dissertations. Paper 11965 http://lib.dr.iastate.edu/etd/11965 (accessed 11 January 2013); (f) Jones, S.B., Holladay, J.E., Valkenburg, C., Stevens, D.J., Walton, C.W., Kinchin, C., Elliott, D.C., and Czernik, S. (2009) *Production of gasoline and diesel from biomass via fast pyrolysis, hydrotreating and hydrocracking: a design case*. U.S. Department of Energy report PNNL-18284 http://www.pnl.gov/main/publications/external/technical_reports/pnnl-18284.pdf (accessed 11 January 2013); (g) Boonpo, J., Udomsap, P., Yoosuk, B., and Sukkasa, S. *Towards commercialization of alternative biofuel: Improving the stability of pyrolysis liquid by physical fractionation*, The Second TSME International Conference on Mechanical Engineering, October 19–21, 2011. Paper ETM04. http://rcme.engr.tu.ac.th/TSME-ICoME%20Web/full_paper_file/ETM/ETM04%201011-1831-1-RV.pdf (accessed 11 January 2013); (h) Xu, J., Jiang, J., and Lu, W. (2010) Preparation of Novolacs using phenolic rich components as partial substitute of phenol from biomass pyrolysis oils. *Bull. Chem. Soc. Ethiopia*, **24**, 251–257; (j) Mullen, C.A., and Boateng, A.A. (2008) Chemical composition of bio-oils produced by fast pyrolysis of two energy crops. *Energy & Fuels*, **22**, 2104–2109; (k) Melligan, F., Kapinski, W., Hayes, M.H.B., and Leahy, J. J. Pyrolysis of Biomass to Produce Bio-Oil, IRC SET 2009 Symposium-Innovation Fuelling the Smart Society, Dublin, September 25, 2009. Available at http://www.carbolea.ul.ie/downloads.php (accessed 11 January 2013); (l) Zhuang, X.L., Zhang, H.X., Yang, J.Z., and Qi, H.Y. (2001) Preparation of levoglucosan by pyrolysis of cellulose and its citric acid fermentation. *Bioresource Technology*, **79**, 63–66; (m) Shafizehd, F., Furneaux, R.H., Cochran, T.G., Scholl, J.P., and Sakai, Y. (1979) Production of levoglucosan and glucose from pyrolysis of cellulosic materials. *J. Appl. Polymer Sci.*, **23**, 3525–3539.

10 Huber, G.W., Iborra, S., and Corma, A. (2006) Synthesis of transportation fuels from biomass: chemistry, catalysts, and engineering. *Chem. Rev.*, **106**, 4044–4098. See page 4064.

11 List of structures compiled from lists of bio-oil components in references 9b, 9j, and 9k.

12 (a) Blommel, P.G., and Cortright, R.D. (2008) Production of conventional liquid fuels from sugars, http://www.biofuelstp.eu/downloads/Virent_Technology_Whitepaper.pdf (accessed 11 January 2013); (b) W Heinen, A.W., Peters, J.A., and van Bekkum, H. (2000) Hydrogenation of fructose on Ru/C catalysts. *Carbohydr. Res.*, **328**, 449–457; (c) Kwak, B., Lee, B., Kim, T., Kim, I., and Lee, S. (2006) Method for Preparing Sugar Alcohols Using Ruthenium Zirconia Catalyst, PCT International Application 2006/093364A1, September 8, 2006; (d) Smith, P.B. (2012) *Carbohydrate Hydrogenolysis*, Chapter 12 in Biobased Molomers, Polymers, and Materials, Smith, P., ed., ACS Symposium Series, American Chemical Society, Washington, DC, 2012; (e) See Table 2, reference 12d; (f) Werpy, T., Frye, J., Zacher, A., and Miller, D.

Hydrogenolysis of 6-Carbon Sugars and Other Organic Compounds, US Patent 6841085B2, January 11, 2005. Compare to conditions in examples to those in reference 12c; (g) Vispute, T.P., Zhang, H., Sanna, A., Xiao, R., and Huber, G.W. (2010) Renewable chemical commodity feedstocks from integrated catalytic processing of pyrolysis oils. *Science*, **330**, 1222–1227.

13 Antos, G.J., and Aitani, A.M. (eds.) (2004) *Catalytic Naphtha Reforming*, 2nd edn, Marcel Dekker, New York.

14 (a) Balat, H., and Kirtay, E. (2010) Hydrogen from biomass-present scenario and future prospects. *Int. J. Hydrogen Energy*, **35**, 7416–7426; (b) Alonso, D.A., Bond J.Q., and Dumesic, J.A. (2010) Catalytic conversion of biomass to fuels. *Green Chemistry*, **12**, 1493–1513. See page 1508.

15 Kuhl, G.H., and Kresge, C.T. (1995) Molecular Sieves in *Encyclopedia of Chemical Technology*, **16**, 916–925, John Wiley & Sons, Inc., New York.

16 See reference 9b and references cited in Section 8.2.2 of this chapter.

17 (a) Cortright, R.D., and Blommel, P.G. *Synthesis of Liquid Fuels and Chemicals from Oxygenated Hydrocarbons*, US Patent 7977517B2, July 12, 2011; (b) Cortright, R.D., and Dumesic, J.A. *Catalyst to Dehydrogenate Paraffin Hydrocarbons*, US Patent 5736478A1, April 7, 1998; (c) Cortright, R.D., and Dumesic, J.A. *Method for catalytically reducing carboxylic acid groups to hydroxyl groups in hydroxycarboxylic acids*, US Patent 6441241B1, August 27, 2002; (d) Cortright, R.D., and Dumesic, J.A. *Low-temperature hydrogen production from oxygenated hydrocarbons*, US Patent 6964758B2, November 15, 2005; (e) Cortright, R.D., and Blommel, P.G. *Synthesis of Liquid Fuels and Chemicals from Oxygenated Hydrocarbons*, US Patent 8053615B2, November 8, 2011; (f) Cortright, R.D., and Blommel, P.G. *Synthesis of Liquid Fuels and Chemicals from Oxygenated Hydrocarbons*, US Patent 8017818B2, September 13, 2011; (g) Cortright, R.D., and Blommel, P.G. *Synthesis of Liquid Fuels and Chemicals from Oxygenated Hydrocarbons*, US Patent Aplication 2011/0245543A1, October 6, 2011; (h) Dumesic, J.A., Simonetti, D.A., and Kunkes, E.L. *Single-Reactor Process for Producing Liquid-Phase Organic Compounds from Biomass*, US Patent 8075642B2, December 13, 2011; (i) Nagaki, D.A., Cortright, R.D., Kamke, L., and Woods, E. *Improved Catalysts for Hydrodeoxygenation of Polyols*, PCT International Publication WO2011/082222A2, 7 July 2011; (j) Blommel, P.G., and Cortright, R.D. (August 25, 2008) *Production of Conventional Liquid Fuels from Sugars*, http://www.biofuelstp.eu/downloads/Virent_Technology_Whitepaper.pdf (accessed 11 January 2013).

18 See reference 17a, Example 6.
19 See reference 17a, Examples 7 and 8.
20 See reference 17a, Example 45.
21 See reference 17a, page 27.
22 See reference 17a; Table 13, Figure 17 and Example 55.
23 Komula, D. (2011) *Completing the puzzle: 100% plant-derived PET*, http://www.virent.com/wordpress/wp-content/uploads/2011/08/Virent-Article-in-Bioplastics-Magazine.pdf (accessed 11 January 2013).

24 (a) Huber, G.W., Cheng, Y., Carlson, T., Vispute, T., Jae, J., and Tompsett, G. *Catalytic Pyrolysis of Solid Biomass and Related Biofuels, Aromatics, and Olefin Componds*, US Patent Application 2009/0227823, September 9, 2009; (b) Huber, G.W., Gaffney A.M., Jae, J., and Cheng, Y. *Systems and Processes for Catalytic Pyrolysis of Biomass and Hydrocarbonaceous Materials for Production of Aromatics With Optional Olefin Recycle, and Catalysts Having Selected Particle Size for Catalytic Pyrolysis*, US Patent Application 2012/0203042, August 9, 2012; (c) Huber, G.W., Cheng, Y., Carlson, T., Vispute, T., Jae, J., and Tompsett, G. *Catalytic Pyrolysis of Solid Biomass and Related Biofuels, Aromatics, and Olefin Componds*, PCT International Publication WO2009/111026A8, September 11, 2009; (d) Carlson, T.R., Vispute, T.P., and Huber, G.W. (2008)

Green gasoline by catalytic fast pyrolysis of solid biomass derived compounds, *Chem. Sus. Chem.*, **1**, 397–400.
25. See reference 24a, page 11.
26. See Example 12 in reference 30a for reactor conditions. Yields approximated from Figure 15.
27. (a) Vispute, T., Zhang, H., Sanna, A., Xiao, R., and Huber, G.W. (2010) Renewable chemical commodity feedstocks from integrated catalytic processing of pyrolysis oils. *Science*, 330, 1222–1227; (b) Renewable Chemicals Digest (November 26, 2010) *UMass Amherst, Anellotech Pioneer Conversion Of Bio-Oil to Chemicals Intermediates*, http://www.biofuelsdigest.com/biotech/2010/11/26/umass-amherst-anellotech-pioneer-conversion-of-bio-oil-to-chemical-intermediates/ (accessed 6 July 2012); (c) New Energy and Fuel (December 3, 2010) *Pyrolysis Oil Gets More Practical Refining*, http://newenergyandfuel.com/ http://newenergyandfuel/com/2010/12/03/pyrolysis-oil-gets-more-practical-refining/ (accessed 11 January 2013).
28. See reference 9b, pages 25–27.
29. Data from reference 27a.
30. Data from reference 27a.
31. Vermeiren, W., Bouvart, F., and Dubut, N. (2011) A Process for the Production of Bio-Naphtha from Complex Mixtures of Natural Occurring Fats & Oils, PCT International Publication WO2011/012440A2, 3 February 2011.
32. Vanrysselberghe, V., and Vermeiren, W. (2011) Use of Free Fatty Acids Produced from Bio-Sourced Oils&Fats as the Feedstock for a Steamcracker, PCT International Publication WO2011/012438A1, 3 February 2011.
33. Data from reference 42b, page 9.
34. (a) Hieronymus, E. (1961) Process for Preparing para-Xylene, US Patent US3002035, September 26, 1961; (b) Manzer, L. (2005) Process for the Preparation of Xylene, PCT International Publication WO2005/054160A1, 16 June 2005; (c) Probst, O., and Hieronymus, E. (1961) Manufacture of Aromatic Hydrocarbons, US Patent 2985693, May 23, 1961; (d) Slaugh, L.H. (1981) Para-Xylene Process and Catalyst, US Patent 4247726, January 27, 1981.
35. Manzer, L.E., Kourtakis, K., Herron, N., McCarron, E.M., and Ver Nooy, P.D. (2003) Process for the Preparation of p-Xylene, US Patent 6600081B2, July 29, 2003.
36. Manzer, L.E., Kourtakis, K., Herron, N., McCarron, E.M., and Ver Nooy, P.D. (2006) Process for the Preparation of p-Xylene, US Patent 7067708B2, June 27, 2006.
37. Hawkins, A.C., Glassner, D.A., Buelter, T., Wade, J., Meinhold, P., Peters, M.W., Gruber, P.R., Evanko, W.A., and Aristidou, A.A. (2009) Methods for the Eonomical Production of Biofuel From Biomass, PCT International Publication WO2009/059253A2, 7 May 2009.
38. (a) Donaldson, G.K., Eliot, A.C., Flint, D., Maggio-Hall, L.A., and Nagarajan, V. (2010) Fermentive Production of Four Carbon Alcohols, US Patent 7851188B2, December 14, 2010; (b) Donaldson, G.K., Eliot, A.C., Flint, D., Maggio-Hall, L.A., and Nagarajan, V. (2011) Fermentive Production of Four Carbon Alcohols, US Patent 7993889B1, August 9, 2011; (c) Donaldson, G.K., Eliot, A.C., Flint, D., Maggio-Hall, L.A., and Nagarajan, V. (2011) Fermentive Production of Four Carbon Alcohols, US Patent Application 2011301388A1, December 8, 2011.
39. Tyagi, R., Lee, Y.-T., Guddat, L.W., and Duggleby, R.G. (2005) Probing the mechanism of the bifunctional enzyme ketol-acid reductoisomerase by site-directed mutagenesis of the active site. *FEBS J.*, **272**, 593–602.
40. Hawkins, A.C., Glassner, D.A., Buelter, T., Wade, J., Meinhold, P., Peters, M.W., Gruber, P.R., Evanko, W.A., and Aristidou, A.A. (2009) Methods for the Eonomical Production of Biofuel From Biomass, US Patent Application 2009/0215137A1, August 27, 2009.
41. Buelter, T., Meinhold, P., Smith, C., Aristodou, A., Dundon, C.A., and Urano, J. (2010) Engineered Microorganisms for the Production of One or More target Compounds, PCT International Publication WO2010/075504A2, 1 July 2010.

42 Anthony, L.C., Huang, L.L., and Ye, R.W. (2010) Production of Isobutanol in Yeast Mitochondria, US Patent Application 2010/129886A1 May 27, 2010.

43 (a) Feldman, R.M.R., Gunawardena, U., Urano, J., Meinhold P., Aristodou, A.A., Dundon, C.A., and Smith, C. (2011) Yeast Organism Producing Isobutanol at a High Yield, US Patent 8017375B2, September 13, 2011; (b) Feldman, R.M.R., Gunawardena, U., Urano, J., Meinhold, P., Aristodou, A.A., Dundon, C.A., and Smith, C. (2011) Yeast Organism Producing Isobutanol at a High Yield, US Patent Application 2011/183392A1, July 28, 2011.

44 Donaldson, G.K., Maggio-Hall, L.A., and Nakamura, C.E. (2009) Deletion Mutants for the Productin of Isobutanol, PCT International Publication WO2009/149240A1, 10 December 2009.

45 Buelter, T., Meinhold, P., Feldman, R.M.R., Eckl, E., and Hawkins, A. (2010) Engineered Microorganisms Capable of Producing Target Compounds Under Anaerobic Conditions, PCT International Prublication WO2010/051527A2, 6 May 2010.

46 Anthony, L.C., Dauner, M., Donaldson, G.K., and Paul, B.J. (2010) Carbon Pathway Optimized Production Hosts for the Produciton of Isobutanol, US Patent Application 2010/120105A1, May 13, 2010.

47 Peters, M.W., Meinhold, P., Buelter, T., and Landwehr, M. (2008) Engineered Microorganisms for Increasing Product Yield in Biotransformation, Related Methods and Systems, PCT International Publication WO2008/013996A2, 31 January 2008.

48 Dundon, C.A., Aristidou, A., Hawkins, A., Lies, D., and Albert, L.H. (2011) Methods of Increasing Dihydroxy Acid Dehydratase Activity to Improve Produciton of Fuels, Chemicals, and Amino Acids, US Patent 8017376B2, September 13, 2011.

49 Peters, M.W., Taylor, J.D., Henton, D.E., and Manzer, L.E. (2011) Integrated Methods of Preparing Renewable Chemicals, US Patent Application 2011/0172475A1, July 14, 2011.

50 Peters, M.W., Taylor, J.D., Jenni, M., Manzer, L.E., and Henton, D.E. (2011) Integrated Process to Selectively Convert Renewable Isobutanol to p-Xylene, US Patent Application 2011/0087000A1, April 14, 2011.

51 Peters, M.W., Taylor, J.D., Jenni, M., Manzer, L.E., and Henton, D.E. (2011) Integrated Process to Selectively Convert Renewable Isobutanol to p-Xylene, PCT International Publication WO2011/044243A1, 14 April 2011.

52 (a) Marliere, P. (2011) Method for Producing an Alkene Comprising The Step of Converting an Alcohol By an Enzymatic Dehydration Step, PCT International Publication WO2011/076691A1, 30 June 2011; (b) Marliere, P. (2011) Method for Producing an Alkene Comprising The Step of Converting an Alcohol By an Enzymatic Dehydration Step, European Patent Application EP2336341A1, 22 June 2011.

53 (a) Marliere, P. (2011) Method for Producing an Alkene Comprising The Step of Converting an Alcohol By an Enzymatic Dehydration Step, PCT International Publication WO2011/076689A1, 30 June 2011; (b) Marliere, P. (2011) Method for Producing an Alkene Comprising The Step of Converting an Alchol By an Enzymatic Dehydration Step, European Patent Application EP2336340A1, 22 June 2011.

54 (a) Marliere, P. (2010) Production D'Alcenes Par Decarboxylation Enzymatique D'Acies 3-Hydroxy-Alcanoiques, PCT International Publication WO2010/001078A2, 7 January 2010; (b) Marliere, P. (2011) Production of Alkenes By Enzymatic Decarboxylation of 3-Hydroxyalkanoic Acids, US Patent Application 2011/0165644A1, July 7, 2011; (c) Marliere, P., Anissimova, M., Chayot, R., and Delcourt, M. (2011) Process for the Production of Isoprenol from Mevalonate Employing a Dihydro-mevalonate

Decarboxylase, PCT International Publication WO2011/076261A1, 30 June 2011.

55. (a) Burow, L.C., Mabbett, A.N., Borras, L., and Blackall, L.L. (2009) Anaerobic central metabolic pathways active during polyhydroxyalkanoate production in uncultured cluster1 defuviicoccus enriched inactivated sludge communities. *FEMS Microbiol. Lett.*, **298**, 79–84; (b) Rohwerder, T., and Müller, R.H. (2010) Biosynthesis of 2-hydroxyisobutyric acid (2-HIBA) from renewable carbon. *Microb. Cell Fact.*, **9**, 13–23.

56. Marliere, P. (2011) Method for the Enzymatic Production of 3-Hydroxy-3-methylbutyric Acid from Acetone and Acetyl-CoA, European Patent Application EP2295593A1, 16 March 2011.

57. Gogerty, D.S., and Bobik, T.A. (2010) Formation of isobutene from 3-hydroxy-3-methylbutyrate by diphosphomevalonate decarboxylase. *Appl. Environ. Microbiol.*, **76** (24), 8004–8010.

58. See Figure 4 in Marliere, P. (2011) Method for the Enzymatic Production of 3-Hydroxy-3-methylbutyric Acid from Acetone and Acetyl-CoA, PCT International Publication WO2011/032934A1, 24 March 2011.

59. Dominguez de Maria, P. (2011) Recent developments in the biotechnological production of hydrocarbons: paving the way for biobased platform chemicals. *Chem. Sus. Chem.*, **4**, 327–329.

60. (a) Bramucci, M.G., McCutchen, C.M., Nagarajan, V., and Thomas, S.M. (2001) Microbial Production of Terephthalic Acid and Isophthalic Acid, US Patent 6187569B1, February 13, 2001; (b) Bramucci, M.G., McCutchen, C.M., Nagarajan, V., and Thomas, S.M. (2002) Terephthalic Acid Producing Proteobacteria, US Patent 6461840B1, October 8, 2002; (c) Bramucci, M.G., McCutchen, C.M., Nagarajan, V., and Thomas, S.M. (2003) Terephthalic Acid Producing Proteobacteria, US Patent Application US2003/170836A1, September 11, 2003.

61. Bramucci, M.G., McCutchen, C.M., Singh, M., Thomas, S.M., Larsen, B.S., Buckholz, J., and Nagarajan, V. (2002) Pure bacterial isolates that convert *p*-xylene to terephthalic acid. *Appl. Microbiol. Biotechnol.*, **58**, 255–259.

62. Morgan, J.A., Lu, Z., and Clark, D.S. (2002) Toward the development of a biocatalytic system for oxidation of *p*-xylene to terephthalic acid: oxidation of 1,4-benzenedimethanol. *J. Mol. Catal., B Enzym.*, **18**, 147–154.

63. Wood, C.D., Shupe, A.M., Wang, Y., Graves, M., Liu, S., and Amidon, T.E. (2008) Biorefinery: conversion of woody biomass to chemicals, energy and materials. *J. Biobased Mater. Bio.*, **2**, 100–120.

64. Stone, W.E., and Lotz, D. (1891) Xylose from maize cobs. *J. Chem. Soc.*, Abstr. **60**, 1001.

65. (a) Miner, C.S., and Brownlee, H.J. (1929) Process of Manufacturing Furfural, US Patent 1735084, November 12, 1929; (b) Brownlee, H.J. (1933) Process of Manufacturing Furfural, US Patent 1919878, July 25, 1933.

66. Adams, R.A., and Voorhees, V.F. (1941) *Organic Syntheses*, Col. vol. 1, 2nd edn, John Wiley & Sons, Inc, pp. 280–282.

67. (a) Haworth, W.N., and Wiggins, L.F. (1950) 5-Hydroxymethyl-2-furfural, US Patent 2498918, February 28, 1950; (b) Haworth, W.N., and Wiggins, L.F. (1948) 5-(Hydroxymethyl)-2-furfural, British Patent GB600871, April 21, 1948; (c) Haworth, W.N., and Jones, W.G.M. (1944) Conversion of sucrose into furan compounds. I. 5-hydroxymethylfurfuraldehyde and some derivatives. *J. Chem. Soc.*, 667–670; (d) Middendorp, J.A. (1919) Hydroxymethylfurfural. *Rec. Trav. Chim. Pays-Bas Belg.*, **38**, 1–71; (e) Verendel, J.J., Church, T.L., and Andersson, P.G. (2011) Catalytic one-pot production of small organics from polysaccharides. *Synthesis*, 1649–1677; (f) Chheda, J.N., and Dumesic, J.A. (2007) An overview of dehydration, aldol-condensation and hydrogenation processes for production of liquid alkanes from biomass-derived carbohydrates. *Catal. Today*, **123**, 59–70; (g) Kuster, B.F.M. (1990) 5-Hydroxymethylfurfural (HMF). A

review focusing on its manufacture. *Starch/Staerke*, **42**, 314–321.

68 Sanborn, A.J. (2008) Processes for the Preparation and Purificaon of Hydroxymethylfuraldehyde and Derivatives, US Patent 7317116B2, January 8, 2008.

69 Haworth, W.N., and Wiggins, L.F. (1948) Improvements Relating to the Manufacture of 5-Hydroxymethyl-2-Furfural, British Patent Specification 600871, April 21, 1948.

70 Werpy, T., and Petersen, G. (eds) (2004) Top Value Added Chemicals from Biomass Volume I-Results of Screening for Potential Candidates from Sugars and Synthesis Gas, Office of the Biomass Program, US Department of Energy.

71 (a) Merat, N., Verdeguer, P., Rigal, L., Gaset, A., and Delmas, M. (1992) Procédé de Fabridation d'Acide 2-5-Durane Dicarboxylique, French Patent Application FR2669634A1, May 29, 1992; (b) Lew, B.W. (1967) Method of Producing Dehydromucic Acid, US Patent 3326944, June 20, 1967.

72 (a) Partenheimer, W., and Poliakoff, M. (2009) The aerobic oxidation of *p*-xylene to terephthalic acid: a classic case of Green Chemistry in action. *Handb. Green Chem.*, **1**, 375–397; (b) Partenheimer, W., and Grushin, V.V. (2001) Synthesis of 2,5-diformylfuran and furan-2,5-dicarboxylic acid by catalytic air-oxidation of 5-hydroxymethylfurfural. Unexpectedly selective aerobic oxidation of benzyl alcohol to benzaldehyde with metal/bromide catalysts. *Adv. Synth. Catal.*, **343**, 102–111; (c) Partenheimer, W. (1995) Methodology and scope of metal/bromide autoxidation of hydrocarbons. *Catal. Today*, **23**, 69–158.

73 (a) Taarning, E. (2008) Value Added Chemicals from Renewables Using Heterogeneous Catalysis. Presentation at BIO 2008, San Diego, June 18, 2008; (b) Klitgaard, S.K., Gorbanov, Y., Taarning, E., and Christensen, C.H. (2009) Renewable chemicals by sustainable oxidations using gold catalysts, in *Experiments in Green and Sustainable Chemistry* (eds. Roesky, H.W., Kennepohl, D.K.), Wiley, pp. 57–63; (c) Taarning, E., Nielsen, I.S., Egeblad, K., Madsen, R., and Christensen, C.H. (2008) Chemicals from renewables: aerobic oxidation of furfural and hydroxymethylfurfural over gold catalysts. *Chem. Sus. Chem.*, **1**, 75–78.

74 Hanke, P.D. (2009) Enzymatic Oxidation of HMF, PCT International Publication WO2009/023174A2 19 February 2009.

75 Ruijssenaars, H.J., Wierckx, N.J.P., Koopman, F.W., Straathof, A.J.J., and Winde, D.J.H. (2011) Polypeptides Having Oxidoreductase Activity and Their Uses, PCT International Publication WO2011/026913A1, 10 March 2011.

76 Roman-Leshkov, Y., Barrett, C.J., Liu, Z.Y., and Dumesic, J.A. (2007) Production of dimethylfuran for liquid fuels from biomass-derived carbohydrates. *Nature*, **447**, 982–985.

77 (a) Dumesic, J.A., Roman-Leshkov, Y., and Chheda, J.N. (2009) Catatlytic Process for Producing Furan Derivatives in a Biphasic Reactor, US Patent 7572925B2, August 11, 2009; (b) Dumesic, J.A., Roman-Leshkov, Y., and Chheda, J.N. (2007) Catatlytic Process for Producing Furan Derivatives in a Biphasic Reactor, PCT International Publication WO2007/146636A1, 21 December 2007.

78 Pedersen, S., Christensen, T.B., Boisen, A., Jugensen, V.W., Hansen, T.S., Kegnaes, S., Riisager, A., Woodley, J.M., Jensen, J.S., and Fu, W. (2011) A Method for Producing Hydroxymethylfurfural, PCT International Publication WO2011/124639A1, 13 October 2011.

79 (a) Gong, W.H. (2008) Terephthalic Acid Composition and Process for the Production Thereof, US Patent 7385081B1, June 10, 2008; (b) Gong, W.H. (2009) Terephthalic Acid Composition and Process for the Production Thereof, PCT International Publicaton WO2009/064515A1, 22 May 2009.

80 Barndvold, T.A. (2010) Carbohydrate Route to para-Xylene and Terephthalic

Acid, PCT International Publication WO2010/151346, 29 December 2010.

81 Shiramizu, M., and Toste, F.D. (2011) On the Diels–Alder approach to solely biomass-derived Polyethylene terephthalate (PET): conversion of 2,5-dimethylfuran and acrolein into p-xylene. *Chem. Eur. J.*, **17**, 12452–12457.

82 de Jong, E., Dam, R., Sipos, L., den Ouden, D., and Gruter, G.-J. (2011) Furandicarboxylic acid (FDCA). A versatile building block for a very interesting class of polyesters. *Biobased Monomers, Polymers, and Materials* Chapter 1, 2012, pp. 1–13. ACS *Symposium Series*, Volume **1105**.

83 (a) Drewitt, J.G.N., and Lincoln, J. (1951) Polyesters from Heterocyclic Components, US Patent 2551731, May 8, 1951; (b) Drewitt, J.G.N., and Lincoln, J. (1949) Improvements in Polymers, British Patent Specification 621971, April 25, 1949.

84 Roerink, F. (2011) Avantium's YXY: Green Materials and Fuels. Presentation at 2nd Annual Biobased Chemicals Summit, San Diego, Feb. 15, 2011.

85 Gruter, G.J.M., and Dautzenberg, F. (2007) Method for the Synthesis of 5-Alkoxymethyl Furfural Ethers and Their Use, PCT International Publication WO2007/104514A2, 20 September 2007.

86 (a) Munoz De Diego, C., Schammel, W.P., Dam, M.A., and Gruter, G.J.M. (2011) Method For The Preparation Of 2,5-Furandicarboxylic Acid And Esters Thereof, PCT International Publicaiton WO2011/043660A2, 14 April 2011; (b) Munoz De Diego, C., Dam, M.A., and Gruter, G.J.M. (2011) Method for the Preparation of 2,5-Furandicarboxylic Acid and for the Preparation of the Dialkyl Ester of 2,5-Furandicarboxylic Acid, PCT International Publication WO2011/043661A1, 14 April 2011.

87 Sipos, L. (2010) A Process for Preparing a Polymer Having a 2,5-Furandicarboxylate Moiety Within the Polymer Backbone and Such (co) Polymers, PCT International Publication WO2010/077133A1, 8 July 2010.

88 (a) Berti, C., Binassi, E., Colonna, M., Fiorini, M., Kannan, G., Karanam, S., Mazzacurati, M., Odeh, I., and Vannini, M. (2010) Biobased Terephthalate Polyesters, US Patent Application 2010/168371A1, July 1, 2010; (b) Berti, C., Binassi, E., Colonna, M., Fiorini, M., Kannan, G., Karanam, S., Mazzacurati, M., and Odeh, I. (2010) Biobased Terephthalate Polyesters, US Patent Application 2010/168372A1, July 1, 2010; (c) Berti, C., Binassi, E., Colonna, M., Fiorini, M., Kannan, G., Karanam, S., Mazzacurati, M., Odeh, I., and Vannini, M. (2010) Biobased Terephthalate Polyesters, US Patent Application 2010/168373A1, July 1, 2010; (d) Berti, C., Binassi, E., Colonna, M., Fiorini, M., Kannan, G., Karanam, S., Mazzacurati, M., and Odeh, I. (2010) Biobased Terephthalate Polyesters, US Patent Application 2010/168461A1, July 1, 2010.

89 (a) Reggel, L., Friedman, S., and Wender, I. (1958) The lithium-ethylenediamine system. II. Isomerization of olefins and dehydrogenation of cyclic dienes. *J. Org. Chem.*, **23** (8), 1136–1139; (b) Pines, H., and Eschinazi, H.E. (1955) Studies in the terpene series. XXIV.1 sodium-catalyzed double bonds migration and dehydrogenation of d-limonene, l-α-phellandrene and of 2,4(8)- and 3,8(9)-p-Menthadiene[2,2a]. *J. Am. Chem. Soc.*, **77** (23), 6314–6321; (c) Mitchell, P.W.D., and Sasser, D.E. (1993) Catalytic Dehydrogention of Cyclic Dienes, European Patent Application EP0522839A2, 13 January 1993.

90 (a) Kyoto Encylopeadia of Genes and genomes (KEGG) http://www.genome.jp/kegg/ (accessed 6 July 2012); (b) University of Minnesota Biocatalysis/Biodegradation Database (UM-BDD) http://umbbd.msi.umn.edu/ (accessed 6 July 2012).

91 (a) Gosset, G. (2009) Production of aromatic compounds in bacteria. *Curr. Opin. Biotechnol.*, **20**, 651–658; (b) Berry, A. (1996) Improving production of aromatic compounds in *Escherichia coli* by metabolic engineering. TIBTECH, **14**, 250–256; (c) Ikeda, M. (2006)

Towards bacterial strains over-producing L-tryptophan and other aromatics by metabolic engineering. *Appl. Microbiol. Biotechnol.*, **69**, 615–626; (d) Bongaerts, J., Kramer, M., Muller, U., Raeven, L., and Wubbolts, M. (2001) Metabolic engineering for microbial production of aromatic amino acids and derived compounds. *Metab. Eng.*, **3**, 289–300.

92 (a) Draths, K.M., and Frost, J.W. (1998) Improving the environment through process changes and product substitutions. *Green Chem.*, 150–165; (b) Snell, K.D., Draths, K.M., and Frost, J.W. (1996) Synthetic modification of the *Escherichia coli* chromosome: enhancing the biocatalytic conversion of glucose into aromatic chemicals. *J. Am. Chem. Soc.*, **118**, 5605–5614; (c) Frost, J.W., and Draths, K.M. (1995) Biocatalytic syntheses of aromatics from D-glucose: renewable microbial sources of aromatic compounds. *Annu. Rev. Microbiol.*, **49**, 557–579; (d) Frost, J.W., and Draths, K.M. (1995) Sweetening chemical manufacture. *Chem. Br.*, **31**, 206–210; (e) Draths, K.M., and Frost, J.W. (1994) Microbial Biocatalysis. Synthesis of Adipic Acid from D-Glucose in ACS Symposium Series, *Benign by Design*, **577**, 32–45; (f) Niu, W., Draths, K.M., and Frost, J.W. (2002) Benzene-free synthesis of adipic acid. *Biotechnol. Prog.*, **18**, 201–211; (g) Draths, K.M., Frost, J.W. (1995) Environmentally compatible synthesis of catechol from D-glucose. *J. Am. Chem. Soc.*, **117**, 2395–2400; (h) Draths, K.M., and Frost, J.W. (1994) Environmentally compatible synthesis of adipic acid from D-glucose. *J. Am. Chem. Soc.*, **116**, 399–400; (i) Li, W., Xie, D., and Frost, J.W. (2005) Benzene-free synthesis of catechol: interfacing microbial and chemical catalysis. *J. Am. Chem. Soc.*, **127** (9), 2874–2882.

93 (a) Krämer, M., Bongaerts, J., Bovenberg, R., Kremer, S., Müller, U., Orf, S., Wubbolts, M., and Raeven, L. (2003) Metabolic engineering for microbial production of shikimic acid. *Metab. Eng.*, **5**, 277–283; (b) Pantaleone, D.P. (2006) Biotransformations: "Green" processes for the synthesis of chiral fine chemicals, in *Handbook of Chiral Chemicals*, 2nd edn (ed. David Ager), CRC Press, Francis & Taylor Group, pp. 359–403.

94 Li, K., Mikola, M.R., Draths, K.M., Worden, R.M., and Frost, J.W. (1999) Fed-batch fermentor synthesis of 3-dehydroshikimic acid using recombinant *Escherichia coli*. *Biotechnol. Bioeng.*, **64**, 61–73.

95 (a) Frost, J.W. (1992) Enhanced Production of Common Aromatic Pathway Compounds, US Patent 5168056, December 1, 1992; (b) Draths, K.M., and Frost, J.W. (1990) Synthesis using plasmid-based biocatalysis: plasmid assembly and 3-deoxy-D-arabino-heptulosonate production. *J. Am. Chem. Soc.*, **112** (5), 1657–1659; (c) Yi, J., Draths, K.M., Li, K., and Frost, J.W. (2003) Altered glucose transport and shikimate pathway product yields in *E. coli*. *Biotechnol. Prog.*, **19**, 1450–1459; (d) Chandran, S.S., Yi, J., Draths, K.M., Von Daeniken, R., Weber, W., and Frost, J.W. (2003) Phosphoenolpyruvate availability and the biosynthesis of shikimic acid. *Biotechnol. Prog.*, **19**, 808–814; (e) Draths, K.M., Pompliano, D.L., Conley, D.L., Frost, J.W., Berry, A., Disbrow, G.L., Staversky, R.J., and Lievense, J.C. (1992) Biocatalytic synthesis of aromatics from D-glucose: the role of transketolase. *J. Am. Chem. Soc.*, **114** (10), 3956–3962; (f) Draths, K.M., and Frost, J.W. (1991) Conversion of D-glucose into catechol: the not-so-common pathway of aromatic biosynthesis. *J. Am. Chem. Soc.*, **113** (24), 9361–9363; (g) Chandran, S.S., Yi, J., Draths, K.M., von Daeniken, R., Weber, W., and Frost, J.W. (2003) Phosphoenolpyruvate availability and the biosynthesis of shikimic acid. *Biotechnol. Prog.*, **19**, 808–814.

96 (a) Ran, N., Draths, K.M., and Frost, J.W. (2004) Creation of a shikimate pathway variant. *J. Am. Chem. Soc.*, **126** (22), 6856–6857; (b) Ran, N., and Frost, J.W. (2007) Directed evolution of 2-keto-3-deoxy-6-phosphogalactonate

aldolase to replace 3-deoxy-d-arabino-heptulosonic acid 7-phosphate synthase. *J. Am. Chem. Soc.*, **129** (19), 6130–6139; (c) Frost, J.W. (2010) Methods and Materials for the Production of Shikimic Acid, US Patent 7790431B2, September 7, 2010; (d) Frost, J.W. (2011) Methods and Materials for the Production of Shikimic Acid, US Patent Application 2011/0045539 February 24, 2011.

97 Jose Luis Baez, F., and Bolıvar, G. (2001) Determination of 3-deoxy-DArabino- heptulosonate 7-phosphate productivity and yield from glucose in *Escherichia coli* devoid of the glucose phosphotransferase transport system. *Biotechnol. Bioeng.*, **73**, 530–535.

98 Escalante, A., Calderón, R., Valdivia, A., de Anda, R., Hernández, G., Ramírez, O.T., Gosset, G., and Bolívar, F. (2010) Metabolic engineering for the production of shikimic acid in an evolved *Escherichia coli* strain lacking the phosphoenolpyruvate: carbohydrate phosphotransferase system. *Microb. Cell Fact.*, **9**, 21–33.

99 Frost, J. (2005) Methods and Materials for the Production of Shikimic Acid, PCT International Publication WO2005/030949A1, 7 April 2005.

100 Dell, K.A., and Frost, J.W. (1993) Identification and removal of impediments to biocatalytic synthesis of aromatics from D-glucose: rate-limiting enzymes in the common pathway of aromatic amino acid biosynthesis. *J. Am. Chem. Soc.*, **115** (24), 11581–11589.

101 (a) Frost, J.W., and Draths, K.M. (1996) Synthesis of Adipic Acid From Biomass-Derived Carbon Sources, US Patent 5487987, January 30, 1996; (b) Frost, J.W., and Draths, K.M. (1997) Bacterial Cell Transformants for Production of cis,cis-Muconic Acid and Catechol, US Patent 5616496, April 1, 1997; (c) Frost, J.W., and Draths, K.M. (1997) Synthesis of Catechol From Biomass-Derived Carbon Sources, US Patent 5629181, May 13, 1997; (d) Frost, J.W., and Draths, K.M. (1993) Biocatalytic Synthesis of Catachol from Glucose, US Patent 5272073, December 21, 1993; (e) Frost, J.W., Snell, K.D., and Frost, K.M. (1998) Deblocking the Common Pathway of Aromatic Amino Acid Synthesis, US Patent 5776736, July 7, 1998.

102 (a) Bui, V., Lau, M.K., and MacRae, D. (2011) Methods for Producing Isomers of Muconic Acid and Muconate Salts, PCT International Publication WO2011/085311, 14 July 2011; (b) Frost, J.W., Miermont, A., Schweitzer, D., and Bui, V. (2010) Preparation of trans,trans-Muconic Acid and trans,trans-Muconates, PCT International Publication WO2010/148049A2, 23 December 2010; (c) Frost, J.W., Miermont, A., Schweitzer, D., and Bui, V. (2010) Cyclohexene 1,4-Carboxylates, PCT International Publication WO2010/148063A2, 23 December 2010; (d) Frost, J.W., Miermont, A., Schweitzer, D., and Bui, V. (2010) Novel Terephthalic and Trimellitic based Acids and Carboxylate Derivatives Thereof, PCT International Publication WO2010/148081A, 23 December 2010; (e) Frost, J.W., Miermont, A., Schweitzer, D., and Bui, V. (2010) Cyclohexane 1,4-Carboxylates, PCT International Publication WO2010/148080A, 23 December 2010; (f) Frost, J.W., Miermont, A., Schweitzer, D., Bui, V., and Paschke, E. (2010) Biobased Polyesters, PCT International Publication WO2010/148070A2, 23 December 2010.

103 Frost, J.W., and Draths, K.M. (1995) Synthesis of Adipic Acid from Biomass-Derived Carbon Sources, PCT International Publication WO 9507996 A1, 23 March 1995.

104 Osterhout, R.E. (2011) Microorganisms and Methods for the Biosynthesis of P-Toluate and Terephthalate, PCT International Publication WO2011/094131A1, 4 August 2011.

105 Burk, M.J., Osterhout, R.E., and Sun, J. (2011) Semi-Synthetic Terephthalic Acid Via Microorganisms That Produce Muconic Acid, PCT International Publicaton WO2011/017560A1, 2 February 2011.

106 (a) Donnelly, M., Millard, C.S., and Stols, L. (1998) Mutant *E. coli* Strain With Increased Succinic Acid Production, US Patent 5770435, June 23, 1998; (b) Guettler, M.V., and Jain, M.K. (1996) Method for Making Succinic Acid, Anaerobiospirillum Succiniciproducens Variants for Use in Process and Methods for Obtaining Variants, US Patent 5521075, May 28, 1996; (c) Datta, R. (1992) Process for the Production of Succinic Acid by Anaerobic Fermentation, US Patent 5143833, September 1, 1992; (d) Guettler, M.V., Jain, M.K., and Soni, B.K. (1996) Process for Making Succinic Acid, Microorganisms for Use in the Process and Methods of Obtaining the Microorganism, US Patent 5504004, April 2, 1996; (e) Yi, J., Kleff, S., and Guettler, M. (2008) Recombinant Microorganisms for Increased Production of Organic Acids, PCT International Publication WO2006/083410A2, 10 August 2008.

107 Frost, J.W., and Frost, K.M. (2003) Biocatalytic Synthesis of Quinic Acid and Conversion to Hydroquinone, US Patent 6600077B1, July 29, 2003.

108 Ran, N., Knop, D.R., Draths, K.M., and Frost, J.W. (2001) Benzene-free synthesis of hydroquinone. *J. Am. Chem. Soc.*, **123** (44), 10927–10934.

109 (a) Frost, J.W., and Frost, K.M. (2003) Biocatalytic Synthesis of Quinic Acid and Conversion to Hydroquinone, US Patent 6620602B2, September 16, 2003; (b) Frost, J.W., and Frost, K.M. (2006) Biocatalytic Synthesis of Quinic Acid and Conversion to Hydroquinone, US Patent 7002047B2, February 21, 2006; (c) Frost, J.W., and Frost, K.M. (2011) Biocatalytic Synthesis of Quinic Acid and Conversion to Hydroquinone, US Patent 8080397B2, December 20, 2011.

110 (a) Frost, J.W. (2002) Biocatalytic Synthesis of Galloid Organics, US Patent 6472190B1, October 29, 2002; (b) Frost, J.W. (2001) Biocatalytic Synthesis of Galloid Organics, PCT International Publication WO0171020A2, 27 September 2001.

111 (a) Kambourakis, S., Draths, K.M., and Frost, J.W. (2000) Synthesis of gallic acid and pyrogallol from glucose: replacing natural product isolation with microbial catalysis. *J. Am. Chem. Soc.*, **122** (37), 9042–9043; (b) Richman, J.E., Chang, Y.-C., Kambourakis, S., Draths, K.M., Almy, E., Snell, K.D., Strasburg, G.M., and Frost, J.W. (1996) Reaction of 3-dehydroshikimic acid with molecular oxygen and hydrogen peroxide: products, mechanism, and associated antioxidant activity. *J. Am. Chem. Soc.*, **118** (46), 11587–11591.

112 Frost, J.W. (2002) Synthesis of Vanillin from a Carbon Source, US Patent 6372461B1, April 16, 2002.

113 (a) Frost, J.W., Frost, K.M., and Knop, D.R. (2002) Biocatalytic Synthesis of Shikimic Acid, US Patent 6472169B1, October 29, 2002; (b) Frost, J.W., Frost, K.M., and Knop, D.R. (2003) Biocatalytic Synthesis of Shikimic Acid, US Patent 6613552B1, September 2, 2003; (c) Frost, J. (2005) Methods and Materials for the Production of Shikimic Acid, PCT International Publication WO2005/030949A1, 4 April 2005.

114 (a) Abrecht, S., Harrington, P., Iding, H., Karpf, M., Trussardi, R., Wirz, B., and Zutter, U. (2004) The synthetic development of the anti-infuenza neuaraminidase inhibitor oseltamivir phosphate (Tamiflu®): a challenge for synthesis and process research. *U. Chimia*, **58** (9), 621–629; (b) Bischofberger, N.W., Kim, C.U., Lew, W., Liu, H., and Williams, M.A. (1998) Carbocyclic Compounds, US Patent 5763483, June 9, 1998.

115 Gibson, J.M., Thomas, P.S., Thomas, J.D., Barker, J.L., Chandran, S.S., Harrup, M.K., Draths, K.M., and Frost, J.W. (2001) Benzene-free synthesis of phenol. *Angew. Chem. Int. Ed.*, **40** (10), 1945–1948.

116 Amaratunga, M., Lobos, J.H., Johnson, B.F., and Williams, E.D. (2000) Genetically Engineered Microorganisms and Method for Producing 4-Hydroxybenzoic Acid, US Patent 6030819, February 29, 2000.

117 Barker, J.L., and Frost, J.W. (2001) Microbial synthesis of p-hydroxybenzoic acid from glucose. *Biotechnol. Bioeng.*, **76**, 376–390.

118 Sun, Z., Ning, Y., Liu, L., Liu, Y., Sun, B., Jiang, W., Yang, C., and Yang, S. (2011) Metabolic engineering of the L-phenylalanine pathway in *Escherichia coli* for the production of S- or R-mandelic acid. *Microb. Cell Fact.*, **10**, 71–84.

119 Sariaslani, F.S. (2007) Development of a combined biological and chemical process for production of industrial aromatics from renewable resources. *Annu. Rev. Microbiol.*, **61**, 51–69.

120 (a) Qi, W.W., Vannelli, T., Breinig, S., Ben-Bassat, A., Gatenby, A.A., Haynie, S.L., and Sariaslani, F.S. (2007) Functional expression of prokaryotic and eukaryotic genes in *Escherichia coli* for conversion of glucose to p-hydroxystyrene. *Metab. Eng.*, **9**, 268–276; (b) Ben-Bassat, A., Qi, W.W., Sariaslani, F.S., Tang, X.-S., and Vannelli, T.M. (2004) Microbial conversion of glucose to para-hydroxystyrene, US Patent 7229806B2, Januray 29, 2004.

121 (a) Ben-Bassat, A., Haynie, S.L., Lowe, D.J., and Huang, L.L. (2008) Method for Preparing Para-Hydroxystyrene by Biocatalytic Decarboxylation of Para-Hydroxycinnamic Acid in a Biphasic Reaction Medium, US Patent 7378261B2, May 27, 2008; (b) Ben-Bassat, A., Qi, W.W., Sariaslani, F.S., Tang, X.-S., and Vannelli, T. (2003) Microbal Conversion of Glucose to Para-Hydroxystyrene, PCT International Publication WO03099233A2, 4 December 2003; (c) Ben-Bassat, A., Breinig, S., Crum, G.A., Huang, L., Altenbaugh, A.L.B., Rizzo, N., Trotman, R.J., Vannelli, T., Sariaslani, F.S., and Haynie, S.L. (2007) Preparation of 4-vinylphenol using pHCA decarboxylase in a two-solvent medium. *Org. Process Res. Dev.*, **11**, 278–285.

122 Verhoef, S., Wierckx, N., Maaike Westerhof, R.G., de Winde, J.H., and Ruijssenaars, H.J. (2009) Bioproduction of p-hydroxystyrene from glucose by the solvent-tolerant bacterium pseudomonas putida S12 in a two-phase water-decanol fermentation. *Appl. Environ. Microbiol.*, **75**, 931–936.

123 McKenna, R., and Nielsen, D.R. (2011) Styrene biosynthesis from glucose by engineered *E. coli. Metab. Eng.*, **13**, 544–554.

124 Pharkya, P. (2011) Microorganisms for the Production of Aniline, PCT International Publication WO2011/050326, 4 April 2011.

125 Hansen, C.A., Dean, A.B., Draths, K.M., and Frost, J.W. (1999) Synthesis of 1,2,3,4-tetrahydroxybenzene: exploiting myo-inositol as a precursor to aromatic chemicals. *J. Am. Chem. Soc.*, **121** (15), 3799–3800.

126 Hansen, C.A., and Frost, J.W. (2002) Deoxygenation of polyhydroxybenzenes: an alternative strategy for the benzene-free synthesis of aromatic chemicals. *J. Am. Chem. Soc.*, **124** (21), 5926–5927.

127 (a) Frost, J.W., and Hansen, C.A. (2004) Synthesis of 1,2,3,4-Tetrahydroxybenzenes and 1,2,3-Trihydroxybenzenes Using Myo-Inositol-1-Phosphate Synthase and Myo-Inositol 2-Dehydrogenase, US Patent 6750049B1, June 15, 2004; (b) Frost, J.W., and Hansen, C.A. (2000) Synthesis of 1,2,3,4-Tetrahydroxybenzenes and 1,2,3-Trihydroxybenzenes Using Myo-Inositol-1-Phosphate Synthase and Myo-Inositol 2-Dehydrogenase, PCT International Publicaton WO0056911A1, September 28, 2000.

128 Das, A., and Khosla, C. (2009) Biosynthesis of aromatic polyketides in bacteria. *Acc. Chem. Res.*, **42**, 631–639.

129 (a) Achkar, J., Xian, M., Zhao, H., and Frost, J.W. (2005) Biosynthesis of phloroglucinol. *J. Am. Chem. Soc.*, **127** (15), 5332–5333; (b) Frost, J.W. (2006) Biosynthesis of Philoroglucinol and Preparation of 1,3-Dihydroxybenzene Therefrom, PCT International Publication WO2006/044290A2, 27 April 2006.

130 Delaunois, B., Cordelier, S., Conreux, A., Clément, C., and Jeandet, P. (2009) Molecular engineering of resveratrol in plants. *Plant Biotechnol. J.*, **7**, 2–12.

131 (a) Giovinazzo, G., Ingrosso, I., Paradiso, A., De Gara, L., and Santino, A. (2012) Resveratrol biosynthesis: plant metabolic engineering for nutritional improvement of food. *Plant Foods Hum. Nutr.*, **67** (3), 191–199; (b) de la Lastra,

C.A., and Villegas, I. (2007) Resveratrol as an antioxidant and pro-oxidant agent: mechanisms and clinical implications. *Biochem. Soc. Trans.*, **35** (Pt 5), 1156–1160.

132 (a) Dai, J., and Mumper, R.J. (2010) Plant phenolics: extraction, analysis and their antioxidant and anticancer propertiesm. *Molecules*, **15** (10), 7313–7352; (b) Korkina, L., Kostyuk, V., De Luca, C., and Pastore, S. (2011) Plant phenylpropanoids as emerging anti-inflammatory agents. *Mini Rev. Med. Chem.*, **11** (10), 823–835.

133 Rosazza, J.P.N., Huang, Z., Dostal, L., Volm, T., and Rousseau, B. (1995) Review: biocatalytic transformations of ferulic acid: an abundant aromatic natural product. *J. Ind. Microbiol.*, **15**, 457–471.

134 (a) Ago, S., and Kikuchi, Y. (1999) Ferulic Acid Decarboxylase, US Patent 5955137, September 21, 1999; (b) Ago, S., and Kikuchi, Y. (2002) Ferulic Acid Decarboxylase, US Patent 6468566B2, October 22, 2002.

135 (a) Lesage-Meessen, L., Delattre, M., Haon, M., and Asther, M. (1996) Method for Obtaining Vanillic Acid and Vanillin by Bioconversion by an Association of Filamentous Microorganisms, PCT International Publication WO9608576A1, 21 March, 1996; (b) Lesage-Meessen, L., Delattre, M., Haon, M., Thibault, J.F., Colonna Ceccaldi, B., Brunerie, P., and Asther, M. (1996) Two-step bioconversion process for vanillin production from ferulic acid combining *Aspergillus niger* and *Pycnoporus cinnabarinus*. *J. Biotechnol.*, **50**, 107–113; (c) Paterson, D. (2010) Vanilla: natural or not? *ChemEdNZ*, **118**, 2–6.

136 Sandborn, L.T., Richter, S.J., and Clemens, H.G. (1936) Process of Making Vanillin, US Patent 2057117, October 13, 1936.

137 Hocking, M.B. (1997) Vanillin: synthetic flavoring from spent sulfite liquor. *J. Chem. Educ.*, **74** (9), 1055–1059.

138 (a) Adlercreutz, H. (2007) Lignans and human health. *Crit. Rev. Clin. Lab. Sci.*, **44** (5–6), 483–525; (b) Saarinen, N.M., Werri, A., Airio, M., Smeds, A., and Mekele, S. (2007) Role of dietary lignans in the reduction of breast cancer risk. *Mol. Nutr. Food Res.*, **51** (7), 857–866; (c) Sok, D.E., Cui, H.S., and Kim, M.R. (2009) Isolation and bioactivities of furfuran type lignan compounds from edible plants. *Recent Pat. Food Nutr. Agric.*, **1** (1), 87–95; (d) Peterson, J., Dwyer, J., Adlercreutz, H., Scalbert, A., Jacques, P., and McCullough, M.L. (2010) Dietary lignans: physiology and potential for cardiovascular disease risk reduction. *Nutr. Rev.*, **68** (10), 571–603; (e) Webb, A.L., and McCullough, M.L. (2005) Dietary lignans: potential role in cancer prevention. *Nutr. Cancer*, **51** (2), 117–131; (f) Begum, S.A., Sahai, M., and Ray, A.B. (2010) Non-conventional lignans: coumarinolignans, flavonolignans, and stilbenolignans. *Fortschr. Chem. Org. Naturst.*, **93**, 1–70.

139 Ahmad, R., Sharma, V.K., Rai, A.K., Shivananda, R.D., and Shivananda, B.G. (2007) Production of lignans in callus culture of *Podophyllum hexandrum*. *Trop. J. Pharm. Res.*, **6** (4), 803–808.

140 Vogt, T. (2010) Phenylpropanoid biosynthesis. *Mol. Plant*, **3**, 2–20.

141 Enkvist, T. (1955) Possibilities of lignin research. *J. Am. Chem. Soc.*, **64** (1), 26–48.

142 Ribbons, D.W. (1987) Chemicals from lignin. *Philos. Trans. R. Soc. London Ser. A*, **321**, 485–494.

143 Goheen, D.W. (1988) Chemicals from lignin. *Org. Chem. Biomass*, **143**, 61.

144 Lora, J.H., and Glasser, W.G. (2002) Recent industrial applications of lignin: a sustainable alternative to nonrenewable materials. *J. Polym. Environ.*, **10**, 39–48.

145 Amthor, J.S. (2003) Efficiency of lignin biosynthesis: a quantitative analysis. *Ann. Bot.*, **91**, 673–695.

146 Holladay, J.E., Bozell, J.J., White, J.F., and Johnson, D. (eds) (October 2007) Top Value-Added Chemicals from Biomass Volume II-Results of Screening for Potential Candidates from Biorefinery Lignin, Prepared for the U.S. Department of Energy under Contract DE-AC05-76RL01830.

147 Gosselink, R.J.A., de Jong, E., Guran, B., and Abächerli, A. (2004)

Co-ordination network for lignin-standardisation, production and applications adapted to market requirements (EUROLIGNIN). *Ind. Crops Prod.*, **2004**, 20121–20129.

148 Bunn, H. (1952) Cost and availability of raw materials. *Ind. Eng. Chem.*, **44**, 2128–2133.

149 Deindoerfer, F.H., Mateles, R.I., and Humphrey, A.E. (1963) Microbiological process report 1961 fermentation process review. *Appl. Microbiol.*, **11**, 273–303.

150 Lipinski, E.S. (1981) Chemicals from biomass: petrochemical substitution options. *Science*, **212**, 1465–1471.

9
Organosolv Biorefining: Creating Higher Value from Biomass
E. Kendall Pye and Michael Rushton

9.1
Introduction

During most of human history, fibrous biomass in the form of wood and other vegetative materials has been the primary source of heat, clothing, paper and construction materials. Only in the last 160 years or so, since the first commercial oil wells were drilled, have oil and natural gas become dominant sources of fuels for heat, power and transportation in the developed world. The rapid advancement of synthetic organic chemistry, which was encouraged greatly by the availability of low cost chemical feedstocks from petroleum, inspired the production of plastics, polymer coatings and synthetic fibers that are now viewed as necessities of our modern lifestyles. Today, less than two centuries after the first commercial applications of oil and natural gas, the industries that rely on petroleum-based fuels, chemicals and materials are being forced to seek alternative resources because of mounting concerns about climate change from fossil carbon use, the predicted decline in availability of low cost crude oil, market disruptions caused by world political events and the frequent rapid and unpredictable feedstock price fluctuations [1]. Clearly, the most readily available, sustainable and sufficiently sized alternative source of feedstocks for transportation fuels and the chemicals industry is vegetative biomass in its various forms that range from trees, agricultural residues, and marine algae. This new direction is being further encouraged in many developed countries by political motives that include the need to stimulate domestic economies, concerns for imbalances of international trade, national security, and the widening raw material costs between crude oil and domestically produced biomass.

One difficulty that must be faced during this desirable transition to a renewable chemicals and fuels economy is that the feedstocks on which much of the current chemicals industry depends are typically hydrocarbons, while vegetation is mostly composed of carbohydrates. To simplify and to minimize disruptions caused by this feedstock transition, it is, therefore, advisable to develop processes that transform fibrous biomass and other plant materials into the same, or similar, chemical intermediates that are used today in our chemicals industry. Additionally,

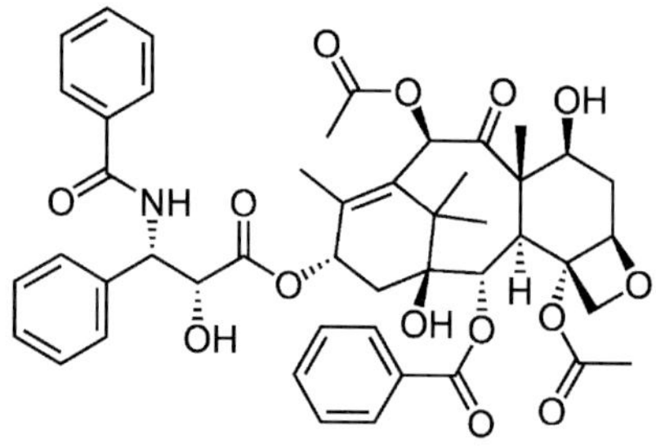

Figure 9.1 Structure of paclitaxel; an example of the chemical synthesis capabilities of plants.

processes should be developed that convert biomass into new chemicals and materials that will replace the presently utilized materials. Both approaches, as well as hybrids of these two approaches, are now undergoing rapid development in many parts of the world [2].

Another factor that should be expected to encourage this transition to biobased chemicals is that the chemical synthesis capabilities of the modern chemicals industry is relatively primitive compared to the chemical synthesis capabilities of plants and microorganisms. It is only necessary to consider the chemical structure of complex biomolecules, such as paclitaxel, the naturally occurring anti-cancer compound produced in the Pacific Yew Tree, shown in Figure 9.1, to recognize that plants have exceptional chemical synthesis capabilities, since they are able to create such compounds starting with only carbon dioxide, water and sunlight. To achieve this synthetic ability with the same high efficiency of biological systems would be an exceptionally difficult challenge for the chemicals industry of today. With the recent rapid advances in genetic engineering and biotechnology it is clear that we are entering an era when genetically modified plants and microorganisms can be cultivated for the production of many of the feedstocks presently required by the modern chemicals industry [3]. The biotechnology industry is already creating genetically modified plants and microorganisms that are tuned to produce desirable chemicals and foods in greater yields than the native species. These feedstock chemicals can supplement and even replace equivalent feedstocks presently produced from petroleum and natural gas. A utopian vision is that new plant and microbial life forms might ultimately become the highly efficient, sustainable and environmentally benign "chemical factories" of the future. A major problem facing this vision is that the current separation and purification technologies used to produce various chemicals from petroleum sources typically involve refinery processes for separating and fractionating liquid and gaseous streams. The recovery and purification of materials from woody biomass will generally involve separation of organic solids for which there is far less experience and developed methodology.

The technologies and the facilities that will perform the functions of separating, converting and purifying the various component chemicals and materials from biomass are now being described as biorefining and biorefineries, respectively.

Despite numerous attempts in the previous two decades to narrow the definition of biorefining [4], this term is now being used to describe any technology that processes plant-based material to create useful chemicals and materials [5]. The range of technologies now being described as biorefining is very large and includes thermochemical processes that gasify or pyrolyze biomass at one extreme, and processes that utilize relatively benign chemical and biological processing conditions at the other. Clearly, the various biorefining processes that now fall within this broad definition range have both advantages and disadvantages, which frequently are functions of either feedstock properties, or product objectives, or both.

9.2
Concepts and Principles of Biorefinery Technologies

9.2.1
Types of Biorefineries for Biomass Processing

Vegetative biomass has been the source of certain industrial chemicals and materials for some time. These include pine tars, rosin and pitch, as well as tannins from tree bark for the processing of leather and various other applications. Furthermore, woody biomass-derived by-products of the chemical pulping industry, such as lignosulfonates from the sulfite pulping process and thiolignin from the kraft pulping process, have a long history of industrial utilization [6]. With the current recognition that society should soon start to seek alternatives to fossil carbon feedstocks, it is clear that wood and woody biomass are the only sources of renewable chemicals and materials able to provide the enormous quantities of materials needed by the world's growing population. Fast growing algae are also potential sources [7], but infrastructure must first be created for their large scale production and recovery. On the other hand, vast quantities of woody biomass in the form of agricultural residues and harvestable trees already exist in many countries [8], together with well-developed systems for their easy collection and recovery. For that reason, most biorefinery technologies currently under development are targeting cellulosic biomass as the feedstock of choice, since these materials are the only biomass sources that presently exist on a scale comparable with the present-day basic chemicals industry. The short-term objective is to develop processes that convert existing cellulosic biomass into the materials and chemicals on which the industry now depends. A longer-term and possibly harder to implement objective is to produce, directly from biomass, new chemical products and materials that can replace existing products now made from fossil carbon sources.

Using these arguments it is the logical role of biorefining technologies, over the near term, to convert plant-based materials into chemicals and materials that are compatible with, or even identical to, those used in the existing chemicals and materials industry. As with oil refining, the primary purpose of biorefining is the separation, modification and purification of chemically dissimilar components of the raw material to yield pure compounds, or fractions of similar materials having greater utility and higher values than the complex raw material itself. Processes used in the refining and fractionation of crude oil are almost exclusively limited to the processing of liquids and gases. Since cellulosic biomass is solid and composed dominantly of cellulose, hemicellulose and lignin, it is, therefore, necessary to develop new processes that are capable of separating, processing and purifying solid materials on a commercial scale, and then converting them to useful products.

9.2.1.1 Biorefineries Employing Thermochemical Treatment of Biomass

Gasification One approach to overcoming the difficulties of solids separation and variations in feedstock composition is to convert woody biomass into either liquids or gases, or combinations of these, which can then be further fractionated and purified by existing processes. They can then be more easily utilized for the synthesis of desired chemicals using typical and existing catalytic processes. A number of biorefinery technologies start with the heating of preferably relatively dry biomass in the presence of very limited amounts of air or oxygen. This causes gasification of biomass that creates a mixture of hydrogen and carbon monoxide (synthesis gas or syngas), together with CO_2 and limited amounts of more complex chemical compounds. Following purification of the gas stream, the syngas can then be readily converted through Fischer–Tropsch catalysis into a number of useful hydrocarbons and, under certain operating conditions, into various aliphatic alcohols [9]. Various inorganic catalysts, mostly based on iron or cobalt, are required to convert the syngas into these products. Generally, the resultant products of this synthesis stage are mixtures that must be further processed and purified in order to create the desired pure products. The general principles of biomass gasification for chemical synthesis are illustrated in Figure 9.2.

An alternative to using Fischer–Tropsch chemistry is to feed the syngas to specialized bacteria that create products, such as ethanol, from the syngas. Several companies are commercializing biorefinery technologies based on these principles [10].

Pyrolysis An alternative thermochemical biorefinery process for converting biomass into useful products is fast pyrolysis, which involves heating the biomass to high temperatures in the absence of oxygen. The resultant products typically include a char, a gas, and an oily liquid [11]. The gas can be burned as fuel to

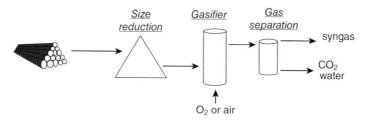

Figure 9.2 Basic process for gasification of biomass for renewable fuels and chemicals.

maintain the temperature of the pyrolysis vessel, while the char has some commercial or fuel value. The mostly desired product is the pyrolysis oil, which is a crude mixture of aromatic and aliphatic compounds. This can be further purified to create higher-value purified fractions, or used as a versatile liquid fuel. Pyrolysis-based biorefinery processes are currently being used to create valuable specialty chemical products, used in such industries as the flavor and fragrance industry [12], but these are relatively small and specialized markets.

A recently reported innovation in biomass pyrolysis is the H2Bioil process in which biomass is heated in the presence of hydrogen to between 400 and 600 °C for typically less than one minute. The fast pyrolysis process involves the presence of hydrodeoxygenation (HDO) or zeolite catalysts which create a bio-oil product containing less oxygen than typical pyrolysis bio-oils. As with most other pyrolysis processes the solid char and the gases can be burned to provide the heat required for pyrolysis [13].

Torrefaction Another form of thermal processing of biomass is torrefaction, in which the thermal treatment conditions employed favor the formation of a solid char, sometimes described as bio-coal [14]. Torrefaction densifies the biomass and makes it more hydrophobic, allowing simpler storage and greater compatibility with conventional solid fuels such as pulverized coal. This char is the only significant product of the process and is used almost exclusively as a solid fuel alternative to coal, with similar low value, and is now frequently used in fuel pellet form.

While biorefineries using thermal processing have some significant advantages, such as final product flexibility and relative indifference to the type of biomass feedstock, they typically decompose and degrade the valuable complex chemical structures in biomass, creating mostly simple materials of relatively low value. Creation of more structurally complex chemicals must then depend on the generally energy intensive and complex chemical processes used to convert the thermochemical products back into more valuable chemicals.

Anaerobic Digestion Another biorefinery process that involves decomposition of biomass into products having much simpler chemical structures is the biological conversion of biomass using the digestive capabilities of certain anaerobic bacteria. The primary product of this technology is a methane-rich gas that can be used as a fuel, or used in the chemicals industry [15]. However, such gas streams provide higher value when the methane is purified by removal of carbon dioxide, which is the second major component of the anaerobic gas stream. Low cost natural gas, now available in many parts of the world from shale, would provide a serious economic competitor to this technology except where regional and possibly remote locations are involved, and where environmental considerations are important.

9.2.1.2 Biorefineries Using Physical and Chemical Pretreatment with Biochemical Processing

Unlike the thermochemical and anaerobic biological digestion processes discussed above, many biorefinery processes under development today aim to preserve much of the valuable complex chemical structures present in the biomass, rather than converting them into basic chemical materials such as syngas, methane or highly chemically modified pyrolysis products.

Such biorefinery processes using woody biomass necessarily start with a feedstock preparation step that creates a reasonably uniform-sized material, such as wood chips, corn stover fiber, corn cobs, sugar cane bagasse, etc. This material may then be steamed to a desired moisture content, or otherwise conditioned before being fed to a treatment stage involving a process such as steam explosion [16], supercritical water treatment [17], ammonia fiber expansion (AFEX) [18], dilute acid pretreatment [19], or a number of variations of these and other treatments. A primary objective of such treatments is to expose physically the cellulosic fibers by separating them from the lignin and hemicellulose that encases them in the natural biomass. By so doing, the cellulose and the hemicellulose becomes accessible to the large, multi-component, cellulolytic enzymes that are then able to hydrolyze efficiently the cellulose to glucose. Alternatively, certain of those treatments that use inorganic acids, or supercritical water, rely on these to hydrolyze the cellulose and hemicellulose in the absence of enzymes [20].

These processes generally preserve most of the chemical structures in the polysaccharides, which typically make up between two thirds and three quarters of biomass material, although some conversion to shorter chain oligomers, or even the monomers, such as the simple pentose and hexose sugars will also occur. These can then be further converted by fermentation or chemical processes into useful chemicals and fuels. In most of these processes, the post-process residual material from the biomass, primarily lignin and some undigested cellulosic fiber, is regarded as a waste product having typically little more than solid fuel value.

Other biorefinery technologies now under development use chemicals such as concentrated sulfuric acid [21] and, in a few cases, supercritical water [22]. There is clearly a very broad range of technologies currently under development to function as process steps in biorefineries, but their commercial success or failure will eventually be determined by economics and proven technical performance.

Organosolv Processes Employing Solvent Extraction A distinctly separate category of the physical and chemical biorefinery technologies is that known as organosolv, or solvent extraction, which employs organic solvents in the treatment stage.

A number of organosolv processes for the production of cellulosic fibers for papermaking were developed over the past century as environmentally benign alternatives to the kraft and sulfite chemical pulping processes [23]. The various organosolv processes utilize different concentrated or aqueous organic solvents at elevated temperatures to remove lignin from the woody biomass in order to create separated cellulosic fibers. However, the conditions employed and the solvents used in most of these processes also initiate chemical reactions, such as hydrolysis, that cause depolymerization of hemicellulose and, generally to a much lesser extent, the cellulose. It is these same process conditions that depolymerize the lignin into smaller fragments that then dissolve in the organic solvent phase. Organosolv delignification processes, therefore, involve not only the separation of the lignin from the biomass, but also perform some chemical modification, not only of the lignin but also of the hemicellulose and cellulose.

Included among the organosolv pulping processes that have been developed are those that utilize formic acid (Fomacell) [24], acetic acid (Acetosolv) [25] and phenol [26]. None of these have reached a full commercial scale of operation at this time. However, some organosolv processes that utilize lower molecular weight aliphatic alcohols, namely methanol, ethanol and butanol, have, in some cases, achieved pilot scale and even commercial scale production. The Organocell pulping process, originally developed by the German pulp and paper company, MD Papier, initially used an aqueous acidic methanol organosolv stage, but later followed this with an alkaline stage [27]. A fully commercial pulp mill (about 400 ton per day of pulp) was constructed and started up at Kelheim in Bavaria, Germany, in the early 1980s but was soon shut down for commercial and technical reasons. One organosolv pulping process, the Alcell process, which used aqueous ethanol as a solvent, is now the primary biomass treatment step of the Lignol biorefinery process [28, 29].

9.3
Catalytic Processes Employed in Biorefining

With the above-described broad range of processes employed in the various forms of biorefining, together with the diverse chemistries involved, it is not surprising that catalytic processes are used extensively in the developing biorefining industry. These catalytic processes range from the classical Fischer–Tropsch processes for the creation of various small aliphatic molecules from synthesis gas, to the sophisticated enzyme bio-catalysts used to efficiently hydrolyze large polysaccharides, such as cellulose and hemicellulose, to their constituent sugars. Additionally, there are numerous acid-catalyzed hydrolytic and dehydration processes and chemistries that occur within the various process unit operations of various biorefineries.

9.3.1
Catalysis in Biorefineries Employing Gasification and Pyrolysis

The type of catalytic systems employed in the various biorefining technologies depends on the characteristics, feedstocks and products of the specific biorefinery process. Biorefineries using gasification processes that destroy the original chemical structures of the biomass, create simple gaseous product streams from which synthesis gas can be recovered and then converted to liquid chemical products using the Fischer–Tropsch metals catalysts of iron, nickel and cobalt. Biomass pyrolysis processes also create relatively chemically simple gas, liquid and char streams that can be processed with inorganic catalytic systems for further upgrading. Or, as described previously, catalysts can be employed, together with hydrogen, within the pyrolysis vessel to create a more chemically reduced bio-oil product profile that has greater energy density and is more compatible with typical hydrocarbon fuels of today.

9.3.2
Catalysts in Anaerobic Digestion Biorefineries

As described earlier, anaerobic biological digestion does not employ high temperature methods, but still destroys most of the chemical structures in biomass. The primary product of anaerobic digestion is a mixed gas stream, rich in methane and CO_2, that can be used directly as a low-value fuel. But, it could also be purified and the resultant methane converted to synthesis gas by steam reforming [30], thus providing a route to higher value liquid fuels or chemical products.

9.3.2.1 Catalysts in Non-Thermochemical Biorefineries
Technologies that process biomass through non-thermochemical processes typically retain, to the maximum extent possible, the valuable and complex chemical structures present in the original biomass. The objective of most biorefineries using physical and chemical treatment processes is generally to open up the structure of the woody biomass to provide access for cellulolytic enzyme systems to the cellulose fiber. This can be done by physical methods, such as steam explosion [16], ammonia fiber expansion (AFEX) [18], or even intense mechanical grinding [31].

The dominant chemistry of organosolv pretreatment processes is acid-catalyzed hydrolysis. This chemistry is primarily responsible for the partial depolymerization of the lignin through mostly acid-catalyzed hydrolysis of the β-O-4 aryl-alkyl ether linkages [32]. The hemicellulose is also partially depolymerized to monosaccharides and oligosaccharides during the organosolv treatment stage by acid-catalyzed hydrolysis of mostly the β-D-(1→4) linkages between the various pentose and hexose moieties in the hemicellulose. However, the efficiency of an organosolv biorefinery is generally determined by having minimal hydrolytic depolymerization of the cellulose that occurs through cleavage of the β-D-(1→4) glycosidic bonds during the organosolv stage. Operating conditions are, therefore, selected that maximize lignin and hemicellulose depolymerization and minimize cellulose depolymerization. On the other hand, some dilute acid biorefinery processes [20] operate under more aggressive treatment conditions designed to hydrolyze extensively the cellulose to glucose. Such aggressive conditions can modify the lignin structure, through such mechanisms as condensation reactions, to an extent that it has less value than the lignin product created under less extreme conditions.

Even under the more moderate operating conditions of an organosolv biorefinery, pentose sugars produced through the hydrolysis of hemicellulose can be readily converted to furfural by acid-catalyzed dehydration reactions. Furthermore, some hydroxymethylfurufal can be produced during this stage from hexose sugars generally derived from the hemicellulose fraction.

Catalysis in the Further Processing of Biorefinery Products Catalytic processes are used extensively in the further processing of the products from biorefinery operations. As previously discussed, Fischer–Tropsch catalytic systems are used to create a range of useful products from the purified syngas created in biorefineries employing biomass gasification. In most other biorefineries that use physical or relatively mild chemical processing, the cellulose fraction is today typically hydrolyzed to the constituent monosaccharide, glucose, using the biocatalyst cellulase enzyme systems. Despite their current relatively high cost, these enzymes provide a greater final yield than previously-utilized acid catalysts, and are also more compatible with sequential and combined saccharification and fermentation processing that creates ethanol or other fermentation products.

The relatively pure lignin fraction produced in organosolv biorefineries can be used directly in polymer and other applications, but it can also be processed further by catalytic hydrocracking followed by thermal hydrodealkylation, similar to technologies used in the petroleum refining industry [33]. These processes create a mixed stream of phenols, catechols, benzene and various hydrocarbons, which can be further refined to produce clean product streams. One such process, using ebullated bed reactor technology, was developed by Hydrocarbon Research, Inc., Lawrenceville, NJ, in the late 1970s as a means of producing aromatic chemicals from renewable materials [34].

9.4
An Organosolv Biorefinery Process for High-Value Products

The organosolv-based Lignol biorefinery process is a relatively mature technology that has been developed from the extensive experiences gained from almost seven year of operations of the Alcell pulping process [35]. This process was operated successfully for that time in a pre-commercial scale facility having a capacity for processing 70 tonnes, dry weight basis, of wood chips. The primary process step is an elevated temperature, aqueous ethanol extraction, operating at a mildly acidic pH that partially hydrolyzes the lignin and hemicellulose of the lignocellulosic biomass feedstock. These materials then become soluble in the liquor. This primary organosolv step is further integrated with systems for the continuous recovery of a high purity organosolv lignin co-product, now called HP-L™ lignin, as well as several other relatively simple chemical co-products created in the process, including furfural, acetic acid and wood extractives, such as phytosterols and terpenoids [36]. Because of the relatively mild conditions employed in the extraction stage and the benign nature of the solvent, many other high-value co-products can be obtained by this technology from specific types of biomass raw materials. For example, betulin, a precursor to betulinic acid, under investigation as an anti-cancer drug [37], can be recovered from the wood and bark of certain tree species, and suberin and triterpenoids can be recovered from the bark of birch trees [38]. While these are minor components by volume, they provide examples of additional revenues that may be recovered from the extraction liquor of the Lignol biorefinery when appropriate. An overview of the operations of a standard Lignol biorefinery is provided in Figure 9.3.

Cellulose Production and Applications. The mostly delignified cellulose fiber that remains after the organosolv extraction stage is readily hydrolyzable by cellulase enzymes to glucose, with high yield. This can then be fermented to ethanol for

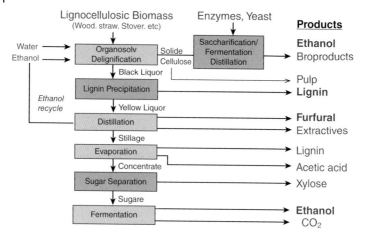

Figure 9.3 Overview of the Lignol biorefinery process.

the motor fuel market, or used in any number of biological or chemical processes being developed for the production of valuable chemicals, such as isobutanol, ethyl acetate, adipic acid and succinic acid. Alternatively, the delignified cellulose fibers can be used in higher value applications, such as the manufacture of papermaking pulp, or dissolving pulp for the production of cellulose ethers and esters, rayon, microcrystalline cellulose (MCC), or nanocrystalline cellulose (NCC). It has been shown that because the lignin has been mostly removed from the cellulose fiber, less cellulase enzyme, which can be inactivated by lignin, is required to saccharify the cellulose. Furthermore, since the feedstock to the saccharification step has a much higher cellulose content than the original biomass, it is easier to create higher glucose concentrations during saccharification. Higher glucose concentrations in the fermentation stage result in higher ethanol concentrations in the beer, which in turn reduces the cost of ethanol recovery. Several other saccharification and fermentation advantages result from the prior removal of the lignin from the cellulosic fiber, all of which improve the process economics.

Lignin Recovery. The aqueous ethanol liquor from the organosolv stage is processed to precipitate the dissolved lignin, which is a lower molecular weight fraction created mostly by simple hydrolysis of the native lignin, generally at the aryl-alkyl ether linkages between two phenyl propane sub-units [32]. It is a very pure material that has many positive attributes and industrial applications, as will be discussed later in this chapter. Under well-defined conditions the lignin precipitates as fine particles, which are then filtered, washed and sent to a dryer. The pure dry lignin powder is then sold into numerous commercial applications creating a major revenue stream for the biorefinery.

Furfural Recovery. Following recovery of the lignin powder cake the filtrate is sent to distillation for recovery and recycle of the process ethanol. Within this stage, furfural is easily recovered and further purified in a separate step, thus becoming an additional revenue source for the biorefinery. Furfural is created from the biomass in the Lignol biorefinery by hydrolysis of the hemicellulose xylan and then dehydration of the resultant xylose and arabinose. The process conditions of the organosolv step favor these reactions, but the yield of furfural is determined mostly by the nature of the biomass feedstock in the process. Hardwoods and annual fibers will provide a significant amount of furfural, but softwoods, which have a much lower xylan content, will produce much less furfural.

Wood Extractives Recovery. Wood and annual fiber biomass contains small quantities of lipophylic chemicals, termed "extractives" in the pulping industry. This very diverse group of chemicals, which usually represents up to 5% of the dry weight of wood, consists of such chemicals as phytosterols, other terpenes and terpenoids, lignans, esters of fatty acids, fatty acids and tannins. Many of these chemicals are currently used in industry and are of high commercial value. Because the Lignol biorefinery process performs a solvent extraction of the woody feedstock, many of these "extractives" are solubilized into the organosolv liquor. They are then readily separated and recovered from the liquor as a hydrophobic fraction, following the recovery of the lignin and the ethanol. They can then be further fractionated and sold as pure products, or sold as a crude mixture to others, thus providing an additional revenue stream to the biorefinery. While the volume of "extractives" recovered from wood is relatively small, typically around 2–3% of the original dry weight, much larger volumes can be recovered when processing certain types of annual fibers that contain significant quantities of waxes. Annual fiber feedstocks, such as Musa, can contain greater than 15% of lipophilic extractives [39]. Some of these extractives are of significant value as fine chemicals used in the pharmaceutical industry. Relevant to this "extractives" product, it has been shown that Lignol's organosolv biorefinery process is able to treat wood from various hardwood and softwood trees, as well as a range of annual plant materials, including bamboo, palm fronds, sugar cane bagasse, corn stover and switchgrass.

Acetic Acid and Hemicellulose Sugars Recovery. Hemicellulose consists primarily of galactoglucomannan (glucomannans) and arabinoglucuronoxylan (xylan) polymers with the type and ratio of each component being different between hardwoods and softwoods [36]. Some of the sugar moieties in hemicellulose are acetylated on the C_2-OH and the C_3-OH groups. As the temperature rises at the start of the organosolv stage these acetyl groups are readily hydrolyzed to form acetic acid, which has the effect of lowering the pH of the aqueous ethanol liquor to some degree. This mild lowering of the pH favors the hydrolysis of the hemicellulose to produce monosaccharides and oligosaccharides that are then soluble in the liquor. Some of these pentoses, the arabinose and xylose, are then converted under the conditions of the organosolv stage into furfural, which is recovered as described previously.

Following recovery of the process ethanol and the furfural, the acetic acid and the unconverted sugars and oligosaccharides remain as soluble components in the ethanol recovery tower stillage, which can be readily concentrated by evaporation. Provided there is sufficient economic incentive, the acetic acid can be recovered by one of several established methods from the evaporator condensate and the concentrated hemicellulose sugars and oligosaccharides can be processed, either chemically, or by enzyme-enhanced fermentation into a number of useful chemicals [40]. This biorefinery process, therefore, recovers, in useful and valuable forms, almost all the components of the original cellulosic biomass.

9.4.1
Guiding Principles of the Lignol Organosolv Biorefinery

This process is a classic example of a biorefinery driven by the following principles;

- *To the maximum extent possible, preserve the valuable complex chemical structures that exist in the lignocellulosic plant material and transform them into valuable chemicals and materials compatible with the needs of existing industry.* By employing this principle, the biorefinery creates a series of higher value co-products from the one feedstock, the combined value of which provides greater revenues compared to similar biorefinery processes producing only one or two products. An added advantage of this strategy is that multiple product streams provide some protection for the economics of the biorefinery. The economic success of other biorefineries, which produce only one product are subject to the market changes for that single product, the price of which might be highly variable over time. For this same reason, modern oil refineries operate most profitably by processing a single feedstock into multiple products – for example, gasoline, aviation fuel, heating oil, diesel fuel, and basic petrochemicals. It was not always that way. Gasoline was originally considered to be a low-value waste stream when the primary objective of the earliest oil refineries was to produce an alternative lantern fuel to whale oil, which was becoming scarce. There is much to learn from history!

- *Integrate the individual process steps to maximize energy efficiency.* This biorefinery technology produces a number of high-value products through several unit processes. Because there are several valuable products being created and recovered simultaneously within the process there is greater opportunity for energy and heat integration, leading to greater process efficiency.

- *Use processes and equipment that have been fully developed for other industries.* The pulp and paper industry has a long history, extending over almost two centuries, of using wood as a raw material. Forestry harvesting practices, debarking equipment, chipping, chip storage methods, chip screening, chip steaming and chip delivery to a digester are all well-established operations in the pulping industry. These same unit processes are used in a Lignol biorefinery, together with distillation, evaporation and solvent recovery processes that are well developed in the petroleum refining and petrochemical industry. Using such well established processes and equipment, wherever possible, reduces the potential of operational difficulties. In this regard, a major advantage of having experience over seven years of semi-commercial operations of the Alcell process has provided great knowledge of the systems and equipment that operate successfully in a biorefinery and those that do not.

9.4.2
Applications and Markets for Organosolv Biorefinery Products

9.4.2.1 Native Lignin – Its Properties and Composition

Lignin is a natural binding material found in the structural parts of plants that serves to strengthen and protect the long cellulose fibers from physical, biological and chemical destruction. To fully appreciate the commercial significance and potential value of lignin products it should be recognized that lignin is one of the most abundant organic polymers on earth, exceeded only by cellulose and chitin [41], constituting about a quarter to a third of the dry mass of wood and stems of plants. Lignin is also the only significant renewable source of aromatic compounds. Aromatic chemicals such as benzene, toluene, xylene, phenol, and cresols, which today are derived from oil and to a much lesser extent from coal, provide a very significant proportion of the feedstocks for the chemicals and materials industries, about 35 million tonnes per year worldwide [42]. Many critically important products, including polystyrene, phenolic resins, polycarbonates, polyurethanes and epoxy resins, are composed of aromatic structures.

Although the specific polymeric structure of lignin is still being debated in the scientific literature [43], it is clear that native lignin is a cross-linked, highly branched, three-dimensional amorphous macromolecule with molecular masses that exceed 10 000 Da. It is relatively hydrophobic and insoluble in hydrocarbons. The degree of polymerization is difficult to measure, and the macromolecule consists of three monolignol monomers, methoxylated to various degrees in the σ-position to the phenolic hydroxyl group: p-coumaryl alcohol, coniferyl alcohol, and sinapyl alcohol. Many investigators believe that it lacks a defined repeating structure since it contains a number of different chemical links that connect these monomeric structures. However, studies show that there is a high proportion of aryl-alkyl ether linkages present in native lignin [32].

9.4.2.2 Lignin from Other Processes

Several forms of lignin obtained from the wood pulping industry, often referred to as technical lignins, have been commercially available for many years. These include lignosulfonates from the various sulfite pulping processes and thiolignin from the kraft pulping process. More recently, lignin obtained from the soda pulping of annual plant woody biomass has also become available commercially [44]. Because of the nature of the delignification chemistry involved in these pulping processes and the method of their recovery, these lignins are generally quite distinct from the pure organosolv lignin derivatives produced in the Lignol biorefinery process. These differences, which include their relatively broad and varied molecular weight distributions, generally poor reactivity, and the presence of contaminants such as ash, sugars, carbohydrates and sulfur, frequently make them unsuitable for many of the applications being developed for the HP-L lignins produced by Lignol's organosolv process.

9.4.2.3 HP-L™ Lignin–Organosolv Lignin from the Lignol Biorefinery

While the lignin products might account for only about one quarter of the original cellulosic biomass dry weight, experience has shown that they can provide more than half the revenues of a typical Lignol biorefinery when used in certain chemicals and materials applications. Support for this proposition comes from an understanding of the characteristics and properties of the lignin co-product from an organosolv biorefinery. Detailed research and development has shown that it is possible to significantly alter both the physical and chemical properties of the lignin through changes in the operating conditions of the biorefinery organosolv stage. Through this knowledge, it is possible to target and enhance certain characteristics of the lignin to create a product with properties best-suited to any one of a number of diverse applications.

Under the selected operating conditions of the biorefinery organosolv stage, certain chemical bonds in the native lignin structure are preferentially hydrolyzed, in particular the β-O-4 alkyl-aryl ether linkage, which creates smaller fragments of the native lignin macromolecule and consequently produces additional aryl and alkyl hydroxy functional groups. As the molecular weight of these lignin fragments becomes smaller through this chemistry, they eventually reach a size that is soluble in the hot aqueous ethanol solvent. This selective solubility and extraction, together with post-processing recovery systems, provides lignin fractions having relatively narrow molecular weight ranges and small polydispersities. These lower molecular weights and smaller polydispersities are important in many of the applications of this product.

9.4.2.4 Lignin Derivatives

Lignin Derivatives is the general term used here to describe the entire generic class of lignin components that can be extracted from lignocellulosic biomass by the organosolv treatment technology. These components contain both polar and non-polar regions, allowing close associations of components that bond physically through hydrogen bonding, Van der Waals interactions, π-bonding, physical entrapment, and organometallic coordination complexes. Lignin derivatives form charge transfer complexes with many substances, increasing the difficulty of separating certain non-lignin materials from dispersed or solubilized lignin. Acid–base chemistry promotes the pH-driven formation of ionic complexes with other substances, such as protein and protein fragments which contain amine groups.

The major class of lignin derivatives created in the organosolv biorefinery shows strong structural and physical resemblance to some significant commercial chemicals, such as the pre-polymer of phenol-formaldehyde resins. One such possible

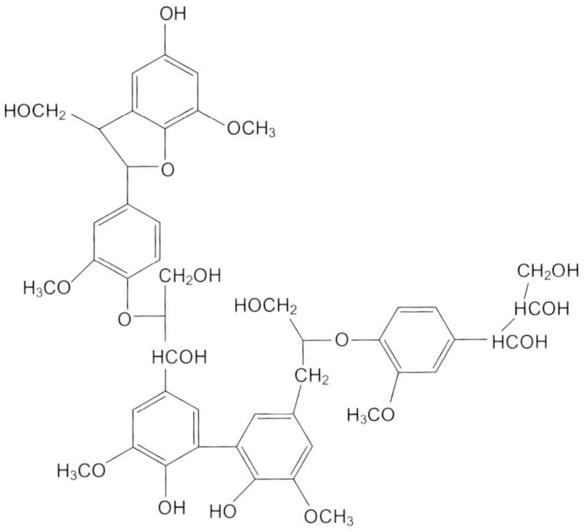

Figure 9.4 Possible representative fragment of HP-L lignin.

HP-L lignin fragment is represented in Figure 9.4. It should be recognized that such a polymer fraction, despite having a low polydispersity, is certainly not composed of chemically identical fragments. However, this product can be described through its average content of aryl and alkyl hydroxy groups, as well as other chemically important groups.

9.4.3
HP-L Lignin Properties

HP-L lignin derivatives have distinct physical and chemical properties and unique functional attributes, making them ideally suited for a range of applications in the chemicals and materials industries. They differ from other commercially available lignin products since they offer the performance features shown in Table 9.1.

9.4.4
Current Applications and Market Opportunities for HP-L Lignin

The chemical and physical attributes of HP-L lignins listed in Table 9.1, and the comparison to those of existing technical lignins, provide direction for identifying potential commercial applications for this product. Lignol has undertaken a major program of market investigation, followed by applications development projects, mostly undertaken jointly with potential end-users of the lignin.

Some of these applications are shown in Table 9.2.

The market-ready opportunities are mostly based on the use of HP-L lignin as a substitute for petrochemicals in the formulations of various materials currently

Table 9.1 Physical and chemical characteristics of HP-L lignins and applications benefits.

Performance characteristics of HP-L lignin	Applications advantages	Comparison with most other technical lignins
High purity, with very low ash, sulfur and carbohydrate content	Many applications require high purity, very low ash content and low content of hydrophilic materials like sugars	Most lignosulfonates and kraft lignins contain significant amounts of sulfur and ash
Highly chemically reactive in polymerization and cross-linking	In polymer applications, especially thermoset applications, cross-linking is very important	Most other technical lignins are already condensed or, in the case of lignosulfonates, derivatized to make water-soluble materials
High temperature stability and melt flow properties	The ability to melt and flow is critical in thermoplastic applications	Solid kraft and soda lignins decompose and sinter before melting making them unsuitable for most thermoplastic applications
Moderate hydrophobicity	HP-L lignins are hydrophobic, making them useful in moisture barrier coatings and reducing swell as panel board binder, but they are soluble in polar solvents	Lignosulfonates are water-soluble and kraft lignin is highly hydrophobic, which limits its applications
Low water content	HP-L lignin is produced as a dry powder having 3%, or lower, water content. This is important in some applications but it also extends shelf life	Lignosulfonates are most often sold as an aqueous solution. Kraft lignin is sold as a dry powder
UV light resistant and antioxidant	Useful for stability of exterior coatings and protection of lubricant and metal drawing properties, etc	Some technical lignins have similar properties, but are less readily utilized because of other properties
Ease of chemical derivitization	HP-L lignin has various functional groups that can be utilized during chemical modification	Chemical modification can be achieved but with greater effort

being used by industry. These kinds of applications are attractive because they pose fewer barriers to entry since the end product and its market already exist.

9.4.4.1 New Product Opportunities for Lignin Derivatives

In addition to developing applications in which HP-L lignin is typically substituting for, or partially replacing, an existing product or material, a number of opportunities have been identified in which a new or improved product is developed from HP-L lignin acting as a feedstock or major component. A number of these opportunities are very attractive because:

Table 9.2 Market-ready opportunities for HP-L lignin.

Market-ready opportunity	End-use
Primary applications	
Foundry and refractory resins	Resins are used to bind sand in the production of molds for metal casting in foundries, typically in the auto industry; HP-L lignin derivatives can replace some of the normal functional components in these resins.
Rigid foam insulation	A liquid resin formulation can be sprayed into cavities or sprayed onto a backing material, after which the liquid forms an expanding foam that subsequently hardens into a rigid insulating material used in the construction and many other industries. HP-L lignin can be a component in certain of the resin formulations.
Wood composite adhesives	Wood composites consist of wood particles, wafers or sheets glued together with adhesives to provide structural integrity and strength. HP-L lignin can be added to certain of these adhesive formulations
Epoxy resins and coatings	Specialized adhesives and coatings are used in the auto industry and many other marine and industrial sectors, where chemical reactions after application lead to rapid hardening and superior bond strength. HP-L lignin can be added to certain of these resin formulations.
Secondary applications	
Friction materials	Resins are used to bind together solid particles that provide the friction properties needed in brakes and clutches in automotive and industrial applications. HP-L lignin has been shown to be an effective component in some of these resin systems.
Paints and coatings	Paints and coatings for a wide range of domestic, commercial and industrial uses can use HP-L lignin as an effective renewable ingredient
Decorative laminates	Laminates used in countertops and other furnishings are a composite of a plastic top layer together with a backing of wood products and fibrous material impregnated with a resin formulation that bonds the layers together. HP-L lignin can be added to certain of the resin formulations
Molding compounds	Traditional phenolic resin materials can be partially replaced or substituted with renewable products such as HP-L lignin.
Carbon products	Carbon powders and fibers produced by graphitization or carbonization of carbon-rich materials, usually petroleum-based; HP-L lignin can be used as one of the feedstocks for carbonization
Filtration media	Filters and adsorbents for separation of contaminants from gaseous or liquid streams can use HP-L lignin as an ingredient
Rubber ingredients	Rubber products such as tires and belts in automotive and industrial applications can use HP-L lignin as an antioxidant and functional ingredient.

Table 9.3 New product opportunities for HP-L lignin.

New product opportunity	End-use
Carbon fiber	Well established in the aerospace industry, carbon fiber consists of strands or fibers of carbon which are molded into shapes and impregnated with resins to form extremely lightweight and strong structures. HP-L lignin is being tested as a precursor for carbon fiber for automotive and industrial sectors.
Adsorbents	HP-L lignin can that soak up and adsorb contaminants from gases and liquids by virtue of its surface properties
Animal feed	HP-L lignin can be added to animal feed both as a pelletizing aid and also to provide nutritional benefits to the animal
Antioxidants	HP-L lignin has significant properties as a natural antioxidant in numerous applications that include additives for lubricating oils and greases, fuels, rubber products such as tires and belts and animal feeds
New plastic materials	HP-L lignin is compounded with conventional thermo-plastics to form new materials with distinct functionality and high renewable content

- They represent large-volume markets where the value of the lignin is sufficiently high to be very attractive
- Once launched commercially, they will provide future market growth for HP-L lignin production and sales

Examples of these longer term new product opportunities are listed below in Table 9.3.

9.4.4.2 Market Drivers for Commercial Use of HP-L Lignin and Other Bio-Products

As stated previously, there are now numerous drivers and incentives for industry to use biobased chemicals and materials from renewable feedstocks. Evidence for this is seen by the number of major international chemicals (e.g., DuPont, BASF) and oil companies (e.g., Shell, BP, Chevron) that are partnering with and investing in smaller biobased technology development companies. These larger, well-established companies are typically selecting technology companies that can produce identical or near-identical chemicals or materials to those presently used by them from alternative, renewable feedstocks. Other larger companies are seeking new products, such as novel biopolymers, that represent new market opportunities.

For the fuels, chemicals and materials industries, biobased products created in biorefineries represent a unique opportunity to obtain the following strategic and financial advantages;

- ✓ Greater confidence in feedstock supply and price compared with the unpredictable supply and world price volatility of feedstocks obtained from crude oil that is subject to international competition and events
- ✓ An ability to satisfy customers demanding products with higher contents of green, renewable materials
- ✓ The ability to lower the carbon footprint of their products as required by current and future government regulation and customer preferences
- ✓ Realize potential cost savings by making products from domestically produced agricultural residues and other low cost feedstocks
- ✓ Benefit from tax credits offered by various governments around the world for use of renewable materials and GHG reduction
- ✓ Qualify for purchase preferences by various governments and their contractors for the use of biobased chemicals and materials
- ✓ Stimulate local industry and domestic employment

These incentives for the use of biorefinery products apply to essentially all biorefineries and are independent of their technology and operating principles. However, those biorefineries, such as the Lignol biorefinery, that operate on the principle of multiple products and maximum retention of chemical structures and value have additional advantages. These include;

- ✓ Increased revenues and profitability from smaller biorefineries requiring less capital to construct and that can be supported by smaller renewable feedstock resource capabilities
- ✓ Releasing biorefinery economics from dependence on a single product in a highly variable market such as liquid fuels and energy
- ✓ Improved carbon footprints and reducing aggregate operating costs by creating more biobased products from less biomass

In reality, these several benefits are highly connected – biorefinery investment returns will be buoyed by HP-L lignin revenues while biofuels production will likely provide the primary reason for the construction of HP-L lignin producing operations.

9.4.4.3 Application Strategies for the Lignol Biorefinery Process

It is now becoming recognized that more value should to be obtained from lignin than its low value as a solid fuel. However very few other biorefinery technologies under development can create a lignin product having the required combined attributes of purity, reactivity, hydrophobicity and thermal properties that encourage its use in the modern chemicals and materials industries. Nevertheless, it is possible to conceive of opportunities to integrate key process steps of the Lignol biorefinery process (Figure 9.5) with certain other types of biorefineries to achieve the desirable goals of multiple saleable products and enhanced revenues.

Lignol Biorefinery Process

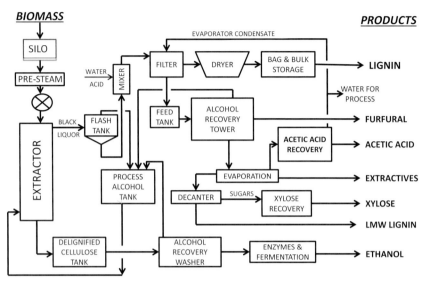

Figure 9.5 Process flow of the Lignol biorefinery process.

Obviously, for technologies that generally destroy the chemical structures of the biomass, this integration is not possible. But, for other technologies that use relatively benign chemical and physical treatment stages, this could be an exciting possibility. The Lignol biorefinery process therefore offers opportunities not only for construction of stand-alone facilities but also for integration with other types of existing and future industrial operations. Within this latter option might be integration with existing forest products industries, such as pulp and paper mills and large sawmill operations, as well as integration with corn ethanol facilities able to harvest stover and cobs along with the corn kernels. Additional opportunities for process integration exist with cane sugar mills and oil palm plantations. For all of these opportunities, sharing of infrastructure, transportation and utilities, together with energy integration possibilities, promise enhanced returns above stand-alone operations.

9.4.4.4 Development of the Lignol Biorefinery Process

The development history of the Lignol biorefinery process exemplifies the various trials and direction changes that many new technologies often undergo before they are successfully commercialized. It also illustrates the various ways that the technology has been successfully deployed in the past and the directions that are still available to it in the future. Unlike certain other types of biorefinery processes, it is a flexible biorefining technology that is not limited to the production of alcohol fuels, or other types of transportation fuels. Additionally, it can be used to produce commodity chemicals and materials, higher value specialty chemicals, and even bio-derived fine chemicals used by the pharmaceutical and fine chemicals industries – often within the same operation. Furthermore, the development history of

9.4 An Organosolv Biorefinery Process for High-Value Products

the Lignol process illustrates how external factors, such as international politics, economic cycles, government policy, and sometimes unrelated serendipitous events often influence the development and commercial acceptance of new technology. It also identifies the high level of operating experience and know-how that has been gained during its many years of development [45].

The core of Lignol's organosolv biorefinery process, which is the organosolv pre-treatment stage, can trace its history through more than three decades and a number of industrial sectors. It started at the University of Pennsylvania, in collaboration with General Electric, in the 1970s as a concept, far ahead of its time as a biofuels process. It then passed into the pulp and paper industry as the Alcell pulping process [35] owned by Repap Enterprises Inc., and was then acquired by Lignol. At each stage, the process was developed towards specific process objectives with additional learning and additions to the technology and intellectual property. Along the way, products were produced, test marketed and gained commercial acceptance; the process was enhanced and adapted for its primary end use.

It was as a direct result of the break-up and sale of Repap, that all of the Alcell technology, know-how, certain patents, and all the seven years of operating records and documentation of the Miramichi Alcell pulp mill, having 70 tonne per day wood processing capacity, became available. One view of this Alcell pulp mill is shown in Figure 9.6. A new company, Lignol Innovations Corporation, (now Lignol Innovations Ltd) formed in British Columbia, acquired these assets in 2002 and re-configured the technology back to its original purpose as a biorefinery. Since that time major advances have been made to the technology as well as to lignin marketing developments.

To further develop the engineering base for a biorefinery plant design, and for technical advancement of the process, as well as to provide large samples of the

Figure 9.6 The Alcell pulp mill at Miramichi, New Brunswick, Canada.

Figure 9.7 Lignol's pilot biorefinery – fermentation area and ethanol distillation and recovery.

various products for market development activities, Lignol has constructed a fully integrated pilot plant at its engineering development center in Burnaby, British Columbia. Two of the unit processes of this pilot plant, which has a nominal processing capacity of one tonne per day, are shown in Figure 9.7 This plant has been instrumental in providing Lignol with the engineering design, operating parameters and projected economics for a fully commercial-scale plant, which is now in the planning stages.

The pilot plant has also been used to investigate several novel operating conditions that create products having new and desirable properties. Several new classes of lignin derivatives have been produced and are being tested in commercial product applications studies. This pilot plant is one of the few such fully integrated pilot plants of this scale available around the world and it provides the company with a unique advanced capability for both process and product development.

9.5
Conclusions

The Lignol biorefinery process is a relatively mature process, involving numerous catalytic steps, that has benefited from over 30 years of development in several different forms. The primary operating principle behind the process is to fractionate the cellulosic biomass feedstock to create multiple products of high value by preserving, to the maximum extent possible, the complex chemical structures that pre-exist in various plant materials. These products can then be supplied to the chemicals and materials industries, either as biobased alternatives to chemicals already used in industry, or as novel feedstocks for production of new products.

The production of multiple products from a single process has numerous economic and operating advantages. It maximizes product yield and value from the original biomass feedstock, and allows the economics of the facility to be less dependent on the current market value of any one product, such as fuel ethanol.

The principle organosolv process step of the technology has operated at a pre-commercial design capacity of 70 tonnes of dry biomass processing capacity per day for almost seven years. In this operation, thousands of tonnes of products were made and mostly sold into commercial markets, thus proving the commercial value of these products. The technology continues to undergo further engineering and process refinement and new applications for its products are being developed in partnership with potential customers. The process, which involves numerous catalytic steps, appears to meet all of the criteria for "green" biobased chemicals and materials production that are being legislated and encouraged by governments and organizations around the world.

References

1 Dale, B.E., and Kim, S. (2006) Biomass refining global impact–the biobased economy of the 21st century, in *Biorefineries–Industrial Processes and Products*, vol. 1 (eds B. Kamm, P.R. Gruber, and M. Kamm), Wiley-VCH Verlag GmbH, Weinheim, pp. 41–66.

2 de Jong, E., van Ree, R., van Tuil, R., and Elbersen, W. (2006) Biorefineries for the chemical industry–a Dutch point of view, in *Biorefineries–Industrial Processes and Products. Status Quo and Future Directions*, vol. 1 (eds B. Kamm, P.R. Gruber, and M. Kamm), Wiley-VCH Verlag GmbH, Weinheim, pp. 85–111.

3 De Luca, V., Salim, V., Atsumi, S.M., and Yu, F. (2012) Mining the biodiversity of plants: a revolution in the making. *Science*, **336**, 1658–1661.

4 NREL (2012) National Renewable Energy Laboratory, http://www.nrel.gov/biomass/biorefinery.html (accessed April 2012).

5 Kamm, B., Kamm, M., Gruber, P.R., and Kromus, S. (2006) Biorefinery systems–an overview, in *Biorefineries–Industrial Processes and Products. Status Quo and Future Directions*, vol. 1 (eds B. Kamm, P.R. Gruber, and M. Kamm), Wiley-VCH Verlag GmbH, Weinheim, pp. 3–40.

6 McCarthy, J.L., and Islam, A. (1999) Lignin chemistry, technology, and utilization: a brief history, in *Lignin: Historical, Biological, and Materials Perspectives*, ACS Symposium Series 742 (eds W.G. Glasser, R.A. Northey, and T.P. Schultz), American Chemical Society, Washington, DC, pp. 2–99.

7 *Algae Biomass Organization* 2012. http://www.algalbiomass.org/ (accessed 28 January 2013).

8 USDA, USDOE (2005) Biomass as Feedstock for a Bioenergy and BioproductsIndustry: The Technical Feasibility of a Billion-Ton Supply. Washington, DC, April 2005.

9 Evans, G., and Smith, C. (2012) Biomass to liquids technology, in *Comprehensive Renewable Energy*, vol. 5 (ed. A. Sayigh), Elsevier, pp. 155–204.

10 Sobolic, J. (2008) Anaerobic Organisms Key to Coskata's Rapid Rise. *Ethanol Producer Magazine*, June 2008.

11 Bridgewater, A.V., and Peacocke, G.V.C. (2000) Fast pyrolysis processes for biomass. *Renew. Sust. Energ. Rev.*, **4**, 1–73.

12 Ramakrishnan, S., and Moeller, P. (2002) Liquid smoke: product of hardwood pyrolysis. *Am. Chem. Soc., Fuel Chem. Div. Preprints*, **47** (1), 366–367.

13. Agrawal, R., Agrawal, M., and Singh, N. (2009) United States Patent 2009/0082604 A1.
14. *integro earth fuels* (2012) http://www.integrofuels.com/index.html (accessed 30 August, 2012).
15. Chynoweth, D.P., and Isaacson, R. (1987) *Anaerobic Digestion of Biomass*, Springer.
16. Sawada, T., and Nakamura, Y. (2001) Low energy steam explosion of plant biomass. *J. Chem. Technol. Biotechnol.*, **76**, 139–146.
17. *Renmatix* (2012) renmatix.com (accessed 2012).
18. Teymouri, F., Laureano-Perez, L., Alizadeh, H., and Dale, B.E. (2004) Ammonia fiber explosion treatment of corn stover. *Appl. Biochem. Biotechnol.*, **113–116**, 951–963.
19. Lee, Y.Y., Lyer, P., and Torget, R.W. (1999) Dilute-acid hydrolysis of lignocellulosic biomass. *Adv. Biochem. Eng. Technol.*, **65**, 93–115.
20. Scholler, H. (1935) France Patent 777,824.
21. Hester, R.D., and Farina, G.E. (2000) Concentrated sulfuric acid hydrolysis of lignocellulosics. US Patent 6063204.
22. Soberg, M. (2011) Supercritical water technology closer to commercial production. *Ethanol Producer Magazine*, Nov 2011.
23. Tjeerdsma, B.F., Zomers, F.H.A., Wilkinson, E.C., and Sierra-Alvarez, R. (1994) Modelling Organosolv Pulping of Hemp. *Holzforschung*, **48**, 415–422.
24. Saake, B. (1995) Production of pulps by the Formacell Process. *Das Pap.*, **49**, 1–7.
25. Niemz, H.H., Berg, A., Granzow, C., Casten, R., and Muladi, S. (1989) Pulping and bleaching by the ACETOSOLV PROCESS. *Das Pap.*, **43**, 102–108.
26. Jimenez, L., de la Torre, M.J., Maestre, F., Ferrer, J.L., and Perez, I. (1997) Organosolv pulping of wheat straw by use of phenol. *Bioresour. Technol.*, **60**, 199–205.
27. Dahlmann, G., and Schroeter, M.C. (1990) The Organocell process: pulping with the environment in mind. *Tappi J.*, **73** (4), 237–240.
28. Arato, C., Pye, E.K., and Gjennestad, G. (2005) The Lignol approach to biorefining of woody biomass to produce ethanol and chemicals. *Appl. Biochem. Biotechnol.*, **121–124**, 871–882.
29. MacLachlan, R., and Pye, E.K. (2007) Biorefining, the future. *Canadian Chemical News*, 12–15.
30. Collodi, G. (2010) Hydrogen production via steam reforming with CO_2 capture. *Chem. Eng. Trans.*, **19**, 37–42.
31. Cadoche, L., and Lopez, G.D. (1989) Assesment of size reduction as a preliminary step in the production of ethanol from lignocellulosic wastes. *Biol. Wastes*, **30**, 153–157.
32. Bose, S.K., and Francis, R.C. (1999) The Role of β-O-4 Cleavage in acidic organosolv pulping of softwoods. *J. Pulp Pap. Sci.*, **25** (12), 425–430.
33. Thring, R.W., Katikaneni, S.P.R., and Bakhshi, N.N. (2000) The production of gasoline range hydrocarbons from Alcell lignin using HZSM-5 catalysts. *Fuel Process. Technol.*, **62**, 17–30.
34. Parkhurst, H.J., Hubers, D.T.A., and Jones, M.W. (1980) Production of Phenol from Lignin. Symposium on Alternative Feedstocks for Petrochemicals, San Francisco: American Chemical Society, Aug 1980.
35. Pye, E.K., and Lora, J.H. (1991) The ALCELL Process: a proven alternative to kraft pulping. *Tappi J.*, **74** (3), 113–118.
36. Alen, R. (2000) Structure and chemical composition of wood, in *Forest Products Chemistry* (ed. P. Stenius), Fapet Oy, Helsinki, pp. 11–57.
37. Pisha, E., Chai, H., Lee, I.-S., Chagwedera, T.E., and Farnsworth, R.N. (1995) Discovery of betulinic acid as a selective inhibitor of human melanoma that functions by induction of apoptosis. *Nat. Med.*, **1**, 1046–1051.
38. Ekman, R., and Eckerman, C. (1985) Aliphatic carboxylic acids from suberin in birch outer bark by hydrolysis, methanolysis and alkali fusion. *Paperi ja Puu-Papper och Tra*, **67**, 255–274.
39. Omotoso, M.A., and Ogunsile, B.O. (2009) Fibre and chemical properties of some Nigerian grown Musa species for pulp production. *Asian J. Mater. Sci.*, **1**, 14–21.
40. Ji, X.J., Huang, H., Nie, Z.K., Xu, Q., and Tsao, G.T. (2012) Fuels and chemicals

from hemicellulose sugars. *Adv. Biochem. Eng. Biotechnol*, **128**, 199–224.

41 *Primex* (2012) http://www.primex.is/Company/The-History-of-Chitin/Chitin/ (accessed 2012).

42 Wikipedia (2012) Aromaticity, http://en.wikipedia.org/wiki/Aromaticity (accessed 2012).

43 Vanholme, R., Demedts, B., Morreel, K., Ralph, J., and Boerjan, W. (2010) Lignin biosynthesis and structure. *Plant Physiol.*, **153**, 895–905.

44 *GreenValue* (2012) http://www.greenvalue-sa.com/industrysitu.html (accessed 2012).

45 Lora, J.H., Powers, J.L., Pye, E.K., and Branch, B. (1993) Recent developments in ALCELL technology – towards a closed cycle mill, Proceedings, *1993 TAPPI Pulping Conference*, Atlanta, GA, November 1–3.

10
Biomass-to-Liquids by the Fischer–Tropsch Process

Erling Rytter, Esther Ochoa-Fernández, and Adil Fahmi

10.1
Basics of Fischer–Tropsch Chemistry and BTL

10.1.1
The FT History and Drivers

Mixtures of carbon monoxide and hydrogen, referred to as synthesis gas or "syngas", can be converted by the Fischer–Tropsch (FT) process to hydrocarbons. Natural gas, biomass or coal is first gasified to syngas, and one therefore distinguishes between gas-to-liquids (GTL), biomass-to-liquids (BTL) and coal-to-liquids (CTL) processes, or in general terms: XTL.

FT synthesis can be classified as a high-temperature FT (HTFT) process operating at 330–370 °C and low-temperature FT (LTFT) at 210–260 °C. The former gives high yield in the naphtha range, containing linear and branched olefins with some aromatics and oxygenate content. The HTFT process is extensively practiced in South Africa, based on precipitated or fused iron catalysts with stability and selectivity promoters, but all plants built or planned after 2000 are of the LTFT type. In this process cobalt- or iron-based catalyst converts syngas mainly to linear long-chain paraffins and some lighter olefins which, after upgrading by hydro-treating and cracking, results in a clean-burning diesel fuel. The produced fuel is virtually free of sulfur, aromatics and nitrogen compounds, and is excellent as a blending stock for conventional diesel. Use of neat XTL diesel in an engine is possible, but faces an issue with the low density and, therefore, lower calorific value and possible need for engine adjustments. In the present BTL overview emphasis will be given to LTFT with cobalt catalyst due to current industry practice and focus on environmentally benign liquid fuels.

Fischer–Tropsch synthesis was discovered by Franz Fischer and Hans Tropsch in the Kaiser Wilhelm (now Max Planck) Institute for coal research in the 1920s. In the following years after the initial patent, filed in Germany on 21 July, 1925 [1, 2], the FT concept was developed further, and the first commercial plant was put on stream by Ruhrchemie in Oberhausen in 1936. During World War II the Ruhrchemie atmospheric fixed-bed technology was practiced for the production

Catalytic Process Development for Renewable Materials, First Edition. Edited by Pieter Imhof and Jan Cornelis van der Waal.
© 2013 Wiley-VCH Verlag GmbH & Co. KGaA. Published 2013 by Wiley-VCH Verlag GmbH & Co. KGaA.

of gasoline and other hydrocarbons in 13 plants (Germany, France, Manchuria and Japan). In Germany this accounted for an overall 9% of the total fuels production and as much as 25% of automotive fuels. The catalyst consisted of a mixture of cobalt, magnesium oxide, thorium oxide (later replaced) and kieselguhr in a 100:5:8:200 ratios. The FT history has been summarized by for example, Schultz [3], Dry [4] and Stranges [5].

In the post-war period a number of pilot FT tests were performed on both fixed, ebulating bed and slurry type reactors. Most well-known is the Kölbel slurry technology, as demonstrated in a 1.5 by 9 m reactor by a Rheinpreussen and Koppers JV between 1953 and 1955 [6]. Sasol started their commercial scale fixed and fluidized-bed operations in 1955. The further technical and commercial development by Sasol has been described in detail [7]. The oil crisis in 1972 and later prompted a strong effort in technology developments of the entire chain, from natural gas to final products. Major US oil and gas companies were very active in patenting technical FT solutions and catalyst formulations. A fixed-bed demo reactor (30 bpd) was operated by Gulf in the early 1980s. Based on the Gulf experiences, Statoil started to develop slurry technology [8]. An extensive slurry development program was also conducted by Exxon in this period and later [9], as well as by ConocoPhillips in the early 2000s. Catalyst developments and a summary of today's technology players and providers, as well as commercial operations are given in subsequent sections.

Historically, the driver for FT-synthesis and CTL has been scarcity of supply of liquid fuels in South Africa due to the UN boycott. In the last decade there has been an opportunity to monetize low-value natural gas by GTL in the Middle East due to the distance from the major markets. Lately, opportunities have emerged in the US due to production of high volumes of shale gas. The resulting pressure on the gas price has currently decoupled the historic link between oil and gas prices in the US, thereby opening a possible window for GTL. In a somewhat longer perspective, XTL has the potential to become a factor that can relieve security of supply of fuels and other hydrocarbons. BTL has the added potential of having a lower carbon footprint and producing renewable fuels. There will, however, be a competition for available biomass for different local uses, and cost reductions as well as improvements in energy efficiency will be required. Hybrid solutions involving BTL are a possibility that deserves further attention.

10.1.2
Reactions

The LTFT process concerns hydro-polymerization of carbon monoxide to give linear alkanes by the overall reaction

$$(2n+1)\,H_2 + n\,CO \rightarrow C_nH_{(2n+2)} + n\,H_2O \quad \Delta H \sim 154 \text{ kJ mol}^{-1} \; (\alpha = 0.95)$$

where n is the chain length of the carbon backbone. The most important additional reactions are α-alkene formation by

$$2n\,H_2 + n\,CO \rightarrow C_nH_{2n} + n\,H_2O$$

and a separate reaction pathway to methane:

$$3\,H_2 + CO \rightarrow CH_4 + H_2O \quad \Delta H = -206 \text{ kJ mol}^{-1}$$

On a schematic level, the feed molecules CO and H_2 are activated on the surface of the FT metal, followed by hydrogenation of carbon and oxygen, chain growth by successively adding $-CH_2$-monomer units, and termination. Alkanes will be formed by hydrogenation of the growing chain, whereas β-hydrogen abstraction leads to α-alkenes. Further hydrogenation of surface $-CH_x$ will give methane. For each carbon unit in the product there will be one water molecule formed. In addition to these main reactions there will be some isomerization to products with short (methyl) side chains, possibly double bond isomerization, in addition to a few percent of oxygenates, mainly alcohols.

A wide range of chain lengths are produced, as determined by the value of chain termination probability relative to chain growth probability. In general, the product slate follows the Anderson–Schultz–Flory (ASF) distribution as expressed by:

$$W_n/n = (1-\alpha)^2 \alpha^{n-1}$$

where W_n is the weight fraction of a chain with a given chain length and α defines the chain growth probability according to

$$\alpha = r_p/(r_p + r_t)$$

Here, r_p and r_t are the reaction rates for propagation and termination, respectively. To minimize the production of light gases, it generally will be desirable to have as high α as possible, defined by the actual catalyst used and the process conditions. The number average molecular weight varies as

$$M_w(\alpha) = 14.026/(1-\alpha) + 2.016$$

resulting in molecular weight averages of 100, 285 and 14 000 for $\alpha = 0.90$, 0.95 and 0.999, respectively.

The H_2/CO usage ratio will be in the range 2.1 to 2.15, depending on α, and to some extent on the selectivity to other products besides alkanes. This ratio is higher than normally obtained by gasification of biomass, and the ratio must therefore be adjusted before synthesis.

Another important characteristic of the FT-reaction is its high exothermicity as given above [10]. The actual enthalpy of reaction varies slightly with the product selectivities, in particular the methane yield. Handling the heat evolved greatly influences the reactor and process designs. Analysis of the FT reactions gives a preferred range of process conditions for LTFT synthesis that include:

- Operating temperature between 210 and 260 °C. At the high end there will be an unfavorable production of light gases including methane. On the other hand, reaction rates will be high and steam produced by the reaction heat will be obtained at a more favorable pressure. Too low temperatures are prohibited by low reaction rates.

- Elevated reaction pressure. Process intensification dictates a reaction pressure of at least 10 bar, and there are commercial references at 25–30 bar. High pressures favor high conversion rates and formation of long-chained hydrocarbons.
- A H_2/CO ratio in the make-up gas to the FT reactor slightly below 2.

10.1.3
Mechanisms and Kinetics

There is still debate about the basic mechanism for the FT-synthesis on cobalt. Proposed mechanisms include hydroxycarbene, alkylidene, CO insertion and carbide. Several groups favor the latter pathway today, but there is discussion on whether the dissociation of CO is dissociative or associative (hydrogen-assisted). For the development of a Langmuir–Hinschelwood type of kinetic expression we will assume a hydrogen-assisted dissociation according to the following simplified scheme:

$$CO + * \leftrightarrow OC*$$

$$H_2 + ** \leftrightarrow 2\,H*$$

$$OC* + H* \leftrightarrow HOC* + *$$

$$HOC* + H* \rightarrow C* + H_2O + *$$

$$C* + x\,H* \leftrightarrow CH_x^* + x*$$

Note that the formation of surface carbide (C*) and water is expected to be rate determining. Further reactions, like chain growth, termination, hydrogenation of olefins and isomerization, will have no influence on the kinetic expression for CO conversion. Several kinetic expressions have been proposed in the literature, but based on certain assumptions and fitting to CSTR data for a Co/Re/alumina catalyst, we have arrived at

$$-r_{CO} = K_1[CO][H_2]/\{1 + K_2[H_2]^a + K_3[CO] + K_4[CO][H_2]^a\}^b$$

as an appropriate expression.

A cycle for FT-synthesis by the carbide mechanism is illustrated in Figure 10.1. Only hydrocarbon formation is shown as only minor amounts of oxygenates will normally be formed. Most of the CH_x monomers generated on the surface of the catalyst will be in the form of CH_2 that is readily incorporated in the chain during polymerization. A smaller portion of the monomer will be hydrogenated all the way to methane. The growing chain can terminate by β-hydrogen abstraction and leave the surface as an olefin or be hydrogenated to an alkane. Olefins can also be hydrogenated in a secondary reaction. There is evidence from experiments at low conversion and small catalyst particle size that the primary product is dominated by olefins, but under standard process conditions the olefin to paraffin ratio can be above 2 for C_3 then diminishing rapidly with chain length [11].

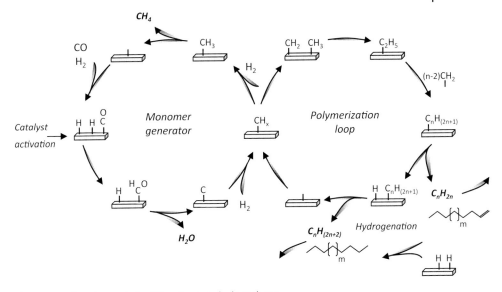

Figure 10.1 Reaction cycle for FT-synthesis to hydrocarbons.

In order to better describe the product selectivities a kinetic model lumping the main product fractions has been found to give a good representation of slurry CSTR data[1]:

$$r_{C5+} = K_1 e^{-(E/RT)} P_{CO} P_{H2} / \{1 + K_2 P_{H2}^a + K_3 P_{CO} + K_4 P_{CO} P_{H2}^a\}^b$$

$$r_{CH4} = K_1 e^{-(E/RT)} P_{CO} P_{H2}^b / \{1 + K_2 P_{H2} + K_3 P_{CO} + K_4 P_{H2O}\}^c$$

$$r_{C2-C4} = K_1 e^{-(E/RT)} P_{CO} P_{H2} / \{1 + K_2 P_{CO}\}^b$$

$$r_{CO2} = K_1 e^{-(E/RT)} P_{CO} P_{H2O}^a / \{1 + K_2 P_{H2} + K_3 P_{CO}\}^b$$

Formation of the main fraction, C_{5+}, will follow the expression for $-r_{CO}$, methane formation will be more sensitive to the hydrogen pressure and will be inhibited by steam, whereas CO_2 will be promoted by steam. The model was developed for a Co/Re on γ-alumina catalyst under conditions with minimal diffusion limitations.

10.1.4
Products

The main products of the LTFT synthesis are linear paraffins. Smaller amounts of oxygenates and olefins can be separated and sold as dedicated products, but normally these are removed during hydroisomerization/hydrocracking to diesel as the main fraction. The upgrading unit can be designed for different products, and an example for diesel optimization is detailed in Section 10.5.5. A special case

---

1) Lian and M. Lysberg, Statoil RD, Trondheim, 2001, personal communication.

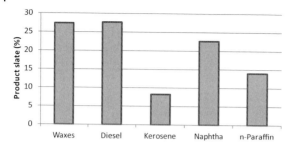

Figure 10.2 Example of XTL product slate.

Table 10.1 Properties of the FT metals.

Property	Nickel	Iron	Cobalt	Ruthenium
FT activity	Moderate	Moderate	Moderate	High
C_{5+} selectivity(α)	Moderate	"High"[a]	High	Very high
Methane formation	High	Moderate	Moderate	Low
WGS activity	Low	High	Low	Low
Deactivation rate	Moderate	High	Moderate	Moderate
Cost (US$/kg)[b]	20	0.4	29	4300

a) Significant oxygenate production in addition.
b) Highly volatile. Prices as of 1 November 2011.

is given by the product slate from the Shell GTL plant in Bintulu, Malaysia, as shown in Figure 10.2 [12]. In this plant focus has been very much on high-end products, like waxes for special applications, naphtha as feedstock for olefin production in steam crackers, n-paraffins in the C_{10}–C_{14} range for production of linear alkyl benzene based detergents, and kerosene. The kerosene can be further upgraded to jet-fuel.

The FT-diesel has a very high cetane number, around 70, and, therefore, excellent ignition properties. It can be used as a blending stock to upgrade low-value refinery gas-oil streams. Appropriate boiling-point range and low-temperature performance without risk of crystallization are obtained by proper hydro-cracking/isomerization of the raw wax product. As the product is free of aromatics and sulfur, exhaust gas particulates and other pollutants will be significantly reduced.

10.1.5
Fischer–Tropsch Metals

Most transition metals are able to hydrogenate CO, including several that favor oxygenates including methanol. However, only four elements are regarded as true FT-catalysts that exhibit sufficient reaction rates [3, 4] and polymerization probabilities, as summarized in Table 10.1.

In general, the activities are moderate with global rates in the range 0.1–1 $kg_{product}/kg_{catalyst}\ h^{-1}$ at typical industrial conditions, meaning that the attainable reaction rates require sizable reactors and catalyst inventory can be a significant cost driver. Within each FT metal family of catalysts the activities and/or selectivities are enhanced by using certain promoters and by securing optimal metal particle dispersion on a porous support. Iron catalysts form a number of oxide and carbide phases during reaction. Cheap fused iron is used by Sasol in fluid-bed operation, whereas precipitated iron is applied in fixed or slurry bed reactors. Methane selectivity remains reasonably low at higher temperatures, and iron is, therefore, used in HTFT synthesis. A specific feature for iron catalysts is the high water-gas-shift (WGS) activity that modifies the H_2/CO ratio according to the reaction

$$CO + H_2O \rightarrow CO_2 + H_2$$

This means that iron can be considered for biomass concepts due to the primary low H_2/CO ratio formed during gasification. If a cobalt catalyst is used the feed composition is adjusted separately by, for example, a dedicated WGS reactor. In both cases CO_2 will have to be removed from the process stream after or before the FT-reactor to avoid accumulation of inerts in recycle streams. It appears that for BTL most concepts have been developed for cobalt catalysts, partly because this catalyst can be operated at higher conversions (enhanced water vapor pressure suppresses the activity of iron catalysts). Nickel gives slightly higher methane yield, particularly in the high temperature range, and is also susceptible to carbonyl formation at elevated pressures. However, nickel might be underrated as an FT metal, and further investigations are suggested. Ruthenium has attractive catalytic properties, however, it is extremely expensive and can be excluded from commercial use.

10.1.6
The Biomass-to-Liquid FT Concept

A conceptual design of a Biomass-to-Liquid (BTL) plant is shown in Figure 10.3. There are three main blocks:

- Biomass harvesting and conditioning for making a suitable feed to the gasification unit. This normally involves milling and/or chipping to a suitable size, drying and, optionally, condensation to a product that more easily can be handled. Pyrolysis to oil and torrefaction to coke are typical processes that can be used.

- The processing block consists of the main XTL steps of making syngas, converting the syngas to paraffinic wax and upgrading the wax to fuels and, optionally, other products.

- After gasification there will be a need for clean-up and purification to very low impurity levels in order not to poison the FT-catalyst. Compared to GTL, this step will be more demanding and there will also be a need to adjust the H_2/CO ratio by, for example, water-gas-shift.

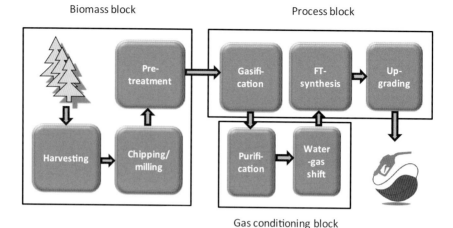

Figure 10.3 The biomass-to-liquids process.

An example of a more detailed process design is described in Section 10.5. There are several factors that will determine the overall design of the BTL plant. The nature of the biomass is important, its composition in terms of cellulose, hemicellulose and lignin, its calorific value and water content, as well as its physical form. Pretreatment of the biomass can be done distributed and must fit with the gasification option selected. Note that there are significant differences between the main gasification options, fluidized-bed and entrained flow, for example, the latter is expected to be more prone to be used at high pressure conditions. As to the FT-synthesis the main choices are between slurry and fixed-bed technology and iron or cobalt catalyst.

10.2
Cobalt Fischer–Tropsch Catalysis

10.2.1
Catalyst Preparation and Activation

A large variety of preparation methods for cobalt based FT-catalysts has been described in the literature, ranging from the original precipitation techniques with kieselguhr, used in the first applications, to advanced gas-phase deposition of cobalt complexes. However, most industrial catalysts are prepared by impregnation on a support. Impregnation methods range from incipient wetness, where the pore volume is impregnated with a cobalt solution, to slurry techniques, including precipitation/deposition where changes in conditions like pH or temperature give a supersaturated solution that preferentially precipitates in the pores.

Sometimes vacuum is used to make sure that the pores will be filled. Additives, typically larger molecular weight oxygenates, can be added to the solution to control the final cobalt particle size and make the distribution of cobalt on the surface more homogeneous [13].

One particular consideration during impregnation is the solubility of the cobalt precursor that may limit the attainable metal loading. For this reason an aqueous cobalt nitrate solution is most convenient. By heating the salt to circa 60 °C it will melt in its own crystal water and, therefore, represent a limiting maximum concentration. Such melt impregnation is possible on an industrial scale while ensuring that all product lines and equipment are temperature controlled. Cobalt loading can be increased by using two or more impregnation steps. Variations in impregnation include using organic solvents and precursors like acetate, acetylacetonate, or carbonate, but industrial manufacture in these directions is not known.

After impregnation the catalyst will be dried and then calcined at a temperature typically around 300 °C. During these steps, first excess solvent will evaporate, then crystal water will be disposed of before the ligands decompose and oxidize, leaving Co_3O_4 on the surface of the support. The details of these steps, for example, heating rates, holding times and composition of the atmosphere, will to a large extent determine both the final cobalt crystal size and how these crystals are dispersed on the surface. Frequently one can observe large aggregates or clusters of cobalt in regions of diameter in the range 50–300 nm, but without necessarily having any detrimental effect on sintering during operation as long as the crystals are sufficiently anchored to the support. It is reasonable to assume that the size of these clusters is linked to droplets that form during the impregnation step. On low surface area supports, like titania, α-alumina and spinels, a more homogeneous cobalt particle distribution is observed.

To obtain an active FT-catalyst, the oxide precursor will be reduced by a hydrogen-rich gas-flow at a temperature between 200 and 450 °C. The degree of reduction will depend on the severity of reduction and, typically, on the support material used. It will be easier to reduce the oxide on a more inert low surface area support than on a material like γ-alumina or high surface area silica. A fraction of unreduced cobalt aluminate, silicate or titanate will occur as a surface layer. Apart from a high degree of reduction, a suitable cobalt particle size should be obtained, as discussed below. This particle size will depend on preparation details and on the formulation, typically on the pore size of the support material [14]. The oxide might also break up into smaller cobalt particles and these can be stabilized by the promoter. Some description of cobalt FT-catalyst formulations and preparations can be found in the reviews by van Steen and Claeys [15] and Khodakov et al. [16].

Note that scale-up to commercial catalyst production is demanding, and little public information on the industrial processes is available. Great care will be taken to obtain a homogeneous catalyst material, but the targeted distribution of cobalt will depend on the actual process. For slurry catalysts with diameters typically in the range 40–120 µm there will be negligible diffusion limitations of the syngas, and full utilization of the available surface area will be desired [11]. On the other hand, a fixed-bed catalyst will be designed as an egg-shell or rim catalyst, where only the outer few hundred micrometers are active. As to the degree of reduction it has been shown by Sasol that the initial value is not critical as the syngas in itself will reduce the catalyst further under the first months of operation [17].

10.2.2
Catalyst Activity

Cobalt in the reduced state is the active component in the H_2/CO conversion. The catalyst activity is largely dependent on the degree of reduction of the metal precursor and the *size of the cobalt particles*, which control the number of catalytically active sites that are available, that is, the dispersion of the metal. It has also been verified that the TOF (turn over frequency, i.e., number of CO insertions per active site and time) is rather constant for Co particles larger than 6–8 nm [18]. The activity falls off rapidly below this threshold and, therefore, more advanced methods for making very high dispersion catalysts have limited relevance for FT-synthesis. As already mentioned, the size of the cobalt particles will depend on several factors, including the cobalt precursor used, the support material, the pore diameter, available surface area, method of impregnation or deposition, drying and calcination conditions, reduction conditions, and so on. It is particularly important not to calcine or reduce the catalyst at too high a temperature to avoid sintering.

In general it can be claimed that to optimize the cobalt particle size is not particularly challenging as long as one is able to control the preparation conditions. However, it is normally regarded necessary to add a metal *promoter* that enhances the degree of reduction and maintains a sufficient dispersion. However, there is no solid evidence that a promoter is able to enhance the reaction rate constants or surface concentration of intermediates. Frequently used promoters are platinum and rhenium. The promoter will add significantly to the cost of the catalyst; Re is today priced at circa 4000 USD/kg and Pt at 55 000 USD/kg. Typical loadings are 0.5 and 0.05 wt%, respectively.

For a given degree of reduction and particle size, the activity per kg catalyst will be proportional to the cobalt *loading*. The loading may vary by as much as 12 to 30 wt% and will be a compromise between several catalyst properties. For instance, a lower surface area and pore volume support will be able to accommodate less cobalt, but might be considerably more attrition resistant. It can also be mentioned that a possible effect of cobalt being in the *fcc* or *hcp* crystallographic phase, with the latter being more active, has been reported [19]. However, studies on the actual configuration of an active cobalt crystal will be needed to be able to correlate activity with atomic arrangement on the surface of a working catalyst. A summary of the most important parameters that dictate initial catalyst activity, as well as selectivity and deactivation can be found in Table 10.2.

Table 10.2 Factors influencing catalytic properties of a freshly reduced catalyst.[a]

Parameter (Increasing-)	Activity	Chain length	Methane selectivity	Deactivation	Comment
Co dispersion	↓	↓	↑	↑	<6–8 nm
Co loading	↑	↔	↔	(↑)	
Promoter	↑	↑	↓	↓	Pt; Re
Support	↑	↑	↔	↓	Critical for attrition
Process parameters					
Temperature	↑	↓	↑	↕	200–280 °C
Pressure	↑	↑	↓	↕	10–40 bar
H_2/CO ratio	↑	↓	↑	↓	0.5–2.5
Conversion	↓	↑	↓	↕	30–70%
Slurry versus fixed-bed reactor					
	↓	↑	↓	↓	

a) Some of the relationships are complex, particularly deactivation, as discussed in the individual sections. ↑: increase; ↓: decrease; ↔: fairly insensitive; ↕: may increase or decrease.

10.2.3
Selectivity

For cobalt FT-catalysis the main selectivity parameter is the polymerization probability α or the more practical C_{5+} fraction. Deviations from the ideal ASF distribution are a low C_2 content and high methane yield. It should be noted that these deviations are not independent of the polymerization probability as C_1 and C_2–C_4 correlate linearly to α or C_{5+} for a large number of FT-catalysts with variations in supports, cobalt particle size, preparation method, and so on [20]. It has also been reported that there can be two or more α-values. Some of the reports, however, are apparently hampered by inadequate analysis. As a first approximation it is sufficient to describe the C_{5+} carbon chain length distribution by a single parameter.

It is imperative to perform investigations of FT-selectivities under controlled process conditions. As can be expected, C_{5+} will decrease with increasing H_2/CO ratio, due to a higher propensity for chain termination by hydrogenation of the growing chain. The effect of temperature is more complex, see the kinetic expressions. There is also a slight increase in C_{5+} with overall pressure. Further, it has been found that the conversion level plays a significant role because the generated water inhibits chain termination. Therefore, comparative catalyst studies must be performed at the same conversion level, a fact often overlooked in the literature. The effect of water on the FT-synthesis has been reviewed recently [21, 22].

There is a significant decrease in C_{5+} selectivity for cobalt particle sizes below 7–8 nm that parallels the drop in activity [13, 18]. One possible explanation is that CO dissociation, and thereby chain growth, becomes more difficult on the cobalt surface geometry of small particles, whereas the termination propensity is unaltered. It appears that there also is a maximum in C_{5+} in the 8–10 nm range [13, 23].

Another catalyst constituent that significantly influences the selectivity is the type of support that is applied. It is well known that that low surface area supports, like titania and α-alumina exhibit much higher C_{5+} values than common high surface silicas or γ-alumina. This difference is evident also when care is taken to make the comparison for constant cobalt particle size [23]. Particularly intriguing is the observation that in the series γ-, δ-, θ-, α-alumina obtained by successive calcination at higher temperatures, the δ- and α-phases have the highest selectivities, and that there is a correlation with the CH_x coverage on the surface [13]. So far it has not been possible to reveal the root cause of the support effect.

Finally, it can be mentioned that the type of promoter influences selectivity. Rhenium consistently improves C_{5+} by around 1% [24]. Platinum, on the other hand, can have an adverse effect, presumably due to an increase in hydrogenation activity.

10.2.4
Activity Loss

A review on deactivation of FT-catalysts has appeared recently [25]. The main causes of deactivation that have been considered are sintering, re-oxidation of cobalt, formation of stable compounds between cobalt and the support, surface reconstruction, formation of carbon species on the cobalt surface, carbidization, and poisoning. The latter is of particular relevance for BTL-FT processes. It should also be mentioned that the chemical environment is challenging with a number of chemical species generated, including significant amounts of water. In addition, the exothermicity of the reaction may lead to hot spots on the catalyst.

No in-depth discussion on this complex topic will be given, but it should be mentioned that there appears to be two main "schools" based on demo slurry operations, one favoring re-oxidation and one favoring poly-carbon formation on the surface as the main long-term deactivation mechanism. Our own experience is in favor of the latter explanation [17], but it should be mentioned that both the catalyst system and the actual process conditions can affect the results. In addition, there will generally be an initial sintering stage. The result is a significant decline in the activity during the first years of operation. In addition comes any catalyst loss and poisoning. All in all, catalyst replacement due to deactivation will contribute significantly to the operational costs of an FT plant. Fortunately, it appears possible to regenerate a deactivated catalyst to a level close to the original activity simply by removing most of the wax, followed by combustion of the remaining carbonaceous species and renewed activation [26]. In fact, indications are that such a procedure will re-disperse the cobalt, possibly even to a level higher than for a freshly reduced catalyst. There is unfortunately no information available on the long-term performance of a regenerated catalyst.

In a study on the effect of biomass relevant impurities there was found [27], by impregnating 400 ppm of the impurity element from nitrate precursors, a poisoning effect in the following order:

$$Na > Ca > K > Mg > P$$

with Mn, Fe and Cl showing minimal effect. The latter is surprising as chlorine causes a 25% reduction in hydrogen chemisorption. The impact of alkali and alkaline earth elements is far stronger than any stoichiometric blocking of surface sites, and presumably is related to the strong electronegativity of the elements. Particular care must be taken to avoid alkali and alkaline earth elements in impregnation fluids and washing water, as well as contaminants in the catalyst support. Sulfur as H_2S or $(CH_3)_2S$ added to the syngas gives a deactivation consistent with stoichiometric blocking of cobalt surface sites as *in situ* measurement of cobalt dispersion by H_2S is consistent with hydrogen-derived dispersion of a fresh catalyst. The effect of ammonia appears to be strongly catalyst dependent, and reports vary from negligible influence to strong negative consequences.

In addition to deactivation, activity loss can stem from loss of catalyst material from the reaction zone due to attrition. When it comes to attrition of catalysts for three-phase slurry bubble column operation, one needs to consider both mechanical abrasion and breakage of catalyst particles, as well as chemical dissolution and any interaction between these mechanisms. Both Sasol [28], GTL.F1 [24], IFP/ENI/Axens [29] and Exxon [30] have developed mechanically robust slurry catalysts. Focus has been both on avoiding chemical attack on the support, as well as on mechanical robustness.

10.2.5
Commercial Formulations

Cobalt FT catalysts can be classified according to the supports and promoters used, and in Table 10.3 some commercial type catalysts have been listed. Certainly, the actual catalyst formulation of the technology providers is proprietary, and the catalysts listed are a selection based partly on presentations and partly on most recent trends in the patent literature. As both the process conditions and reactor environment vary considerably in the reports, it is extremely challenging to compare reported activities and selectivities.

All catalyst systems in Table 10.3 [37], except the Shell fixed-bed catalyst, are designed for slurry reactor operations. It appears that Shell favors a titania-based support over their previous zirconia-modified silica system. Promoters are either Mn or V, the latter claimed to lower CO_2 yield [38]. Titania has relatively large pores and moderate surface areas, but is known to give high selectivities to C_{5+}

Table 10.3 Indicative formulation of commercial type cobalt catalysts.[a]

Technology provider	Support/modifier	Promoter	Verification scale (bbl/d)	Reference
Sasol	γ-Alumina/Si (TEOS)	Platinum	16 000	[32]
Shell	Titania	Mn; V	6 000	[33]
GTL.F1	Ni-aluminate/α-Alumina	Rhenium	1 000	[34]
ENI/IFP/Axens	γ-Alumina/SiO$_2$; spinel		20	[30]
Nippon oil	Silica/Zirconia	Ruthenium	500	[35]
Syntroleum	γ-Alumina/Si (TEOS); La	Ruthenium	80	[36]
Exxon Mobil	Titania/γ-Alumina	Rhenium	200	[31]
Conoco-Phillips	γ-Alumina/Boron	Ru/Pt/Re	400	[37]

a) Deduced from open literature and patents. Actual commercial formulation may vary.

products. Fixed-bed catalysts contain binders or flow improvers, like citric acid, added prior to the forming step. In addition to Shell, it should be mentioned that BP are promoting their fixed-bed technology and claim that a CO_2 resistant support is vital [39].

The platinum-promoted Sasol catalyst on γ-alumina (Purolox SCCa-2/150 or -5/150: pore volume 0.5 ml g^{-1}; surface area 150 m^2 g^{-1}) is modified by impregnation with tetra-ethoxy-silane (TEOS) followed by calcination to give a surface concentration of about 2.5 Si atoms/nm^2. The GTL.F1 catalyst is based on a larger pore diameter γ-alumina that has been modified by nickel and high temperature fired to give a nickel-aluminate (spinel)/α-alumina mixture. The pore properties resemble titania-based supports, but with very high attrition resistance. Also the ENI/IFP catalyst has a support of modified γ-alumina, but probably strengthened by silanation and calcination, giving a final SiO$_2$ content of 6–7 wt%. Other modification methods have also been described in earlier patents, including formation of spinel compounds. It is unclear whether the formulation contains a reduction or other type of promoter. In their slurry developments, it appears that Nippon has reverted to the previous Shell support formulation for fixed-bed, but using ruthenium as a promoter. No information on attrition resistance has been revealed. Syntroleum has been using a catalyst similar to Sasol's, but again with ruthenium-promoted cobalt. Both ExxonMobil, who pioneered titania as a support with alumina as binder, and Conoco-Phillips have terminated their developments in FT-technology.

Caution should be taken when comparing performance. For example, selectivities will vary with the actual gas composition, including the conversion level. Activities are generally lower for a given catalyst operated in slurry compared to fixed-bed in spite of limited apparent diffusion limitations. The origin of this effect is not understood. Particularly low activities have been reported by Nippon and ENI/IFP, indicating large cobalt crystals in the working catalyst.

10.3
Fischer–Tropsch Reactors

10.3.1
Reactor Selection

Apart from the choice of gasification technology, selection of the type of FT-reactor is the main factor influencing the overall design of a BTL process. Comparing BTL to GTL, the main difference will be in the size of the plant. In contrast to a world scale 30 000–160 000 bpd GTL plant, the restricted biomass availability and logistics of a BTL project will probably limit the size to 2000–10 000 bpd. Still, both fixed and slurry type reactors will be possible, but the favorable scalability of the slurry bubble column reactor will not be as important. A comparison of properties of the main reactor types for LTFT is given in Table 10.4.

10.3.2
Tubular Fixed-Bed

Due to the necessity to control the heat evolved during the reaction, the fixed-bed FT reactor is designed as a multi-tube heat-exchange type of reactor, where catalyst pellets are loaded into the tube bundles and the shell contains boiling water. It should be noted, however, that for HTFT synthesis both circulating fluidized-bed and fixed-fluidized bed reactors are operated (Sasol Synthol and advanced Synthol reactors). In this type of reactor there will be a certain temperature profile. In order to minimize this effect several measures are taken. Once-through CO conversion will be limited to 30–35%. The tube diameter is typically 3–5 cm and the size of the catalyst pellets or extrudates will be relatively large for the diameter, in the range of 1–3 mm. It has been shown that above circa 200 µm particle size, the

Table 10.4 Estimated typical properties for different reactor types for low-temperature FT-synthesis.[a]

Reactor	Conversion per pass (CO %)	Max capacity per reactor (bbl/day)	Characteristics
Tubular fixed-bed	30–35	6,000	≤30 000 tubes with catalysts pellets or extrudates
Slurry bubble column	55–65	25,000	Internal heat exchanger and product filter
Micro-channel	65–75	500	Metal block with <2 mm diameter channels

a) Based on open literature and patents for commercial LTFT synthesis with cobalt catalysts.

higher effective H_2/CO ratio in the inner part of the pellets will significantly reduce the C_{5+} yield [11]. Therefore, an egg-shell type of catalyst design is preferred, where only the outer parts are impregnated with active metal. The design will be a compromise between diffusion limitations and pressure drop. The latter will limit the superficial gas velocity that can be applied.

The H_2/CO ratio will depend on several factors, but generally it is expected that the make-up gas will have a ratio slightly below 2. With recycle in the FT-section the feed ratio to the reactor will be significantly lower and there will be a gradual reduction from the inlet to the outlet. Compared to a slurry reactor, the average gas composition will be richer in hydrogen for a given feed ratio and, together with less efficient temperature control, the operating temperature is lower. A consequence is lower reaction rates, but this will at least in part be compensated by a lower average partial pressure of produced water, and thereby a higher syngas pressure.

A distinct advantage of the tubular fixed-bed reactor is a well-proven commercial design. Several tens of thousands of tubes can be incorporated within the reactor shell. Scale-up is comparatively easy, and optimization can be done in a single tube laboratory reactor. Operational experiences as to issues with possible fouling or attrition, and resulting difficulties with unloading the tubes, are not known, but it can be expected that an experienced operator is able to control these factors. Minimizing catalyst deactivation or being able to perform *in situ* regeneration is critical in order to reduce catalyst consumption and avoid an extensive unloading–reloading sequence. With a suitable catalyst formulation, catalyst attrition will be low. The liquid product is inherently separated from the catalyst and any need for removing residual particles and metal components will be low.

10.3.3
Slurry Bubble Column

A slurry bubble column reactor is illustrated in Figure 10.4. The catalyst particles are suspended in the liquid hydrocarbon product of the FT-process and synthesis gas is bubbled through the slurry. Depending on the density of the catalyst particles, their diameter and the superficial gas velocity, there is likely to be a profile of solid concentration diminishing from bottom to top of the reactor. Gaseous components will leave from the top of the reactor. The higher boiling products will have to be removed from the reactor as liquid, and separated from the catalyst/slurry. Several methods for this purpose have been patented, both *in situ* and *ex situ* techniques. Broadly they can be classified as employing filters, settling devices, magnetic separation and hydrocyclones. Polishing of the produced wax down to ppm level of solids will make sure that no clogging or interference with the hydrotreating columns occurs.

Settling of the catalyst should be avoided as overheating, and consequently catalyst deactivation will occur. The FT reactor operates preferentially in the churn turbulent flow regime for best distribution of catalyst particles as well as minimizing mass & heat transport restrictions. In the churn turbulent flow regime there

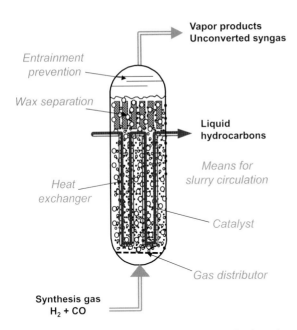

Figure 10.4 Slurry bubble column with critical technology elements.

is a mixture of smaller and larger bubbles that undergo frequent break-up and coalescence. It is beyond the scope of this chapter to discuss operation of the slurry bubble column reactor in detail, but further information can be found in the book on FT-technology by Steinberg and Dry [40]. Some important factors for laboratory scale reactors are to avoid slug flow at small column diameters or high solids concentrations on the one hand, and homogeneous flow at low gas velocities on the other. Other important factors to consider are the physical properties of the suspension and design of the gas distributor.

The flow pattern may also be influenced by the reactor internals. Apart from the gas distributor, the most important is the internal heat exchanger. The latter is removing reaction heat by boiling water in separate heat exchanger tubes. In this way a near isothermal reactor can be obtained. A demister or hydrocyclones can be located in the reactor head to catch liquid droplets that follow the gas stream. Note that mechanical agitation is not normally needed.

To be able to operate a slurry bubble column in a safe and optimal way, it is important to have an elaborate reactor model. Important output of such a model will be axial and radial particle concentrations, pressure drop, gas and solids hold-up, gas bubble size and distribution, gas/liquid interphase, wall heat transfer coefficients, as well as the reactant and product concentrations/pressures, reaction rates and product distribution. A gas/liquid/particle film mass transport and diffusion model is normally not needed due to the slow nature of the FT-reaction, the dynamics of the flow regime and the small size of the catalyst particles being in the range 20–200 μm.

The reason for selection of a slurry bubble column reactor for FT-synthesis is a high heat transfer coefficient and isothermal conditions, as well as economy of scale. Continuous catalyst regeneration of a slip stream is a viable option. Main development areas are in minimizing catalyst particle attrition, and filtration of the products. Another aspect is efficient liquid and gas back-mixing. This leads to high selectivity as the high exit water concentration is beneficial, but on the other hand the reactant concentration will be lower than the average of a fixed-bed reactor, resulting in comparatively lower global rates. The conversion is expected to be significantly higher than for fixed-bed. Conversion will be limited by a feasible height of the reactor, but there will also be an upper conversion limit above which the catalyst may partially re-oxidize. High conversion, and thereby a significant water vapor pressure, will induce an unfavorable water-gas-shift activity. Naturally, extensive recycle of syngas in the FT-section of the plant is necessary to obtain a high overall CO conversion.

10.4
Biomass Pretreatment and Gasification

10.4.1
Pretreatment of Biomass

Size reduction by chipping and/or milling of the biomass will normally be needed. Furthermore, drying of the biomass down to 7–20 wt% moisture is carried out for better operation and process efficiency. A dry biomass will reduce the energy demand of the gasification step, but will simultaneously lower the hydrogen content of the product gas. Size reduction and drying technologies will not be discussed in detail except for noting the need for very small particles, typically in the 0.1 to 1 mm range, to be able to feed entrained flow gasifiers. Milling of wood to this size is very energy demanding and, therefore, some further pre-processing of the biomass is preferred. Most relevant pretreatment technologies are torrefaction [41, 42] and fast pyrolysis [43].

By torrefaction the biomass is roasted under inert or oxygen-lean conditions at 200–300 °C and atmospheric pressure. The biomass particle heating rate is relatively low at <50 °C min^{-1} and the residence time is typically less than 1 h. Torrefied biomass has a coal-like texture that makes it easily pulverized. Moreover, the inert surface of the particles reduces their agglomeration during high pressure feeding to the gasifier (e.g., 40 bar). Such agglomeration may lead to plugging of the lock-hopper and to instabilities in the fluidization regime (a fluidization vessel is needed as part of the feed system). Typically the electricity input for size reduction to 100 μm torrefied (1% moisture) particles is 0.01 kW$_e$ per kW$_{th}$ energy content, and this can be compared to circa 0.04 kW$_e$ for milling dried wood chips [44]. The volumetric energy content will, by the torrefaction, increase from 10 to 16 GJ m^{-3}. Typically 70% of the mass after torrefaction is retained as solid product, but this contains as much as 85–90% of the energy content of biomass. The rest will be in the form of torrefaction gases (a mixture of H_2O, CO_2, light alcohols and acids [41]). Note that sulfur, chlorine and alkali content will not be reduced significantly and will thus be carried over to the gasifier. The milled product is highly reactive (auto-ignition) and must be kept under inert conditions. Torrefaction is a process that can be integrated on site with the rest of the BTL plant or located near the forest as distributed "reactors", thereby reducing the cost of biomass transportation. However, efficient local utilization of the torrefaction off-gas can be a challenge.

In spite of its attractiveness, torrefaction has not been commercialized at large scale. Pechiney pioneered operation of a 12 000 t a^{-1} torrefied wood demonstration plant in France during the 1980s for metallurgical application [44]. During 2010/2011 three projects came on stream in Europe, the largest being the Topell Energy plant in the Netherlands of 60 000 t a^{-1} capacity using a "torbed" technology from Torftech (UK) [45]. Other technologies include oscillating belt and rotary drum reactors. Torrefaction as a pretreatment process for gasification must be regarded as in the early stage of development. Potential issues are lack of feedstock flexibility as agricultural residues appear more challenging than wood chips, and process operation, integration and product quality of the solid residue must be validated. Safety during storage and transport must also be carefully assessed.

By pyrolysis the biomass is thermally degraded under inert conditions or using under-stoichiometric amounts of oxygen. The temperature is 500–600 °C and elevated pressures can be applied. This treatment leads to non-condensable gases like H_2O, CO, CO_2 and CH_4, pyrolysis oil (also called bio-oil), and char. Practically all the biomass ash components will be retained in the char. The fraction of condensable gas (bio-oil) will vary considerably with operating conditions and the biomass quality, ranging from 70 wt% yield for dry wood to 50 wt% for dry straw. The high ash content of the latter will favor catalytic decomposition of many organics to gas. At the Metso R&D Center in Tampere, Finland, a consortium of Metso, UPM, Fortum and VTT has produced more than 70 t of bio-oil from sawdust and forest residues in campaigns up to 200 h [46]. There are several variations of the technology, but the most convenient one is fast pyrolysis that favors high yield of bio-oil. Bio-oil and char could be mixed into a slurry with typically 90% of the energy of the biomass feed [47]. Feeding this slurry to the gasifier is conventional technology. Other design options include feeding pyrolysis gas and char (after grinding) separately to the gasifier, as was practiced by Choren [48]. However, this design leads to direct integration between pyrolysis and gasification, which increases process complexity. In addition, processes where only the bio-oil is sent to the gasifier while the char is used internally in the pyrolysis section as a heat source have been foreseen by BTG, and Ensyn [49, 50]. Note that fast pyrolysis, like torrefaction, has not been demonstrated at commercial scale.

ECN has calculated the energy efficiency (η) for production of syngas from wood for these different pretreatment/feeding options [51]:

- Wood is pulverized to 100 μm and a coal-like feeding system is used (η: 59%)
- Use of torrefaction as pretreatment (η: 74 to 75%)
- Use of flash pyrolysis as pretreatment (η: 76%)
- Pre-gasification at low temperature using a high pressure fluidized bed reactor (η: 84 to 85%)
- Wood is pulverized to 1 mm and a dedicated biomass feeding system is used (η: 80 to 84%)

The efficiencies are for syngas at 40 bar and $H_2/CO = 2$. From their results it can be observed that the options including pulverization to 1 mm (assumed 99% conversion) and pre-gasification at low temperatures are most energy efficient. However, a higher level process simulation with elaborate recycling may alter these values. Further, the technology readiness of the options must be taken into account, and neither a high-pressure fluidized-bed pretreatment nor feeding of larger biomass particles are state-of-the art.

A suggested alternative is to have a fluidized-bed gasifier operated at moderate temperatures up-front of the entrained bed. This solution will give improved overall energy efficiency, but introduces several process integration challenges [51], including matching pressures and flows. Hydrothermal carbonization has also been reported as a pretreatment technology. Such treatment consists in decomposition and dehydration of the biomass, which combined with water drainage and drying results in a coal-like product [52].

10.4.2
Biomass Gasification

The main reactor configurations for gasification of biomass are shown in Figure 10.5. It is emphasized that there are a number of modifications and variations to these main types of gasifiers.

As indicated in Figure 10.5, the fixed-bed reactor illustrates the reactions taking place during gasification: drying, pyrolysis/devolatilization, combustion (oxidation) and reduction. In the devolatilization and pyrolysis section volatile organic material including tar is formed. Depending on the design of the reactor a smaller or larger fraction of the organic vapor will be combusted. Important design parameters are the actual oxygen inlet position and the flow pattern. Preferably, as little tar as possible should leave the reactor. In the gas equilibration section, the following reduction reactions will take place:

		ΔH (kJ mol^{-1})
$C + H_2O \rightarrow CO + H_2$	Water gas reaction	+118
$C + CO_2 \rightarrow 2\,CO$	Boudouard reaction	+160
$CO + H_2O \rightarrow CO_2 + H_2$	Water-gas-shift	−42
$CO + 3\,H_2 \rightarrow CH_4 + H_2O$	Methanation	−206
$C_nH_m + n\,H_2O \rightarrow n\,CO + (n + m/2)\,H_2$	Steam reforming of higher hydrocarbons	
$C_nH_m + n\,CO_2 \rightarrow 2n\,CO + m/2\,H_2$	Dry reforming of higher hydrocarbons	

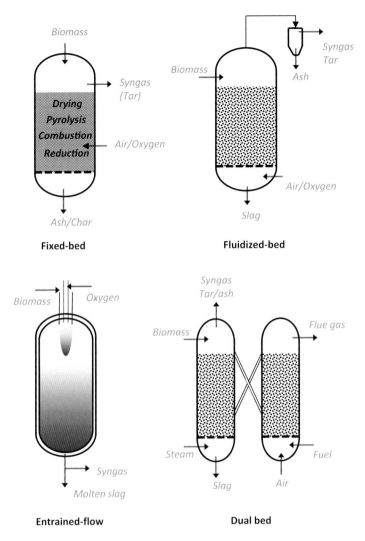

Figure 10.5 Reactor types for gasification of biomass.

As these reactions proceed, the net effect is a reduction in the temperature obtained by the combustion. Sufficient oxygen needs to be supplied for total conversion of char. Apart from the feed itself, the gas composition will depend on the equilibrium composition at the exit and the approach to this equilibrium. To a large extent the gas composition is determined by the water-gas-shift reaction. As a first approximation it can be assumed that during cooling of the product gas this reaction is in equilibrium around 700 °C and that the composition will be frozen at lower temperatures due to slow reaction rates, unless a catalyst is used.

Table 10.5 Properties of biomass gasification reactors.

	Fixed-bed	Fluidized-bed/ Circulating-bed	Dual-bed	Entrained-flow	Plasma
Feed properties	Less critical	Less critical	Less critical	Very fine particles/liquid	Very flexible
Oxidation agent	Air/O_2	Air or steam/oxygen	Air	Oxygen	None
Temperature (°C)	750–900	750–900	750–900	1200–1800	2000–5000
Pressure (bar)	1–5	1–10	1–10	10–60	$\geq$1–(60)
Tars	High to low	Intermediate	Intermediate	Zero	Zero
Size ($MW_{th\text{-biomass}}$)	<20	<100	<50	>100	>100

Methanation will only be in equilibrium at temperatures >1200 °C. Fixed-bed reactors can be designed as updraft (counter-current) or downdraft (co-current) flow of biomass versus product gas.

Fluidized-bed reactors operate at higher gas velocities and normally contain an inert fluidization agent, typically sand. In the circulated fluidized bed design, sand particles, char and ash can be transported out of the top of the reactor and reintroduced to the bottom part. The dual fluidized-bed reactor is a further development where the combustion reaction is performed in a dedicated reactor, thereby eliminating the need to use pure oxygen. Indeed, for FT-applications it is optimal not to carry excessive nitrogen through the synthesis sections.

Whereas the fixed and fluidized-bed reactors operate in the temperature range 750–900 °C and at moderate pressures, the entrained-flow reactors can reach 1500 °C or higher at elevated pressures of typically 40 bar. At these high temperatures all tar components and char will be eliminated and the methane yield is negligible. In the slagging type of gasifiers, ash will melt to a slag that is removed from the bottom of the reactor.

Some further details of the main reactor types are given below, including the more recently developed plasma reactor. General properties for different reactor types are given in Table 10.5. Some values, such as size and pressure ranges, are estimated from general reactor properties as experience from biomass applications does not exist.

10.4.3
Entrained-Flow Gasifier

Slagging gasifiers are operated at pressures typically between 10 and 60 bar, and high temperatures, typically between 1200 and 1800 °C. Under these conditions the gasification process takes place rapidly with high carbon conversion, and the product gas is low in tar and methane content. The entrained-flow slagging gasifier is the preferred option for large capacity BTL. However, experimental work has shown that at temperatures between 1300 and 1500 °C the biomass ash does not always melt completely [51], which is for example the case for willow. Therefore, in order to obtain a good melt, addition of a fluxing material, like silica, is necessary, resulting in slag (ash + silica) mass flows of at least 6% of the fuel flow.

Challenges in feeding the biomass encompass the need for small particle size, pressurizing and continuous operation. Pulverization of biomass requires dedicated milling equipment, especially since milling to particle sizes as low as 100 μm might be necessary if coal-like feeding systems are used for pressurizing (lock-hoppers) and injection (fluidization vessels). Pulverization of biomass to 100 μm will require of the order of 4% of the energy content in biomass. Several pretreatment options designed to minimize this problem have been summarized above, notably torrefaction and pyrolysis. It should be noted that use of a dedicated biomass feeding system which allows feeding of larger particles under pressure will be a distinct advantage, as described above, and different solutions are under development, for example, high-pressure pistons feeders.

10.4.4
Fluidized-Bed Gasifier

The bed material in a fluidized-bed gasifier behaves as a highly turbulent fluid leading to fast and efficient mixing of the fuel with the gasification agent. Typical reactor temperatures are 750 to 900 °C, resulting in a product gas with high amounts of methane, as well as significant concentrations of tars and light hydrocarbons. Operational pressure is normally atmospheric or moderately pressurized and scale of operation is typically 50 MW$_{th}$, significantly smaller than for entrained-flow gasifiers. These conditions are suitable for power generation, but less optimal for BTL. It is mainly the restricted temperature required to avoid melting of ash and resulting high methane yield that limits the suitability.

There exist two main types of fluidized bed gasifiers: stationary bubbling-bed and circulating-bed. In stationary systems, the gasification agent enters the reactor at the bottom with sufficient velocity to set the bed material in bubbling fluidized state, but low enough to minimize biomass carry-over to the top of the reactor. A cyclone is used to separate ash and char from the product gas leaving the reactor. The main advantages of this system are the good mixing and uniform conditions in the reactor, together with high flexibility as to changes in humidity and ash content. On the other hand, there will be lower conversion than for high-temperature systems, and the high concentration of tar requires appropriate gas cleaning. In addition, this system is not suitable for biomass with low ash melting temperature since this would result in lumping of the bed material with ash [53].

A circulating system is similar to the stationary with the difference that now the bed material is carried away from the reactor, separated in a cyclone, and then recycled back to the reactor. One benefit of the circulating system is easier scale-up capabilities due to higher throughput in each individual section. An interesting variation of the circulating fluidized-bed reactor is using two fluidized-beds, where the circulating bed material is used as heat carrier. In this set-up the gasification is performed in one fluidized-bed and the bed material is transferred into the second bed, where the necessary gasification heat is generated by combustion of a part of the processed material, like char, or of external fuel. The heated-up bed material is recycled back to the gasification bed. The main advantage of this system is that gasification and combustion can be optimized separately, and it is thus feasible to use air instead of pure oxygen, which can be a significant cost saving [53].

10.4.5
Plasma Gasifier

In a plasma gasifier, typically generated using electricity, biomass is decomposed at temperatures above 2000 °C into its basic atomic components. The temperature is calculated electron temperatures. For this purpose, electricity is fed to a torch, which has two electrodes, creating an arc. Plasma gasification does not combust the biomass as incinerators do. It converts the organic constituents into a fuel gas that still contains the entire chemical and heat energy of the feed, and it converts the inorganic parts into a molten salt that after the reactor solidifies to an inert vitrified glass. There are no tars [54].

The plasma reactor can process large amounts of any type of biomass, including high moisture levels. In addition, the feed system is also simpler than for an entrained flow gasifier due to operation at slightly negative pressure. The downside is that compressors are needed to carry the gas out from the reactor and attain the operating pressure of the FT-unit. Each reactor can typically process up to 20 tons per hour (tph) compared to 3 tph for typical fluidized-bed gasifiers [54]. Pressurized systems are under development, but are at a very early stage.

Plasma gasification exhibits several advantages, including high total carbon conversion, nearly hydrocarbon-free syngas, emission reduction (you still have CO_2, S-, N-, and fly ash will still be present), fuel flexibility and scaling-up characteristics. On the downside are high capital costs, unproven stability and reliability, and the use of high value energy for the conversion process. Sufficient documentation of mass and energy balances of operating systems is still lacking. However, technology players claim that plasma is becoming more and more competitive, especially for municipal solid waste.

10.4.6
Gasification Pilot and Demonstration Projects

A summary of pilot and demonstration projects for biomass gasification is given in Table 10.6. Projects that have not reached the construction stage have not been included. This includes, among others, the GAYA project coordinated by GDF Suez using a dual-bed gasifier, the catalytic dual-bed concept by RTI/University of Utah, and the ClearFuels concept that resembles an indirectly-fired steam

10.4 Biomass Pretreatment and Gasification

Table 10.6 Biomass gasification pilot and demonstration projects relevant for BTL.

Technology provider/operator	Location	Feed/pretreatment	Gasification conditions	Capacity (feed)	Status
Entrained-flow gasifiers					
Chemrec AB/Weyerhauser	North Carolina USA	Black liquor	Air, atm.	300 tpd	1996–present Commercial
Chemrec AB	Piteå Sweden		Oxygen, HP 1050 °C	20 tpd	2005–present
Choren	Freiberg Germany	Wood chips++ LT gasification	Chemical quench	10.5 tph 35 MW_{th}	Operated 2010–2011
KTI/Lurgi	Karlsruhe Germany	Straw, wood/pyrolysis	1400 °C	0.5 tph	Construction (op. 2012)
Uhde Total	Mardyck/Dunkirk France	Wood Torrefaction	30–40 bar 1200–1600 °C 3–5 s		Demo start-up in 2012/2013
Fluidized-bed gasifiers					
Vienna Univ. of Technology	Güssing Austria	Wood chips, forest residues, municipal solid waste and others Chipping and drying	Dual bed Steam/air, atm. 850 °C	2.2 tph (8 MW_{th})	Commercial since 2002 4 plants
Cutec	Clausthal-Zellerfeld		Circulating bed Air/O_2, atm. ≤950 °C	2–20 $m^3 h^{-1}$	1990–present
ECN	Petten Netherlands		Dual bed Steam/air, atm.	1800 tpa	2008–present
Enerkem	Westbury Canada		Bubbling bed <10 bar, 700 °C	12 000 tpa	2009–present 100.000 tpa constr.
GTI/AndritzCarbona	Chicago Illinois		Bubbling bed Air/O_2, HP	2–5 MW_{th} (100–200 MW_{th})	present Com. plant in Des Plains is on hold
Foster Wheeler Neste/Stora Enso	Varkaus Finland		Circulating bed 850–900 °C	–	2009–present
Foster Wheeler Sydkraft	Värnamo Sweden		18 bar 950–1000 °C	6 MW_{el} + 9 MW_{th} (product)	1997–2002

(Continued)

Table 10.6 (Continued)

Technology provider/operator	Location	Feed/pretreatment	Gasification conditions	Capacity (feed)	Status
Moving-bed					
Blue Tower	India Japan Germany	Molasses, MSW, wood ++	Steam 900 °C	3 + 30 MW$_{th}$ 3 MW$_{th}$ 13 MW$_{th}$	Com. + constr. Commissioning Construction
Plasma gasifiers					
Westinghouse	Utashinai Japan	MSW+ 4% coke	Moving-bed Atm., ~1600 °C	300–500 tpd	2003–present
Westinghouse Coskata	Madison Pennsylvania	Wood++ Very flexible	Moving-bed Atm., ~1600 °C	–	2009–present
Steam reforming					
Range Fuels	Soperton Georgia	Mixed feed Pyrolysis (devolatilization)	HP	300 000 tpa ethanol	2010–present

HP = high pressure; LT = low temperature; GTI = Gas Technology Institute; atm = atmospheric pressure; tpa = tons per annum; tpd = tons per day.

reformer. Note also that there are numerous CHP (combined heat and power) plants that are less relevant for the BTL-FT setting.

10.4.6.1 Entrained-Flow Gasifiers

Entrained-flow gasifiers are well known for coal feeds, including the major technology companies and suppliers ConocoPhillips, Lurgi, Shell, Uhde, Siemens, GE, Chevron-Texaco, KBR and Mitsubishi. It is interesting that the trend goes toward preference for top-fired oxygen-blown reactors with single outlet for gas and slag [55]. Some experience exists for co-feeding coal and biomass (up to 30 wt%), but the experience with dedicated biomass entrained-flow gasifiers is scarce.

The Chemrec technology is based on gasifying black liquor from the pulp and paper industry and has been demonstrated on a larger scale at atmospheric pressure with air. More recently, the technology was developed for oxygen and high pressure. Black liquor and oxygen, fed to the top of the reactor, have a residence time of about 5 s at a temperature slightly above 1000 °C. After water quench, the raw syngas is cooled in a counter-current condenser where the 70% water content of the gas is effective in scrubbing off fine particulate material [56].

The Choren Carbo-V concept uses pyrolysis as pretreatment and the gasification reactor is made of a combustion down-flow chamber and a reduction up-flow chamber. Biomass is pre-dried to moisture content of 15–20% before it enters a low-temperature pyrolysis reactor with air or oxygen operated at 400–500 °C. The product gas containing tars is oxidized by oxygen in the entrained-flow combustion chamber leading to high temperatures (1400 °C). Slag is removed from the bottom of the reactor. The combustion gases are mixed with char from the pyrolysis step in a separate up-flow chamber where the char is converted to syngas, resulting in a temperature reduction to circa 900 °C. This chemical quenching concept results in a tar-free raw syngas [48].

There is also a first pyrolysis step in the Bioliq technology of Karlsruhe Institute of Technology (KTI), but this is operated under inert conditions at 500 °C, producing a thick oily liquid containing solid coke particles. The produced "biosyncrude" is gasified by oxygen in the entrained-flow unit at 1400 °C [57].

The Uhde entrained-flow gasifier PRENFLO-PDQ is adapted to a variety of biomass types by installation of torrefaction pretreatment. After internal water quench the outlet temperature is 200–250 °C. Carbon conversion is reported as >99% [58]. The technology has been extensively tested in a coal-based IGCC unit in Puertollano, Spain.

10.4.6.2 Fluidized-Bed Gasifiers

A dual fluidized-bed with olivine as heat transport medium is used in the Güssing bioenergy plant. Wood chips are gasified with steam at 850 °C in one chamber. Only 10% of the steam is used for gasification, the rest is needed for fluidization purposes. Part of the generated char, together with supplementary fuel, is combusted in the other chamber to reach a temperature of 920 °C. Note that a nitrogen-free syngas can be reached without using pure oxygen. Total efficiency of electricity and heat to the grid is 75% [53]. Recent lab scale experiments indicate that baffles in the gasifiers increase residence time, and thereby secure a tar-free product. The ECN pilot plant is similar to the Güssing one in that a dual fluid-bed is applied with air combusting char in a separate vessel. Sand is used as heat transport medium. However, the gasification takes place in a riser placed inside a bubbling-bed combustor, thereby securing excellent heat utilization [59].

Both Enerkem and GTI-Carbona are using a bubbling fluidized-bed [60]. At the Gas Technology Institute (GTI) in Chicago a pilot facility is used as testing platform for gasification and clean-up in a project with UPM Kymmene and Andritz [61]. The unit has been modified to operate on biomass instead of coal, it is oxygen blown at high pressure and coupled to tar reforming and gas clean-up. Actual operating temperature and pressure have not been released.

The Blue Tower concept can be considered as a further development of the dual-bed concept [62]. A solid heat carrier is heated by combustion of char and transported through a three-stage moving-bed arrangement. The biomass is first pyrolyzed by steam into a gas phase and char, the pyrolysis gas is then further reformed at 950 °C by the hot heat carrier to give a high hydrogen concentration product. The concept aim is a decentralized gasification market, and it is unclear whether the concept offers flexibility for scale-up above the 10–30 MW_{th} range.

10.4.6.3 Plasma Gasifiers

In the down-flow moving bed Coscata plant, heat to the gasification is provided by Westinghouse plasma torches. The torches are located in the bottom of the reactor and are therefore firing into a coke-rich bed. Tars are cracked in the upper part of the reactor, and carbon conversion is claimed to be 100%. Any kind of biomass can be used with practically no pretreatment [63]. The reactor design is somewhat different in the Solena Plasma Gasification Vitrification technology where the temperature is said to be around 5000 °C. Several large-scale BTL projects based on the latter technology have been announced, but no final project sanction appears to have been obtained so far.

10.4.6.4 Steam Reforming

Range Fuels has a significant commercial plant where the mixed biomass first goes through a pyrolysis-like treatment called devolatilization, before the gas is further converted by steam reforming to syngas. The facility was recently purchased by the bio-ethanol company LanzaTech [64]. It should also be mentioned that a novel development by Cortus of their WoodRoll technology involves low-pressure steam reforming of char from pyrolysis pretreatment of the biomass. Combustion of the pyrolysis gas is used for heating the reformer as well as for drying. A 500 kW pilot unit started operation in 2011 in Stockholm. It has been reported that no hydrocarbon byproducts are formed during operation at 1120 °C, thereby significantly simplifying gas clean-up. Due to advanced heat integration a thermal efficiency to syngas of 80% is claimed [65]. In another development Primus Green Energy is gasifying biomass by superheated steam [66].

10.4.7
Syngas Composition

The lower heating value of syngas from gasification of biomass varies from 4 to 18 MJ Nm^{-3}, figures that can be compared to circa 36 MJ Nm^{-3} for natural gas. Trends from entrained-flow to fluidized-bed gasification follow the expected trend of lowering the temperature; that is higher methane, hydrocarbons, tars, and higher hydrogen to carbon monoxide ratio. Using steam for gasification naturally increases the H_2/CO ratio significantly. Note that there are considerable variations between the individual technologies and tests summarized in Table 10.7.

The data from Güssing are after removal of tar that is recycled to the combustion chamber. The gas composition fits with water-gas-shift reaction being in equilibrium at 850 °C, but the methane yield is significantly higher than the equilibrium value.

The hydrogen to carbon monoxide ratio after gasification will be widely different, depending on the raw material in hand and on the gasification technology. Compared to the circa 2 ratio for GTL, a BTL raw syngas can have H_2/CO in the vicinity of 1.0 or lower. To optimize the FT product yield, the optimal H_2/CO ratio in the make-up syngas fed to the FT section with cobalt catalyst should be slightly below the stoichiometric usage ratio in order to maximize C_{5+} yield. The ratio will be further reduced during FT-reaction and as unconverted syngas is recycled internally in the FT-section. It is a frequent misconception in the open literature that the H_2/CO ratio should be 2.1 to match the overall utilization ratio.

Table 10.7 Examples of raw dry syngas compositions (vol%).

Technology	Conditions	CO	H_2	H_2/CO	CO_2	CH_4	C_2-C_5	H_2S	N_2	Comment/reference
Entrained-flow										
Chemrec	Air	8–12	10–15		15–17	0.2–1			55–65	[57, 68]
	O_2	30	43	1.4	27	1		1.4	–	25–40 ppm COS
Choren	O_2	35–40	35–40		Balance	Small			Small	[69]
KTI/Lurgi	Enr.air	50	27	0.54	14				6	[70]
UHDE		>85 (CO + H_2)			6–8	<0.1				
Fluidized-bed										
Cutec	Enr.air	22	31.6	1.4	33.6	7.9	1.8		3	[71]
ECN	Air	29	31	1.1	20	14	5			[60] 45 g m^{-3} tar
Enerkem	O_2	20–24	20–24		30–35	8–12	10–20			[72]
Foster Wheeler	Air	15.5–17.5	10–12		14–17	5–7			45–50	Värnamo plant [73]
Güssing	15 wt% H_2O	29.1	37.7	1.30	19.6	10.4	3.2		0.5–0.7	Tar free [54]
Fixed-bed or moving bed										
Cortus	Steam	14.9	70.3	4.7	22.9	1.8				[74]
		16.2	60.6	3.6	21.8	1.4				
Blue tower	Steam	20	50	2.5	25	5				[75]
Plasma										
Westinghouse	23 wt% H_2O	67.5	26.6	0.39	5.9					N_2 free basis [76]

10.5
Biomass-to-Liquids Process Concepts

10.5.1
Example of Process Flow-Sheet

A BTL flow-sheet for a concept with entrained-flow gasification and pyrolysis pretreatment is shown in Figure 10.6. We are simulating generic pyrolysis and gasification technologies rather than specific ones from technology providers, although the concept resembles the Choren design. The gasification takes place

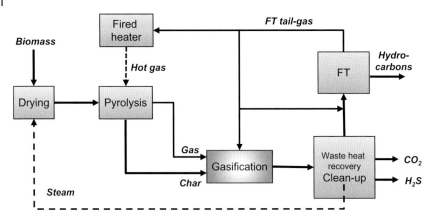

Figure 10.6 Conceptual flow-sheet for BTL-FT with entrained-flow gasification and pyrolysis pretreatment.

at an elevated pressure of 35 bar, which is advantageous for equipment size (costs), thermodynamics and kinetics. Gas from pyrolysis is injected into the gasification reactor through fuel nozzles and mixed with oxygen and possibly steam. Char is fed through a lock-hopper and a fluidization vessel. Note that all the heat needed for pyrolysis comes from the FT-reactor tail gas, whereas drying is supported by steam from raw syngas cooling within the clean-up block. Not shown is the air separation unit (ASU) providing oxygen for the gasification. Steam generated by cooling the FT-reactor is used for electricity production, mainly providing energy for the ASU.

The FT tail-gas directed to the fired-heater also serves as a purge for recycled streams in order not to accumulate inerts. Recycle around the FT reactor is needed to secure a 90+% conversion of the syngas. Additional tail-gas can be directed to the gasification unit. The hydrocarbons produced will vary in chain length, but normally will be optimized to give a high wax fraction. This wax will be upgraded to the desired products as described later. For the upgrading, a small PSA unit will be installed at an appropriate location to extract a minor amount of hydrogen from syngas.

10.5.2
Gas Conditioning and Clean-Up

Gas conditioning and clean-up encompasses a water-gas-shift reactor that increases the H_2/CO ratio to the desired synthesis value at the expense of additional CO_2 produced, hydrolysis to convert HCN to NH_3 and COS to H_2S, as well as removal of CO_2, NH_3 and H_2S. Most CO_2 will be removed to limit the amount of inerts entering the FT-section. Some of the CO_2 will be recycled to the solid feeding system. The water-gas-shift is operated at a high temperature with sulfur-resistant catalysts. H_2S will be further converted to elemental sulfur for deposition.

Table 10.8 Typical dry raw gas syngas composition for the above entrained-flow concept based on wood, and provisional FT feed specifications.

Compound	Raw syngas composition	FT make-up gas
H_2	19 mol%	61 mol%
CO	58 mol%	32 mol%
CO_2	20 mol%	<10 mol%
CH_4	0 mol%	Total inerts
N_2	4 mol%	
NO_x	0 ppm	10 ppb
NH_3	20 ppm	1 ppm
HCN	0 ppm	10 ppb
H_2S	170 ppm	<10 ppb
COS	20 ppm	Sulfur

Tars, particularly the significant amounts made by the lower temperature fluidized-bed gasification technologies, will have to be removed prior to FT-synthesis. Thermal or catalytic cracking or scrubbing are the main options available. Tar can, in this case, represent a significant carbon and energy fraction that needs to be integrated into the overall concept. Dust and soot are removed by filters and cyclones/venturi scrubbers. Organic impurities that will have to be removed are typically BTX compounds (benzene, toluene and xylenes), whereas inorganic components comprise NH_3, HCN, H_2S, COS, HCl, and volatile metals, like alkali metals. As described in Section 10.2.4 a very pure syngas is needed for the FT-operation. An example of raw syngas composition and composition needed for the FT-operation is given in Table 10.8. Further specifications will apply to metal content, in particular alkali and alkaline earth metals which each should be below 10 ppb.

10.5.3
BTL Mass and Energy Balance

Process simulations show that the two main factors that influence the energy efficiency of a BTL plant are the selected pretreatment and gasification configurations. Processes based on fluidized-bed gasifiers are less efficient than entrained-flow due to lower biomass conversion (95 vs 99%), and more hydrocarbons (light hydrocarbons, BTX, tar) in raw syngas that require more severe clean-up. Some published simulation results are given in Table 10.9 together with some case studies carried out by Statoil [76]. We have used Statoil 2 with torrefaction as a base case, implementing technology and cost data from vendors. Note that there is a significant difference between the calculated efficiencies in the studies, and it is challenging to pinpoint the actual differences in the concepts and process assumptions. However, analyses of the discrepancy show that a major factor is in the feed systems. For example, NREL fluidized-bed and entrained-flow systems

Table 10.9 Energy and carbon efficiencies of BTL-FT processes.

	Arlanda 1 (Arlanda 2)	NREL 1	NREL 2	Statoil 1	Statoil 2	Statoil 3
Process conditions						
Feed LHV (MW$_{th}$)	611 (298)	417	417	790	559	493
Biomass	Grot and logs	Corn stover	Corn stover	Hard wood chips	Hard wood chips	Hard wood chips
Drying/humidity	50→20%	25→10%	25→10%	35%	35→15%	35→7%
Pretreatment[a]	–	–	–	–	Torrefaction	Pyrolysis
Gasifier feed[a]	30–50 mm	6 mm	1 mm	3–5 cm	0.1 mm	Liquid
Gasifier type	Bubbling fluidized-bed	Bubbling fluidized-bed	Entrained-flow	Circulating fluidized-bed	Entrained-flow	Entrained-flow
Gasifier design	Carbona	GTI	Texaco	Foster Wheeler/VTT	Siemens	Siemens
Pressure (bar)/temperature (°C)	10–20 850–950	28 870	28 1300	10 865	35 1500	35 1500
Syngas conditioning[b]	Tar reformer	Tar reformer	MDEA	Tar reformer	Water wash	Water wash
CO$_2$ removal	Rectisol	MDEA		Water wash		
FT-technology	Slurry	Reactor n.a.	Reactor n.a.	Slurry	Slurry	Slurry
	Co-catalyst	Co-catalyst	Co-catalyst	Co-catalyst	Co-catalyst	Co-catalyst
FT-conditions[c]	n.a.	25 bar 200 °C 40% conv.	25 bar 200 °C 40% conv.	20 bar 226 °C 60% conv.	20 bar 226 °C 60% conv.	20 bar 226 °C 60% conv.

10.5 Biomass-to-Liquids Process Concepts

Efficiencies

Efficiency reported	44 (46)	43	53	26	29	37
Efficiency recalc.	45 (46)	43	54	27	27	33
Carbon eff. reported	n.a.	39	50	–	–	42
Carbon eff. recalc.	46 (47)	39	50	24	33	38

Indicative fuel production costs

Year	2Q 2009	2007	2007	2010	2007	2010
Biomass ($/ton)	134	75	75	83	83	83
Diesel fuel ($/l)	0.7 (Jet fuel: 1.2)	1.4	1.2	n.a.	2.1	n.a.

GTI = Gas Technology Institute; VTT = Finnish VTT research center; FT diesel: LHV 44 MJ kg^{-1}, density 0.77 kg l^{-1}; Biomass: LHV 18 MJ kg^{-1}; n.a.: not available in report.

a) Chipping and/or milling will be used at different stages.
b) Syngas conditioning will contain various steps including cooling, filtering, hydrolysis, CO_2/H_2S absorption, water-gas-shift and various guard-beds including ZnO and activated carbon.
c) conv. = CO conversion once through. Total conversion: >90%.

assume that 1 mm biomass particles can be fed with a lock-hopper system without appreciable leakage of inert gases, as it is assumed that only 5% of the CO_2 needed for pressurization and fluidization leaks into the gasifier [77]. This feeding system must be regarded as non-proven, but still represents a potential for technology development. By contrast, a proven feeding system for 0.1 mm torrefied wood is used in the Statoil 2 case, with a CO_2 leak to the gasifier equivalent to as much as 50% of the CO_2 needed for pressurization and fluidization. The Arlanda studies report feeding of biomass particles as large as 30 to 50 mm via a lock-hopper as further size reduction does not seem to have been included [78]. The robustness of this configuration should be evaluated further. The Statoil 3 concept is based on hypothetical pyrolysis-entrained flow integrated reactors. The concept requires severe biomass drying, but gives the best efficiency in the Statoil studies since less char (i.e., CO_2) is fed to the gasifier and less inerts will be carried over to the process units. This implies that less oxygen will be needed for heating to the gasification temperature and power for compression will be minimized.

For comparison purposes, we have performed a recalculation of the energy efficiencies on a common basis for LHV of the biomass and FT-products, and used a common efficiency of 40% for generation of electricity from biomass sources. The new efficiencies, named "efficiency recalculated" in Table 10.9, are defined as:

$$\text{efficiency calculated} = \text{LHV FT-products}/(\text{LHV biomass} - \text{El. export}/0.4)$$

Reported and calculated efficiencies are rather close in the Arlanda and NREL studies. This is also true for the carbon efficiencies defined here as ratios of LHVs, not taking into account electricity import or export. The entrained flow Statoil values deserve a special comment. In contrast to the other studies the net efficiency is lower than the carbon efficiency. This is due to a net electricity import instead of export. The reason is that more FT off-gas can be recycled to the gasifier, typically 91 versus 60% for a fluidized-bed, due to lower contents of inerts, mainly methane and CO_2. Therefore, less fuel is directed to the power station.

By direct comparison within the same study we can conclude that *entrained flow is more efficient than fluidized-bed*, in particular concerning carbon efficiency. The origin of the lower efficiency of fluidized-bed gasification is due to lower carbon conversion and to the energy needed for tar and light hydrocarbons reforming. To sum up, energy efficiency from biomass to liquids approaching 35% will be possible using proven entrained-flow technology with pyrolysis as pretreatment, and with technology developments it might be possible to improve on this figure by 10–15% by direct and confined feeding of biomass particles.

10.5.4
CO_2 Management

Life cycle analysis of CO_2 and other greenhouse gas (GHG) emissions is a challenging task and different groups have come up with a large variation in the cal-

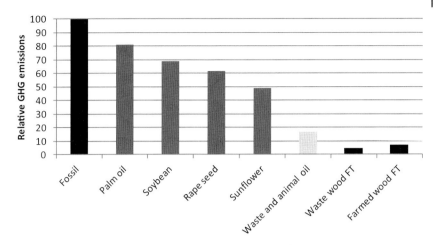

Figure 10.7 Relative greenhouse gas emissions from diesel biofuels.

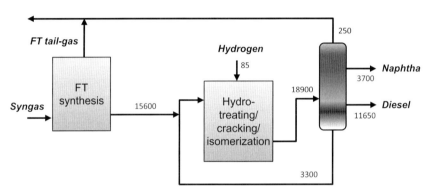

Figure 10.8 Upgrading section of a 3000 bpd BTL plant. Flows are in kg h^{-1}.

culated numbers. Official numbers from the European Commission for the use of several biodiesel types are compared in Figure 10.7 [79]. The numbers encompass average effects of cultivation of raw materials, land-use change, processing, transport and distribution, but manufacture of equipment is not included. Together with the special case where waste oil or animal oil is used as raw material for biodiesel, the FT-route comes out very favorably. Still, large amounts of CO_2 will be emitted from a BTL plant and energy optimization will be needed. There are also feasible options to capture CO_2 from high-pressure streams, meaning that storage or use of CO_2 can result in a negative GHG impact.

10.5.5
Upgrading and Products

Figure 10.8 shows a general lay-out of the upgrading of FT hydrocarbons for maximum diesel production. FT hydrocarbons, mostly wax, are first hydrotreated to remove oxygenates and to saturate olefins, then the longer chains are cracked and isomerized to give the appropriate product slate and quality. The hydrotreatment can be conducted using two reactors or beds with specific catalysts. The process can be optimized for other products, like base oil or the intermediate boiling point range kerosene.

One part of the BTL products that is of particular interest is the kerosene fraction that may be optimized to fulfill the Jet A/A-1 specifications ASTM D1655/D7566 or UK DefStan 91-91 for drop-in aviation fuel. Apart from hydrotreated vegetable oil (HVO) from oil seed plants, the biomass-derived FT-product is the only fuel able to reduce greenhouse gas emissions of aircraft on a short to medium term. Blends of up to 50% with FT-BTL jet-fuel have been certified. A number of criteria as to thermal stability, cold-properties, viscosity, and combustion properties will have to be fulfilled. Around ten flight tests on large commercial or military planes and blends of up to 50% FT jet-fuel have been performed [80], but all were based on natural gas (GTL) or coal (CTL) as feed. The first commercial flight was conducted by Qatar Airways on an Airbus A340-600 flight from London Gatwick to Doha on 12 October 2009. HVO fuel was certified as late as in June 2011, but has received a lot of attention due to its availability, and a number of test flights have already been completed. However, for large-scale utilization HVO will have challenges with possible competition for land use and a less efficient GHG reduction impact.

10.5.6
Production Cost

An example of breakdown of capital expenses is shown in Figure 10.9 for the Statoil 2 case in Table 10.9. Note that the syngas production, including gasification,

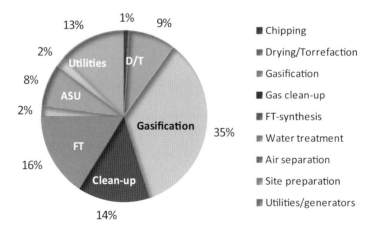

Figure 10.9 Breakdown of investment costs (CAPEX) for BTL with torrefaction and entrained flow gasification.

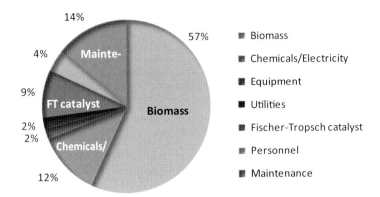

Figure 10.10 Operation costs (OPEX) for a BTL plant with torrefaction and entrained flow gasification.

air separation (ASU) and clean-up represents as much as ~60% of the total. Although main equipment cost is based on best available data from suppliers, caution should be taken at this level of cost estimation as to total cost of installation, including piping, instrumentation, and so on. Further costs including engineering, commissioning and start-up, owner's costs, contingency, financing, and so on will also be significant. In the breakdown of operation costs in Figure 10.10 the biomass as delivered is the main expense. Note also that the FT catalyst cost is significant.

Direct comparison of production costs given in Table 10.9 is difficult due to differences in costing methodologies. At present there is no commercial BTL plant that can serve as a baseline. Nevertheless, we expect that BTL fuels will be at least twice as expensive as refinery diesel using best available technologies. However, technology developments and maturity may bring the costs down to a competitive level in a high oil price scenario.

10.6
BTL Pilot and Demonstration Plants

There are a limited number of BTL-FT projects that have gone into operation beyond the laboratory stage, although there have been ample announcements. Projects are summarized in Table 10.10.

The FT-part of the Choren plant is based on a license from Shell, and an alpha plant has been operated successfully. Further plans for a commercial 640 MW$_{th}$ unit at Schwedt, Germany, are on hold due to the bankruptcy of Choren and possible issues with the FT-license. The Choren technology has been acquired by Linde GmbH. The only other BTL pilot project that has been in operation is the Oxford Catalyst (Velocys) micro-channel reactor coupled to the Güssing gasification plant. Reports are that the tests have been completed according to plans.

Table 10.10 BTL-FT projects.

FT technology provider Operator	Location	Gasification type/ technology	FT-technology	FT liquids (bpd)	Status
Shell Choren	Freiberg Germany	Entrained-flow Choren	Fixed-bed	1.3	Alpha-plant 2003–2011
Shell Choren	Freiberg Germany	Entrained-flow Choren	Fixed-bed	330	Gasification 2010–2011. On hold
GTL.F1	Bure Saudron France	Entrained-flow Choren	Slurry bubble column	560	Basic engineering
Axens French JV	Mardyck/ Dunkirk France	Entrained-flow Uhde	Slurry bubble column	0.1	Demo gasifier 2012–2017
Oxford Catalysts SGC Energia	Güssing Austria	Dual fluidized-bed Vienna U. of Tec.	Micro-channel fixed-bed	1	Demonstration 2010–2011
KTI/Statoil KTI	Karlsruhe Germany	Entrained-flow KTI	Micro-channel fixed-bed	15	Start-up 2012
Neste/Stora Enso	Varkhaus Finland	n.a.	n.a	–	
Flambeau River	Park Falls Wisconsin	Steam reforming TRI	n.a.	1200	Announced/delayed

TRI = ThermoChem Recovery International.

A joint project, BioTfueL, has been announced in France between the partners Total, IFP, Axens, the French Atomic and Alternative Energy Commission and Sofiproteol, with Uhde providing its PRENFLO gasification technology as described above [81]. The Gasel FT technology is based on a slurry bubble column reactor with catalyst developed in cooperation between ENI, IFP and Axens, and demonstrated in a 20 bpd pilot plant. The next phase will be demonstration of the gasification, but only with laboratory scale FT for quality control. This plant is expected to come on stream in 2013.

Also in France, the French Atomic and Alternative Energy Commission has announced the construction of a FT BTL plant in Bure Saudron [82]. The slurry FT-technology will be provided by GTL.F1. Design and operation will be based on the nominal 1000 bpd FT demonstration unit as operated in Mossel Bay, South Africa, from 2004–2009 by partners PetroSA, Statoil and Lurgi. An interesting feature of the process design is that external hydrogen will be added to the FT-synthesis stage to adjust the H_2/CO ratio. This hydrogen will be obtained by electrolysis of water using off-peak cheap electricity. Technology for gasification was originally expected to be provided by Choren, but the status today is uncertain. The Lurgi branch of Air Liquide is coordinating the engineering work.

Other active BTL players are the technology providers Rentech for FT-technology and Solena in plasma gasification, and several moderately sized projects have been proposed. However, no final decisions to go ahead have been announced.

10.7
XTL Energy and Carbon Efficiencies

A comparison of the total thermal process efficiencies for XTL plants is given in Figure 10.11. Data are for lower heating values (LHV) and will vary considerably with type of biomass or coal. The feedstock is assumed to be available at the plant without any energy penalty. Calculations are from open literature with uncertainty bars representing spread in the reports. Electricity and heat import or export has not been included.

Calculations of the CO_2 emissions of the XTL plants for a given production of FT-liquids result in much more distinct differences, see Figure 10.12. The processes compared are torrefaction/entrained-flow (BTL), entrained-flow (CTL) and autothermal reforming (GTL). The reason for the large variations is that, on top of the energy efficiency in the plant itself, one has to take into account that the energy content of the starting materials varies considerably. A GTL plant will have an overall carbon efficiency of ~75%, whereas 2/3 of the carbon of a

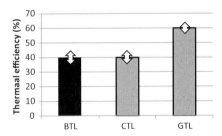

Figure 10.11 Thermal energy efficiencies of XTL FT plants.

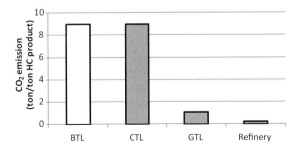

Figure 10.12 CO_2 emissions per FT-liquids produced for XTL plants compared to refinery fuel.

BTL plant goes into CO_2 in the gasification step alone. Using the hot syngas to partially provide heat for the reformer section of a GTL plant (gas-heated reformer) can increase the efficiency to ~81%. A similar concept can be visualized for biomass conversion, although the more aggressive gas composition is an additional challenge.

If CO_2 is removed from the purge gas of a GTL plant the emitted CO_2 can be reduced to 60%, and if we take out CO_2 after the reformer only 54% remains. These modifications will, however, be at the expense of power export, and the latter case will be in energy balance. In a BTL plant significant amounts of CO_2 will be captured after the gasification, with possibilities for subsequent compression, transport, and storage. Such a concept will theoretically impose a negative GHG footprint.

10.8
BTL Summary and Outlook

At present there is no commercial scale BTL-FT plant in operation. Two semi-commercial plants in the 300–600 bpd range have reached the construction or engineering phase, but completion of the projects remains to be seen. There are also numerous announcements of BTL-FT projects from pilot to commercial scale. One interesting trend is that airlines and other aviation companies have caught interest in BTL to secure a biofuel fraction of their fuel pool. Several tests have been performed with up to 50% blending of FT-kerosene with standard jet fuel and this blend has been approved as aviation fuel. Alternative blending with hydrogenated bio-oil appears less attractive in the longer term. However, there is no doubt that, in a 10–20 year perspective, introduction of BTL-FT products will be dictated by overall regulations and will require tax exemption or other incentives to become profitable.

For a BTL project there are several important decisions to be made regarding technology and overall concept selection. Apart from the biomass availability and logistics, the main decision will be the gasification technology, as well as the associated pretreatment and purification of the raw syngas. There are some key points characterizing the different gasification technologies:

- Entrained-flow, and possibly bubbling-bed and plasma gasifiers, is suited for higher pressures.
- Dual-bed gasifiers can use air without getting appreciable amounts of nitrogen into the syngas.
- Entrained-flow reactors need pretreated biomass in the form of pyrolysis or torrefaction.
- Entrained-flow and plasma reactors operate at higher temperatures and give low tar.
- Plasma reactors can gasify raw biomass as long as it can be fed to the reactor.
- The very high temperature and use of electricity are challenges for the energy efficiency of plasma reactors.

All these reactors are, in principle, suitable for BTL applications. For larger scale operations elevated pressures, preferably 20–30 bar, will be needed. This also means that an air separation unit will have to be installed to generate oxygen for the combustion, except when a dual system with a heat transfer medium is applied.

A commercial BTL plant will be at least an order of magnitude smaller than a GTL or CTL plant, but still efficient scale-up is important, and it is highly likely that a pressurized gasifier will be selected and that carrying a nitrogen load in the system should be avoided. At present this lends itself toward using an oxygen-blown entrained-flow gasifier. However, the technologies are immature and there is a significant technology development of alternative gasifiers, and new more efficient solutions might be demonstrated. In particular, we will point out the importance of developing feed systems for entrained flow that limit carry-over of fluidization gas.

The energy efficiency of BTL-FT is relatively low. Although the different process simulations that have been conducted give varying results, it is likely that, all factors included, we are looking at the 35–45% range. The main reason for the limited efficiency is the low energy content of the biomass and the need to adjust the inherently low H_2/CO ratio to fit with the product stoichiometry. A possible option is to add the required hydrogen from other sources, for example, by making use of cheap electricity at off-peak periods to electrolyze water.

References

1 Fischer, F., and Tropsch, H. (1930) US 1,746,464.
2 Fischer, F., and Tropsch, H. (1930) *Brennstoff-Chem.*, **11**, 489.
3 Schultz, H. (1999) *Appl. Catal. A: Gen.*, **186**, 3.
4 Dry, M.E. (2002) *Catal. Today*, **3–4**, 227.
5 Stranges, A.N. (2007) *Stud. Surf. Sci. Catal.*, **163**, 1.
6 Kölbel, H., and Ackermann, P. (1956) *Chem. Ing. Tech.*, **28**, 381.
7 Steynberg, A., Espinoza, R.L., Jager, B., and Vosloo, A.C. (1999) *Appl. Catal., A*, **186**, 41.
8 Schanke, D., Rytter, E., and Jaer, F.O. (2004) *Stud. Surf. Sci. Catal.*, **147**, 43.
9 Eisenberg, B., Fiato, R.A., Mauldin, C.H., Say, G.R., and Soled, R.L. (1998) *Stud. Surf. Sci. Catal.*, **119**, 943.
10 Chaumette, P., Verdon, C., and Boucot, P. (1995) *Top. Catal.*, **2**, 301.
11 Rytter, E., Eri, S., Skagseth, T.H., Schanke, D., Bergene, E., Myrstad, R., and Lindvåg, A. (2007) *Ind. Eng. Chem. Res.*, **46**, 9032.
12 Kort, G.D. (2010) World XTL10 Summit, London.
13 Borg, Ø., Dietzel, P.D.C., Spjelkavik, A.I., Tveten, E.Z., Walmsley, J.C., Diplas, S., Eri, S., Holmen, A., and Rytter, E. (2008) *J. Catal.*, **259**, 161.
14 Borg, Ø., Eri, S., Blekkan, E.A., Storsæter, S., Wigum, H., Rytter, E., and Holmen, A. (2007) *J. Catal.*, **248**, 89.
15 van Steen, E., and Claeys, M. (2008) *Chem. Eng. Technol.*, **31**, 655.
16 Khodakov, A.Y., Chu, W., and Fongarland, P. (2007) *Chem. Rev.*, **107**, 1692.
17 Moodley, D.J., van de Loosdrecht, J., Saib, A.M., Overett, M.J., Datye, A.K., and Niemantsverdriet, J.W. (2009) *Appl. Catal.*, **354**, 102.
18 Bezemer, G.L., Bitter, J.H., Kuipers, H.P.C.E., Oosterbeek, H., Holewijn, J.E., Xu, X., Kapteijn, F., van Dillen, A.J., and de Jong, K.P. (2006) *J. Am. Chem. Soc.*, **128**, 3956.

19 Ducreux, O., Rebours, B., Lynch, J., Roy-Auberger, M., and Bazin, D. (2009) *Oil Gas Sci. Technol. Rev. IFP*, **64**, 49.

20 Lögdberg, S., Lualdi, M., Järås, S., Walmsley, J.C., Blekkan, E.A., Rytter, E., and Holmen, A. (2010) *J. Catal.*, **274**, 84.

21 Blekkan, E.A., Borg, Ø., Frøseth, V., and Holmen, A. (2007) *Catalysis*, **2007**, 13.

22 Das, T.K., Zhan, X., Li, J., Jacobs, G., Dry, M.E., and Davis, B.H. (2007) *Stud. Surf. Sci. Catal.*, **163**, 289.

23 Rane, S.P., Borg, Ø., Yang, J., Rytter, E., and Holmen, A. (2010) *Appl. Catal.*, **388**, 160.

24 Rytter, E., Eri, S., Schanke, D., Wigum, H., Skagseth, T.H., Borg, Ø., and Bergene, E. (2011) *Top. Catal.*, **54**, 801.

25 Tsakoumis, N.E., Rønning, M., Borg, Ø., Rytter, E., and Holmen, A. (2010) *Catal. Today*, **154**, 162.

26 Weststrate, C., Hauman, M., Moodley, D., Saib, A., Steen, E., and Niemantsverdriet, J. (2011) *Top. Catal.*, **54**, 811.

27 Borg, Ø., Hammer, N., Enger, B.C., Myrstad, R., Lindvåg, O.A., Eri, S., Skagseth, T.H., and Rytter, E. (2011) *J. Catal.*, **279**, 163.

28 van Berge, P.J., van de Loosdrecht, J., and Barradas, S. (2005) Method of treating an untreated catalyst support and forming a catalyst precursor and catalyst from the teated support, patent US6875720 B2, Sasol Technology.

29 Bellussi, G., Carluccio, L.C., Zenarro, R., and del Piero, G. (2007) Process for the preparation of Fischer-Tropsch catalysts with a high mechanical, thermal and chemical stability, PCT patent WO2007009680 A1, ENI/IFP.

30 Behrmann, W.C., Davis, S.M., and Mauldin, C.H. (1992) Catalyst preparation, patent EU535790 A1.

31 van Berge, P.J., and Barradas, S. (2008) Catalysts, patent US7365040 B2, Sasol Technology.

32 Dogterom, R.J., Mesters, C.M.A.M., and Reynhout, M.J. (2007) Process for preparing a hydrocarbon synthesis catalyst, PCT patent WO2007068731 A1, Shell.

33 Rytter, E., Skagseth, T.H., Wigum, H., and Sincadu, N. (2005) Enhanced strength of alumina based Co Fischer-Tropsch catalyst, PCT patent WO 2005072866, Statoil.

34 Ikeda, M., Waku, T., and Aoki, N. (2005) Catalyst for Fischer-Tropsch synthesis and method for producing hydrocarbon, PCT patent WO2005099897 A1, Nippon Oil.

35 Inga, J., Kennedy, P., and Leviness, S. (2005) Fischer-Tropsch process in the presence of nitrogen contaminants, PCT patent WO2005071044, Syntroleum.

36 Srinivasan, N., Espinoza, R.L., Coy, K.L., and Jothimurugesan, K. (2004) Fischer-Tropsch catalysts using multiple precursors, patent US6822008 B2, ConocoPhillips.

37 Eri, S. (2012) Statoil RD. Personal Communication.

38 Creyghton, E.J., Mesters, C.M.A.M., Reynhout, M.J., and Verbist, G.L.M.M. (2008) Process for preparing a catalyst, PCT patent WO2008071640 A2, Shell.

39 Font Freide, J.J.H.M., Collins, J.P., Nay, B., and Sharp, C. (2007) *Stud. Surf. Sci. Catal.*, **163**, 37.

40 Steynberg, A.P., Dry, M.E., Davis, B.H., and Breman, B.B. (2004) *Stud. Surf. Sci. Catal.*, **152**, 64.

41 Prins, M.J. (2005) Thermodynamic analysis of biomass gasification and torrefaction. Ph.D. thesis, Eindhoven University of Technology, http://www.alexandria.tue.nl/extra2/200510705.pdf (accessed January 2013).

42 Bergman, P.C.A., Prins, M.J., Boersma, A.R., Ptasinski, K.J., Kiel, J.H.A., and Janssen, F.J.J.G. (2005) Torrefaction for entrained flow gasification of biomass, Energy research Center of the Netherlands, report ECN-C-05-067, http://www.ecn.nl/docs/library/report/2005/c05067.pdf (accessed January 2013).

43 Bridgwater, A.V. (2012) *Biomass Bioeng.*, **38**, 68.

44 Bergman, P.C.A., Boersma, A.R., Zwart, R.W.R., and Kiel, J.H.A. (2005) Torrefaction for biomass co-firing in existing coal-fired power stations, Energy research Center of the Netherlands, report ECN-C-05-013, http://www.ecn.nl/docs/library/report/2005/c05013.pdf (accessed January 2013).

45 Torftech (2010) http://www.torftech.com/news.php (accessed January 2013).
46 Fuel oil from forest residues, www.metso.com (accessed January 2013).
47 Trippe, F., Fröhling, M., and Schultmann, F. (2010) *Waste Biomass Valor.*, **1**, 415–430.
48 Choren (2008) http://213.133.109.5/video/energy1tv/Jan%20NEU/Konferenz/Wirtschaft/FNR_BTL_08/T1/2a_Tech/4_Blades.pdf (accessed January 2013).
49 BTG (2011) www.btgworld.com (accessed January 2013).
50 Ensyn (2011) www.ensyn.com (accessed January 2013).
51 van der Drift, A., Boerrigter, H., Coda, B., Cieplik, M.K., and Hemmes, K. (2004) Entrained Flow Gasification of Biomass: Ash Behaviour, Feeding Issues, and System Analyses, Energy research Center of the Netherlands, report ECN-C-04-039, http://www.ecn.nl/docs/library/report/2004/c04039.pdf (accessed January 2013).
52 Thomas, S., and Wittmann, T. (2011) Biomass Upgrading with the CarboREN-Technology to Facilitate Gasification. International Seminar on Gasification 2011–Gas Quality, CHP and New Concepts, 6–7 October 2011, Malmö.
53 Bacovsky, D., Dallos, M., and Wörgetter, M. (2010) Status of 2nd Generation Biofuels Demonstration Facilities in June 2010, IEA Bioenergy Task 39, report T39-P1b 27.
54 Discussion on Plasma Gasification, www.recoveredenergy.com (accessed November 2011).
55 Gasification Technologies Conference (2011) http://www.gasification.org/page_1.asp?a=96&b=4 (accessed January 2013).
56 Lindblom, M., and Landälv, I. (2007) Chemrec's Atmospheric and Pressurised BLG (Black Liquour Gasification) Technology–Status and Future Plans, www.chemrec.se (accessed November 2011).
57 Bioliq (2011) Karlsruher Institut für Technologie, www.bioliq.de (accessed November 2011).
58 Uhde (2010) PRENFLO Gasification, http://www.thyssenkrupp-uhde.de/fileadmin/documents/brochures/uhde_brochures_pdf_en_11.pdf (accessed January 2013).
59 van der Meijden, C.M. (2010) Development of the MILENA Gasification Technology for the production of Bio-SNG, Energy research Center of the Netherlands, report ECN-B-10-016, http://www.ecn.nl/docs/library/report/2010/b10016.pdf (accessed January 2013).
60 Enerkem (2011) http://www.enerkem.com/en/home.html (accessed January 2013).
61 Andritz (2011) http://www.andritz.com/gasification (accessed January 2013).
62 Mühlen, H.J. (2011) Blue tower concept–gasification concept for wastes, http://www.sgc.se/gasification2011/Resources/Heinz_Jurgen_Muhlen.pdf (accessed January 2013).
63 Coskata (2011) http://www.coskata.com/facilities (accessed January 2013).
64 (2012) *Chem. Eng. News*, **90–3**, 8.
65 Cortus (2011) www.cortus.se (accessed November 2011).
66 Primus Green Energy (2011) www.primusge.com (accessed January 2013).
67 Lindblom, M. (2003) An Overview of Chemrec Process Concepts, presentation at: Colloquium on Black Liquor Combustion and Gasification, May 13–16, 2003, Park City, Utah.
68 Vogels, J. (2010) An Industrial Scale Hydrogen Production from Biomass via CHOREN's Unique Carbo-V-Process. 18th Proceedings of the World Hydrogen Energy Conference, May 16–21, 2010.
69 Arnold, U., Döring, M., Dahmen, N., and Dinjus, E. (2009) Fuel Production from Biomass-Derived Syngas within the Bioliq-Process, presentation at: IEA Bioenergy Agreement Task 33: Thermal Gasification of Biomass, First Semi-annual Task Meeting 2009, Karlsruhe, Germany.
70 Maly, M., Claussen, M., Schindler, M., Vodegel, S., and Carlowitz, O. (2005) The Biomass-to-Liquid Process at CUTEC: Optimisation of the Fischer-Tropsch Synthesis with Carbon Dioxide-rich Synthesis Gas, presentation at: Renewable Resources and Biorefineries Conference, September 19–21, 2005, Ghent, Belgium.
71 Chornet, E. (2009) Gasification of Heterogeneous Biomass Residues and Catalytic Synthesis of Alcohols,

presentation at: Gasification 2009–Gas Clean-up and Gas Treatment, October 22–23, 2009, Stockholm, Sweden.

72 Waldheim, L. (2011) Värnamo Gasification Project, presentation at: International Seminar on Gasification 2011–Gas Quality, CHP and New Concepts. 6–7 October 2011, Malmö, Sweden.

73 Ljunggren, R. (2011) A Novel Three-Stage Gasification Technology, presentation at the International Seminar on Gasification 2011–Gas Quality, CHP and New Concepts, 6–7 October 2011, Malmö, Sweden.

74 Concord Blue Energy (2012) www.the-bluetower.com (accessed January 2013).

75 Westinghouse Plasma Corporation (2012) www.westinghouse-plasma.com (accessed January 2013).

76 Fahmi, A. (2007–2010) Statoil.

77 Swanson, R.M., Satrio, J.A., and Brown, R.C. (2010) Techno-Economic Analysis of Biofuels Production Based on Gasification, NREL Technical Report TP-6A20-46587, November 2010.

78 Ekbom, T., Hjerpe, C., Hagström, M., and Hermann, F. (2009) Pilot Study of Bio-jet A-1 Fuel Production for Stockholm-Arlanda Airport, report VARMEFORSK SYS08-831, 15 November, 2009.

79 European Commission (2010) Scheme for certifying sustainable biofuels, Decision of 10 June 2010, Annex V to Directive 2009/28/EC. http://eur-lex.europa.eu/LexUriServ/LexUriServ.do?uri=OJ:L:2010:151:0019:0041:EN:PDF (accessed January 2013).

80 CSIRO (May 2011) Flight path to Sustainable Aviation, report, http://www.csiro.au/files/files/p10rv.pdf (accessed January 2013).

81 IFP BioTFuel (2013) http://www.ifpenergiesnouvelles.com/developpement-industriel/carburants-alternatifs/carburants-alternatifs-demain (accessed January 2013).

82 Air Liquide (2009) Bure Saudron project, http://www.airliquide.com/en/france-launch-of-a-second-generation-biofuel-project.html (accessed January 2013).

11
Catalytic Transformation of Extractives

Päivi Mäki-Arvela, Irina L. Simakova, Tapio Salmi, and Dmitry Yu. Murzin

11.1
Introduction

Fine and specialty chemicals from wood are produced from various compounds, such as carbohydrates and extractives. This topic was comprehensively reviewed as recently as in 2007 in ref. [1]. In this chapter the main emphasis is on industrial utilization of extractives, and especially terpenes and terpenoids. In general, extractives from wood consist of several types of compounds, such as terpenes and terpenoids, fats and waxes, phenolic compounds, and other substrates, for example alkanes and water-soluble compounds such as mono- and disaccharides [2]. Industrially the most important wood-derived feedstocks composed of extractives are tall oil and turpentine.

Tall oil is a side product from the Kraft pulping process. Crude tall oil (CTO) components are separated via fractional distillation to two main fractions, tall oil rosin (TOR) and distilled tall oil (DTO). The content of resin and fatty acids in CTO is 85%. Fatty acids are important compounds used for production of various chemicals [1], for example conjugated fatty acids. In addition to fatty acids, CTO contains about 10–15% neutral, unsaponifiable compounds, especially sterols [3]. The total amount of sterols in the unsaponifiable fraction is more than 95% and their distribution is given in Table 11.1 [20].

Turpentine is composed of volatile compounds from oleoresins in the softwood fraction. The main compounds in turpentine originated from the US are α-pinene (75–85%) and β-pinene (3%), limonene (5–15%), and camphene (4–15%) [21], together with myrcene, β-phellandrene, trans-β-terpineol, 4-allylanisole, and isoborneol (Figure 11.1) [22], although there are large variations in, for example, the amounts of α- and β-pinene, depending on the softwood species. The amounts of different monoterpenes from *Pinus Ponderosa*, which is common in the US [23], *Picea abies* common in Europe [24], and in the typical Russian *Pinus Silvestris* [25] are given in Table 11.2. Turpentine is produced either via chemical pulping from the Kraft process or directly via distillation of the resin extrudates from living trees. The most common monoterpenes, α-pinene, β-pinene and limonene, are produced according to [36] annually in the quantitites 18 000, 12 000 and 30 000 t,

Catalytic Process Development for Renewable Materials, First Edition. Edited by Pieter Imhof and Jan Cornelis van der Waal.
© 2013 Wiley-VCH Verlag GmbH & Co. KGaA. Published 2013 by Wiley-VCH Verlag GmbH & Co. KGaA.

Table 11.1 Isomerization of terpenes and their derivatives.

Compound	Product	Conventional method	Catalytic method
α-pinene	Camphene	Via bornyl chloride [4] Boron trifluoride [5]	Sulfated zirconia [6] Zeolites [7] Au/Al$_2$O$_3$ [4, 8] Amberlyst 35 [9]
α-pinene oxide	Campholenic aldehyde	ZnBr$_2$, ZnCl$_2$ [10]	Continuous reactor [11–13]
α-pinene	β-pinene	Fractional distillation [14]	Pd/C, Ru/C, Rh/C, Pt/C, Ir/C [15, 16]
β-pinene oxide	Myrtanal		Sn-Beta, Zr-Beta [17]
β-pinene oxide	Perillyl alcohol		HNO$_3$ treated ion exchange resins [18]
Pinane-2-ol	Linalool		Carbon block [19]

respectively. Some other web-based sources indicate higher (even twofold higher) production capacities .The production of fine chemicals is in the midst of changing the conventional industrial synthesis methods, that is, from utilization of homogeneous catalysts, stoichiometric oxidants, strong acids and bases, as well as metal salts and batch operations, toward the use of solid catalysts and continuous processes. This change is needed since the E-factor [37], which reflects the ratio of waste to product, is extremely high for fine chemicals production and cannot be accepted in future according to the green chemistry principles [38]. Heterogeneous

11.1 Introduction

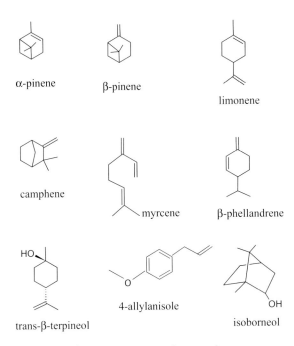

Figure 11.1 The main components in turpentine.

Table 11.2 Oxidation of terpenes.

Compound	Product	Conventional method	Catalytic method
α-pinene	Verbenol	CrO$_3$ [26, 27]	Cr-aluminophosphate [28] Autooxidation by O$_2$ followed by Pd/C [29–31]
α-pinene	Myrtenal	CrO$_3$ [26]	
α-pinene	Verbenone		Cr-MCM-41 [32] Cr-aluminophosphate [28] autooxidation by O$_2$ followed by Pd/C [29–31]

(Continued)

11 Catalytic Transformation of Extractives

Table 11.2 (Continued)

Compound	Product	Conventional method	Catalytic method
β-pinene	Pinocarvone	Insect pheromones [33, 34]	Cr-MCM-41 [32] Cr-aluminophosphate [28]
β-pinene	Pinocarveol		Cr-MCM-41 [28]
β-pinene	Myrtenol		Cr-MCM-41 [32]
β-pinene	Myrtanal		Cr-MCM-41 [32]
(+)-limonene	Carvone		Cr-MCM-41 [32]
Limonene	Trans carveol		Cr-MCM-41 [32]
Limonene	Perillyl alcohol	SeO_2, CrO_3 [35]	

catalysts are often cheaper than the corresponding homogeneous catalysts composed of, for example, complex ligands. They also facilitate easy separation and reuse of the catalyst. Furthermore, catalyst regeneration is in many cases possible, especially for such solid catalysts as zeolites.

The continuous reactor technology facilitates large production capacities. If the synthesis routes involve complex reaction kinetics, it is easy to adjust the contact time of the feed with the catalyst in continuous operation. In addition to continuous operation, process intensification has also been realized by applying one-pot technology [39], which implies the use of a bifunctional catalyst in one reactor pot. This catalyst is able to catalyze, for example, both oxidation and the consecutive isomerization reaction, as is the case in the synthesis of campholenic aldehyde from α-pinene.

The aim of this chapter is to describe catalytic process development for renewable materials, especially using different terpene and sterol fractions as a raw material. In particular, more recent work is discussed, since catalytic transformation of terpenes was reviewed in 2004 [40]. Special attention is put on the comparison between conventional and novel technologies. The latter comprises the use of heterogeneous catalysts and new process design.

11.2
Fine and Special Chemicals from Crude Tall Oil Compounds

Crude tall oil is composed of fatty and resin acids, as well as extractives. It is conventionally mainly used in paints, inks, glues, detergents, and as additives in cosmetics. Value-added fine chemicals, such as health promoting food additives, can be synthesized from unsaponifiable extractives in the CTO fraction. In this section the catalytic hydrogenation of sterols is demonstrated on both lab and pilot scales.

11.2.1
Sitosterol Hydrogenation and Its Application in Food as a Cholesterol-Suppressing Agent

Cholesterol, which is present in food and can be adsorbed on artery walls, increases the risk for coronary heart disease. Cholesterol absorption can be suppressed by a proper selection of diet [41], for example by using phystosterols in the food, a fact which has been known since 1950 [42]. Cholesterol absorption is dependent also on hereditary factors. Especially, Finns have a high presence of lipoprotein, e4 allele, which in turn increases the possibility for them to have a too high cholesterol level in their body [43]. Lipoproteins, which affect cholesterol absorption, are different types, for example e2, e3 and e4. These alleles combine in pairs randomly. When the sum of the subindices in the alleles is higher, more cholesterol is absorbed. Since fatty acid esters of sitosterol dissolve easily in the fat part of food, they are used in food applications [44]. Sitostanol is a more efficient cholesterol-suppressing agent than sitosterol [43, 45]. Thus hydrogenation of

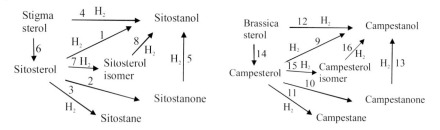

Figure 11.2 Reaction scheme for hydrogenation of sterols [48].

sitosterol is needed and it has been intensively studied over several Pd supported catalysts [46, 47]. Since sitosterol hydrogenation is currently industrially applied, kinetic modeling and a scale-up study are reviewed here.

Generally speaking, hydrogenation of plant sterols yields the respective stanols by hydrogenating the carbon–carbon double bond and the C=O bond. Some by-products can be formed during the hydrogenation: primary ketones and the unsaturated compounds. A proposed network for the hydrogenation of sitosterol and campesterol is sketched in Figure 11.2 [48].

In general, it is possible to identify 12 compounds in the reaction scheme, however, the campesterol isomer species is not always taken into account due to its low content. Sitosterol and campesterol were hydrogenated to sitostanol and campestanol in a batch reactor over 5% Pd on active carbon catalysts (catalyst 1 from Degussa with specific surface area $1010\,m^2 g^{-1}$, and catalyst 2, also from Degussa, with surface area $800\,m^2 g^{-1}$) in an organic solvent (isopropanol). The reactor volume was 100 ml and the liquid volume in the reactor was 500 ml [48].

A mixture of plant sterols (sitosterol, campesterol, stigmasterol and brassicasterol) was used as a raw material. 18 experiments were performed under different conditions, varying the temperature (70–120 °C) and hydrogen pressure (4–45 bar) over a Pd catalyst. It was verified that experiments were conducted in the kinetic regime, free from the influence of external and internal diffusion. The tests showed that a stirring speed of 1000 rpm is sufficient for the reactor to operate in the kinetic regime. Samples were taken during the reaction and at the same time the temperature and pressure were registered. The reaction rate was essentially the same, independent of the hydrogen pressure. Increased temperature resulted, as expected, in higher reaction rates. The influence of the vapor pressure of the solvent (1-propanol) had to be taken into account in the calculations, since it lowers the hydrogen partial pressure in the reactor. In the kinetics and reactor modeling the recorded total pressure in the reactor was thus corrected by subtracting the vapor pressure of 1-propanol, in order to get the hydrogen pressure. The solubility of hydrogen in sitosterol/1-propanol solution was measured.

The kinetic model was based on the following hypothesis: molecularly adsorbed hydrogen was assumed to react on the catalyst surface with adsorbed organic species. Preliminary analysis of the reaction rate dependence on hydrogen concentration revealed that discrimination between molecular and dissociative hydrogen adsorption cannot be performed in a meaningful way, therefore molecular hydrogen adsorption was preferred as, in principle, it gives a possibility to describe first order in hydrogen frequently observed in hydrogenation of carbon–carbon double bonds.

The surface hydrogenation step was assumed to be rate limiting, whereas adsorption and desorption steps were regarded as rapid quasi-equilibria. Finally the rate equations were obtained

$$r_j = \frac{k'_j K_A K_{H_2} c_j c_{H_2}}{\left(1 + \sum K_i c_i\right)\left(1 + K_{H_2} c_{H_2}\right)} \tag{11.1}$$

where index i denotes the components (A, sitosterol; B, sitostanol; C, sitostanone; D, sitostane; E, campesterol; F, campestanol; G, campestanone; H, campestane; I, brassicasterol; J, stigmasterol; K, sitosterol isomer) and index j denotes the reactions (1...16, Figure 11.2).

The influence of temperature on the rate and adsorption coefficients can be accounted for with the Arrhenius and van t'Hoff equations. The rate constants ($k_{0,j}$), adsorption constants ($K_{0,i}$) the activation energies (ΔE_A) and adsorption enthalpies were estimated from the laboratory experiment data by nonlinear regression.

The mass balance for a liquid-phase component (i) in a batch reactor was written as $dc_i/dt = \rho_B r_i$, where r_i is the reaction rate and ρ_B is the catalyst bulk density ($\rho_B = m_{cat}/V_{liq}$). The mass balance is valid in the kinetic regime, that is, in the absence of mass transfer limitations. Hydrogen is present both in the gas and the liquid phases, thus, the mass balance is written as: $dc_H/dt = N_H a_v + r_H \rho_B$, where the flux $N_H a_v$ between the gas and liquid phase is $N_H a_v = k_1 a_v (P_G/RT/K - c_H)$. The gas–liquid equilibrium constant K is obtained from gas–liquid solubility data. The proposed kinetic model and the equations of the batch reactor were linked to the MODEST software [49], the code was compiled and used for the parameter estimation of the data obtained in the experimental work. The rate constants, adsorption coefficients, heats of adsorption and the activation energies included in the kinetic model were estimated from the laboratory experiments by nonlinear regression with the Levenberg–Marquardt method. The reactor mass balances were solved as a subtask to the parameter estimation with the backward difference method. The parameter estimation routine minimizes the objective function, the sum of square error (Q) which is defined as $Q = \sum (y_{exp} - y_{est})^2 w$ where y_{exp} is the component concentration obtained from experiments, and y_{est} is the component concentration predicted by the model, w is the weight factor for the experimental point. The weight factor was set to 1 for all experimental points.

The parameter estimation was performed by applying the estimation procedure directly to the data. The program was executed many times; it has the built-in option to use the last result as a starting condition for the next estimation. Different sets of initial values for the calculation of the model parameters were also tried, finally permitting the convergence of the numerical procedure. The number of components and reactions in the estimation were 11 and 14, respectively, giving 14 rate constants, 14 activation energies, 12 adsorption coefficients, and 12 adsorption energies, altogether 52 adjustable parameters.

Some initial parameter estimations with all 52 parameters revealed that there are too many parameters, resulting in large parameter errors. In order to reduce the number of parameters some assumptions were made, in particular due to the small temperature difference ($\Delta T = 50\,°C$) in the experiments, the possible variations for heat of adsorption of hydrogen with temperature were neglected. Furthermore, the adsorption coefficients for sitosterol–campesterol, sitostanol–campestanol, sitonanone–campestanone, sitostane-campestane and stigmasterol–brassicasterol were set equal to each other, which is a reasonable assumption taking into account their structural similarities. With these modifications the number of parameters was reduced to 36. In order to reduce further the number of parameters some assumptions were made after initial estimations with all parameters. The highest degree of explanation (R^2) after these simplifications was 99.8%.

The kinetic model described very well the concentrations of the reactants and the main products, stanols and sterols, although for very minor by-products, stanes and stanones, the fits were not as good as for the main products. The developed kinetic model was used to predict the plant-size reactor behavior. A characteristic feature for a plant-size reactor is the presence of hydrogen mass-transfer resistance. In the simulations only the gas–liquid mass transfer coefficients were fitted.

Experimental data were collected from a plant reactor of $8\,m^3$ with $25\,kg$ of catalyst. The plant reactor was simulated with the values of kinetic parameters obtained from the laboratory reactor. The simulation of the plant reactor gave the values of the gas–liquid mass transfer coefficient ($k_l a$) equal to $0.34\ 1\,min^{-1}$. For efficiently stirred tank reactors the typical values are circa $3–12\ 1\,min^{-1}$ [50], demonstrating the low efficiency of gas–liquid mass transfer in the plant reactor. The results are displayed in Figure 11.3, demonstrating successful application of the adopted approach in process scale-up.

The model can predict the formation of the main products very well (Figure 11.3). Figure 11.4 shows the calculated concentration of hydrogen in the reaction mixture in the plant conditions. During the first 100 min the hydrogen concentra-

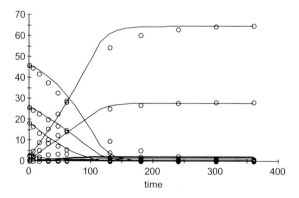

Figure 11.3 Simulation of a plant reactor, 4 bar, 70 °C based on laboratory data. Experimental points "o" from plant reactor and simulated lines [48].

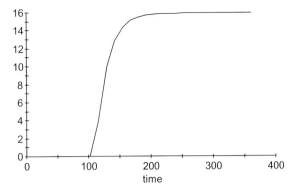

Figure 11.4 Simulation of a plant reactor. Concentration of hydrogen (mol m^{-3}) in the liquid phase as a function of time (min) [48].

tion is very low because all hydrogen in the liquid phase is immediately consumed by the reaction and thus the gas–liquid mass transfer rate controls the overall reaction rate. After 100 min of reaction less hydrogen is consumed and the hydrogen concentration approaches the saturation level determined by the gas–liquid equilibrium constant.

11.3
Fine and Special Chemicals from Turpentine Compounds

The main industrially important compounds in turpentine are α- and β-pinene, limonene, and 3-carene. There are several reactions for monoterpenes and their derivatives which are of industrial relevance, such as isomerization, oxidation, epoxidation, esterification, etherification, hydrogenation and hydration. In isomerization, typically, metal chlorides have been applied as catalysts, whereas oxidation reactions have been performed with chromic acid as an oxidant. In esterification and etherification reactions mineral acids were conventionally used as catalysts. The different transformations of monoterpenes are reported below, both with conventional and novel methods.

11.3.1
Isomerization of Monoterpenes and Their Derivatives

In the isomerization of monoterpenes and their epoxides, typically, Lewis acidic metal chlorides [10] or even strong mineral acids [4] have been used as catalysts (Table 11.3). These systems are environmentally extremely harmful and not following the principles of green chemistry, thus they should preferably not be applied on a large scale. The important reactions found for isomerization of monoterpenes and their epoxides are isomerization of α-pinene to camphene [4, 8, 9, 54–56],

Table 11.3 Epoxidation of terpenes.

Compound	Product	Conventional method	Catalytic method
α-pinene	α-pinene oxide	Peracetic acid [51]	Gold nanoparticles [52]
α-pinene	β-pinene oxide	Peracetic acid [35, 51]	
limonene	1,2-epoxylimonene	Peracids	[53] Ti-MCM-41
Carveol	1,2-epoxycarveol		[53] Ti-MCM-41
α-terpineol	1,2-epoxy-α-terpineol		[53]

α-pinene to β-pinene [15, 16], α-pinene oxide to campholenic aldehyde [10, 11] and β-pinene oxide to myrtanal, myrtenol and perillyl alcohol [17]. The products have applications both as fragrances and in the pharmaceutical industry.

α-Pinene has been used as a raw material for the synthesis of perfumes [25], and in the food and pharmaceutical industries [18, 25, 57]. The desired product in the isomerization of α-pinene is camphene, which is used as an intermediate for the production of camphor, isobornyl acetate and isoborneol [58]. Camphene is either esterified or etherified to isobornylacetate or alkoxycamphane and the formed product is further transformed to camphor (Figure 11.5). Isoborneol and its acetate are used as fragrances, whereas camphor is used in pharmaceutical applications.

Conventionally, camphene is produced via transformation of α-pinene to (+)-bornyl chloride by reacting it with dry HCl, followed by base-catalyzed dehalogenation of bornyl chloride to camphene [4]. Alternatively, α-pinene can be isomerized

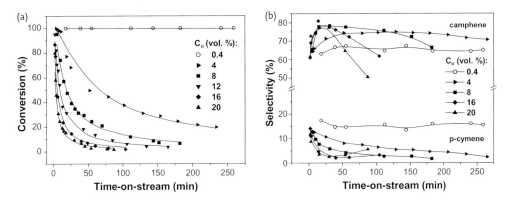

α-Pinene   Camphene   Isobornyl acetate   Isoborneol   Camphor

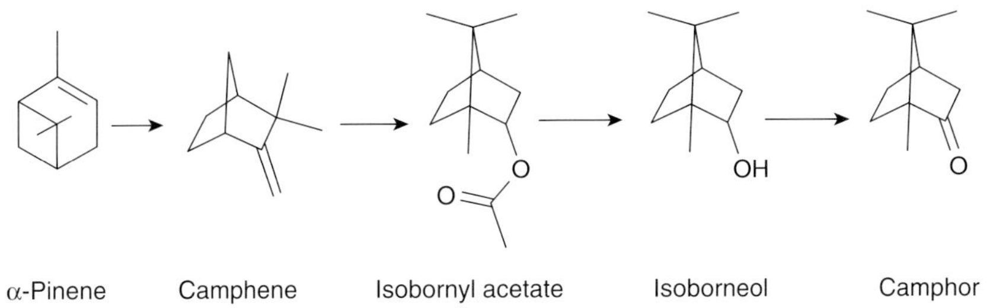

Figure 11.5 Transformation of α-pinene to camphor via isomerization, esterification, hydrolysis of the ester followed by dehydrogenation of isoborneol [40].

Figure 11.6 Time dependence of α-pinene conversion (a, symbols – experiments; line – calculated) and selectivity (b) over the 2.2 wt. % Au/γ-Al$_2$O$_3$ catalyst at different initial concentration of α-pinene (C_0). Reaction conditions: 200 °C, contact time 0.33 s [8].

in the gas phase to camphene using borophosphoric acid as a catalyst [55, 59]. An environmentally more benign method currently used for camphene production is the application of an acid-treated TiO$_2$ as a catalyst [56, 60], but the drawback of this method is low rate and poor selectivity [18]. Selectivity for camphene over amorphous TiO$_2$ catalyst was enhanced by synthesizing camphene in a continuous reactor system, in which the first reactor was operated at 160 °C up to 97% conversion. Thereafter, the residual α-pinene was isomerized at a lower temperature than 160 °C in order to avoid the formation of tricyclene [61].

Several heterogeneous catalysts, other than amorphous TiO$_2$, have been applied as catalysts for isomerization of α-pinene to camphene, such as Au/Al$_2$O$_3$ [54], iron supported sulfated zirconia [6] and zeolites [7].

Contrary to a conventional method of α-pinene to camphene liquid-phase transformation over acid-hydrated TiO$_2$, Au/Al$_2$O$_3$ catalyst was found to afford conversion up to 99% in continuous α-pinene vapor phase isomerization with selectivity of 60–80%, making this catalyst very promising from an industrial viewpoint (Figure 11.6). However, deactivation of the gold catalyst, observed during the reaction, might be a serious obstacle for real industrial implementation [8].

The catalyst deactivation dynamics during α-pinene isomerization [8] was considered based on a so-called "separable" deactivation model. The reaction rate under catalyst deactivation was shown to consist of two terms, one depending only on the main reaction kinetics and the other responsible for deactivation:

$$W(C, T, t) = k(T) \cdot C \cdot a(C_o, t) \tag{11.2}$$

where $a(C_o, t)$ is the empirical expression, also called relative activity or function of deactivation, which is dependent on the initial α-pinene concentration and time-on-stream. Finally, for predicting the α-pinene conversion at different temperatures and initial concentrations the following expression was obtained [8]:

$$X = 1 - e^{-k \cdot a \cdot \tau} \tag{11.3}$$

The deactivation function values for different initial concentrations were calculated from the experimental data using Eq. (11.3). The deactivation function values change from 1 to some stationary values. Activity functions for catalyst deactivation are represented in many textbooks [62]. A hyperbolic function was used, giving an adequate description of deactivation function dependence on time-on-stream (Figure 11.7).

On the basis of experimental observations the following deactivation function was proposed [8]:

$$a = \frac{1}{1 + K_t(C_0) \cdot t} \tag{11.4}$$

where $K_t(C_0)$ is the parameter of deactivation, which changes with the initial concentration of α-pinene. The parameter of deactivation dependence on α-pinene concentration is well described by a parabolic function $K_t = BC_0^2$, where $B = 64.0 \pm 8.3$ (Figure 11.8).

Due to the industrial importance of deactivation, various kinetic models which account for deactivation have been advanced. Probably the most frequently used

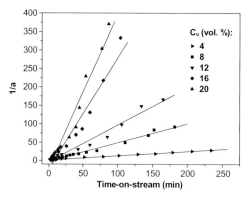

Figure 11.7 Dependence of deactivation function on time-on-stream at different initial concentration of α-pinene (C_0). Reaction conditions: 200 °C, contact time 0.33 s [8].

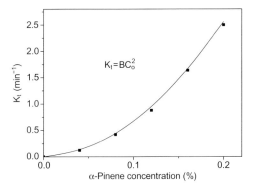

Figure 11.8 Dependence of deactivation rate constant on the α-pinene concentration. Reaction conditions: 200 °C, contact time 0.33 s [8].

approach is based on different empirical and semi-empirical equations. However, increasingly, the nature of deactivation is considered as a constituent part of the reaction scheme. This approach provides more possibilities for elucidating deactivation mechanisms, which should be an essential part of any new catalyst development. For example, often the deactivation function is expressed in terms of time. This is not the true variable, as it can lead to incomplete predictions. More correctly, the deactivation function has to be expressed in terms of the deactivating agent itself: the coke precursor or the poison, which means that the amount of coke (or poison) on the catalyst site should be known. The determination of a rate equation for the formation of the coke precursor is thus an integral part of the kinetic study of the process. In the case of α-pinene isomerization the following scheme was applied [8] to explain the catalyst deactivation from a mechanistic point of view:

$$\text{α-pinene} + * \longrightarrow \text{α-pinene*} \longrightarrow \text{products} + * \quad (11.5)$$
$$\text{α-pinene*} + \text{α-pinene*} \longrightarrow \text{deposit}$$

where * is the surface site. Taking into account the first order in α-pinene, which means its low coverage, and defining relative activity as a function of sites not occupied by deposit, an activity function was obtained.

$$a = \frac{1}{1 + k_d K_p^2 C_0^2 t} \quad (11.6)$$

where k_d is the deactivation constant. Equation (11.6) provides a mechanistic explanation for the utilization of the deactivation function in Eq. (11.4) and for the parabolic expression for the lumped constant K_t, which is defined as $K_t = BC_0^2 = k_d K_p^2 C_0^2$.

The high sintering stability of Au/Al$_2$O$_3$ catalyst found in [8] provided a possibility of restoring its activity in α-pinene isomerization by burning the hydrocarbon deposits on the catalyst surface. Regenerated Au/γ-Al$_2$O$_3$ catalyst exhibited the same conversion profile (and selectivity) to camphene (Figure 11.9) confirming complete regeneration of catalyst, contrary to the results obtained over Fe–Mn-promoted sulfated zirconium oxide [63] and H$_3$PW$_{12}$O$_{40}$/TiO$_2$ [64] when only partial restoration of activity was achieved.

Industrial implementation of α-pinene to camphene isomerization with periodical regeneration may be envisaged, as presented in Figure 11.10.

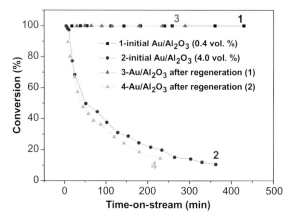

Figure 11.9 Catalyst activity before and after regeneration, temperature 200 °C, contact time 0.33 s: (1) and (3) α-pinene concentration 0.4 vol%, fresh and spent Au/Al$_2$O$_3$, correspondingly, (2) and (4) α-pinene concentration 4.0 vol%, fresh and spent Au/Al$_2$O$_3$, correspondingly [8].

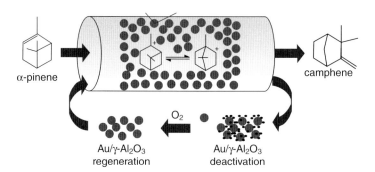

Figure 11.10 Scheme of continuous α-pinene to camphene isomerization over Au/Al$_2$O$_3$ [8].

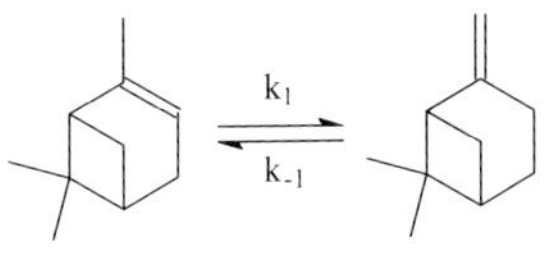

Figure 11.11 Isomerization of α-pinene to β-pinene.

In addition to isomerization to camphene, as discussed above, α-pinene can be isomerized to β-pinene (Figure 11.11) [14]. Although the equilibrium yield of β-pinene is only 4% [25], the process is industrially feasible via fractional distillation. The isomerization of α-pinene is reported to take place in the presence of acid catalysts such as 23 mol% benzoic acid at 155 °C [65] and titanic acids [66]. As a result a mixture of various undesirable by-products other than β-pinene is formed, which hinders significantly efficient separation of α-pinene from β-pinene. The isomerization of α-pinene to β-pinene is observed in the basic solution, for example in the presence of dimethyl sulfoxide with potassium tert-butoxide at 55 °C [67], and much more reactive potassium 3-aminopropylamide in 3-aminopropylamine at 0 °C [68] to give yields of β-pinene of 0.5%. Base-catalyzed isomerization avoids possible rearrangements of the four-member pinene cycle. The disadvantage of this reaction is an unfavorable reactant/catalyst mass ratio of only 1.04–1.4 g β-pinene/g catalyst. Solid base-catalyzed isomerization of α-pinene is carried out over CaO, SrO, MgO evacuated at 900–1300 °C, and superbases consisting of an alkali metal hydroxide and the alkali metal itself supported on γ-Al_2O_3 in the temperature range 25–150 °C [69]. However, very efficient superbase catalysts underwent deactivation. Another β-pinene preparation method is comprised of four steps involving a reductive fission of alkene phosphonates with lithium hydride, leading to β-pinene with an overall yield of 25% [70]. Two-step isomerization of α- to β-pinene via allylstannane afforded a yield of β-pinene of 50% [71]; while a process via hydroboration with diborane, thermal isomerization of the intermediate organoborane, and substitution either with a high-boiling olefin or benzaldehyde gave β-pinene in 51% yield [72]. Schlosser allylic metalation of α-pinene with n-butyllithium and potassium tert-butoxide, with subsequent treatment of the potassium salt with trimethyl borate and hydrolysis, resulted in a 92% yield of β-pinene [73].

Isomerization of α-pinene to β-pinene over Pd/C, Ru/C, Rh/C, Pt/C, Ir/C catalysts was studied in [15, 16] showing a possibility of β-pinene production by α-pinene isomerization over Group VIII metals. Selective catalytic systems on the basis of Pd and Rh metals were suggested. Optimal reaction conditions for β-pinene production in the presence of hydrogen were found when the maximum yield of β-pinene was equal to 2.8 wt.% at 100 °C and 11 bar of hydrogen. It was found that the reaction rate of α-pinene isomerization is first order in α-pinene concentration and 0.5 order in hydrogen pressure for Pd/C, Ru/C, Ir/C and 1.0 order for Rh/C, Pt/C catalysts. Obviously, it is concerned with different double bond isomerization mechanisms under the investigated conditions. The Gibbs energy of the reaction and the composition of the equilibrium mixture of pinene isomers were estimated in the temperature range 20–90 °C over the most active Pd/C catalyst. At 25 °C the α-/β-pinene ratio is 98.7/1.3 and at 100 °C is equal 97.2/2.8.

Campholenic aldehyde is the desired product in the isomerization of α-pinene oxide. Reaction routes for both the oxidation of α-pinene and the epoxidation of α-pinene oxide are shown in Figure 11.12. This aldehyde is used as a sandalwood-like fragrance [75] and synthesized with high selectivities (85%) using Lewis acidic $ZnCl_2$ or $ZnBr_2$ [50] causing, however, disposal problems.

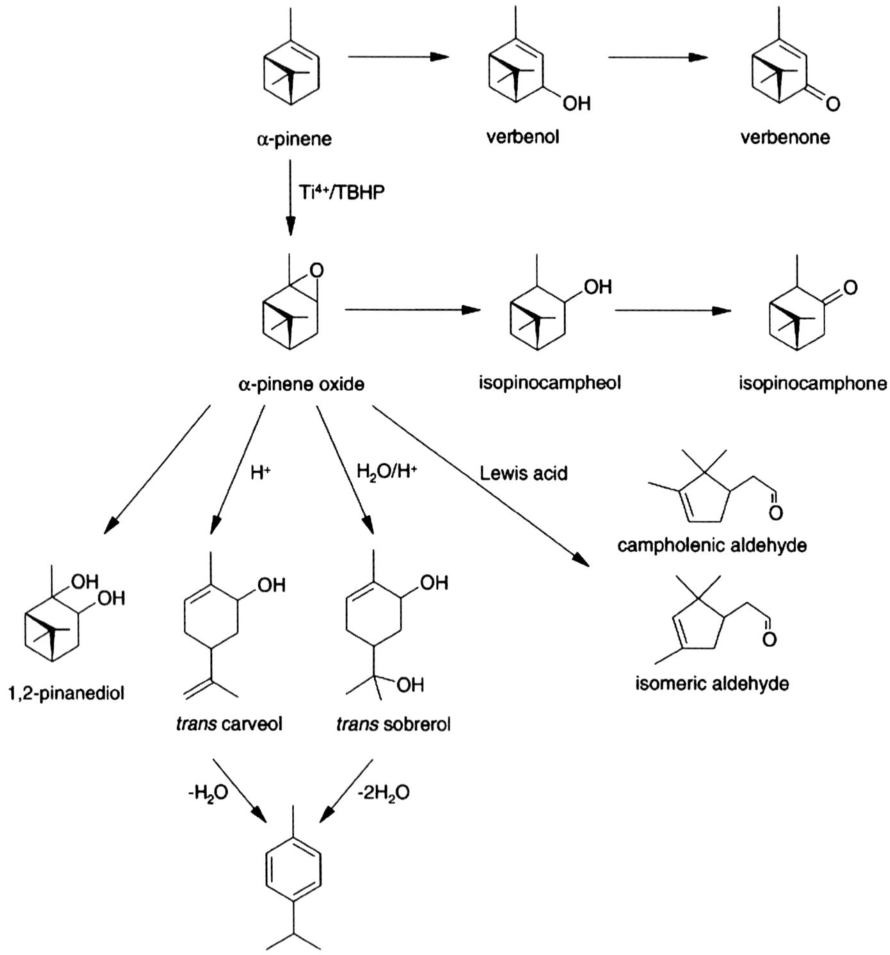

Figure 11.12 The reaction network in catalytic oxidation and epoxidation of α-pinene and isomerization and hydrolysis of α-pinene oxide [74].

Campholenic aldehyde has also been produced in the gas phase in a continuous tubular reactor containing a catalyst [12]. In addition to a tubular reactor, a spinning disc reactor was applied for transformation of α-pinene to campholenic aldehyde [11] over immobilized zinc triflate supported on silica. Since the synthesis of campholenic aldehyde is a complex reaction (Figure 11.12), it is important to shorten the residence time in a spinning disc reactor. At the same time the product selectivity could be increased. When comparing the batch operation with a continuous spinning disc reactor, the results from the isomerization of α-pinene oxide at 45 °C in 1,2-dichloroethane as a solvent were very promising in terms of reaction time, feed and conversion. These key parameters were for batch and continuous operations 3000 and 1 s, 1 and 60 g h^{-1}, 50 and 85% conversion, respectively [11]. Only selectivity towards campholenic aldehyde was slightly lower under the studied conditions, namely 80% compared to 75%. Selectivity towards campholenic aldehyde decreased with increasing contact time, being 75% with 0.2 s contact time and decreasing to less than 15% with 1 s contact time. Furthermore, the catalyst separation step can be avoided in the continuous operation. Moreover, it was stated that heat transfer was improved in this reactor. A schematic picture of the spinning disc reactor is shown in Figure 11.13.

One example of process intensification, as proposed by [39] is to use only one reactor pot together with bifunctional or several catalysts (Figure 11.14). One-pot synthesis for producing campholenic aldehyde directly from α-pinene has been proposed in [61] over bifunctional catalysts, which contain both a metal function for the oxidation reaction together with an organic oxidant, for example tert-butyl hydroxyperoxide and Lewis acidic sites, which are present for example in mesoporous hexagonal molecular sieve (HMS) catalyst [74]. An important step in achieving high yields of campholenic aldehyde was to use a drying agent, MgSO$_4$, in the reactor in the absence of oxygen in order to avoid radical oxidation. Molecular oxygen was stated to be able to initiate radical auto-oxidation of terpenes.

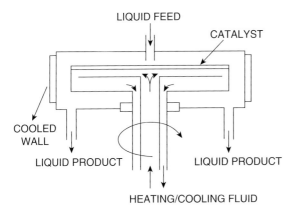

Figure 11.13 A schematic picture of a spinning disc reactor [11].

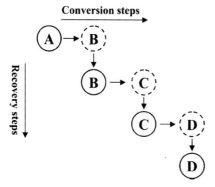

Figure 11.14 A schematic picture of one-pot methodology [39].

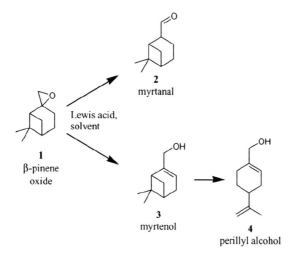

Figure 11.15 Isomerization of β-pinene oxide [17].

Isomerization of β-pinene oxide is typically performed over homogeneous Lewis acid catalysts, analogously to the case of α-pinene oxide. Solid Lewis acidic catalysts, such as Zr-beta zeolite have been active and selective catalysts in the isomerization of β-pinene oxide [17]. The main products in this reaction are myrtanal, myrtenol and perillyl alcohol (Figure 11.15). The product distribution can be adjusted by selecting the catalyst properties as follows. Perillyl alcohol formation is catalyzed by Brønsted acid sites, whereas Lewis acid sites enhance formation of myrtanal. Suitable heterogeneous catalysts for synthesis of perillyl alcohol are ion-exchanged resins treated with nitric acid [76]. Perillyl alcohol was also synthesized by oxidizing limonene to perillyl aldehyde with toxic homogeneous catalysts, followed by reduction to the corresponding alcohol. Perillyl alcohol exhibits antimicrobial effects [77] being the main component in lavender oil [35].

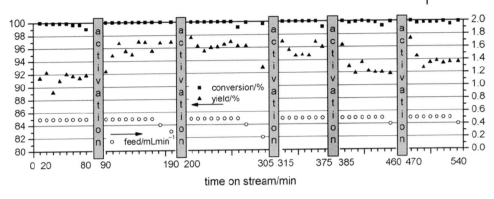

Figure 11.16 Continuous process for isomerization of β-pinene oxide to myrtanal [17].

Relevant from the industrial point of view, a continuous fixed bed flow reactor was investigated in β-pinene oxide isomerization to myrtanal [17]. Zr-beta zeolite was applied as a catalyst, being very active and selective towards myrtanal. When the catalyst activity was decreased after 80 min time-on-stream, it was possible to reactivate the catalyst under air flow at 500 °C, after which the reaction was continued again for 80 min. Several cycles of reaction–reactivation were demonstrated (Figure 11.16). Furthermore, a comparative investigation of batch and continuous operation was performed using acetonitrile as a solvent at 80 °C [17]. The batch experiments were started with 600 mg of β-pinene epoxide and, since the amount of the recovered and reactivated catalyst decreased in each run, a total of 37.2 g of epoxide was converted to myrtanal, corresponding to complete conversion in each batch after 2 h. The TONs, mol of converted epoxide per mol of metal, for batch and continuous operation were calculated, giving the values, 3878 and 2130, respectively. Thus it can be concluded that 1.8-fold TON was achieved in the batch mode compared to continuous operation. The corresponding selectivities for myrtanol in batch and continuous mode were 95% and 92%, respectively. The zeolite crystals remained quite stable in both reactors and no leaching of Zr occurred.

11.3.2
Oxidation of Monoterpenes

The chemistry behind terpene oxidation was far from green at the beginning of the nineteenth century, since typically chromic acid and chromium trioxide were used as stoichiometric oxidants [26, 78]. For example α-pinene is oxidized to myrtenal and verbenone (Table 11.4) [26]. Furthermore, cis-3-pinene-2-ol is a fraction separated from the oxidation of α-pinene with chromic acid. This unsaturated alcohol was further oxidized to verbenone using chromic acid mixture [78]. As mentioned above, α-pinene is a valuable raw material used for the synthesis of camphene by isomerization, as well as for oxidation and epoxidation [29–31].

Table 11.4 Hydration of terpenes.

Compound	Product	Conventional method	Catalytic method
α-pinene	α-terpineol	Hydration with mineral acids to terpin hydrate, followed by dehydration to α-terpineol [79–81]	
α-pinene oxide	trans-sobrerol	Hydrolysis of α-pinene oxide in acidic solution [82]	
Limonene	α-terpineol	Homogeneous acids	Hydration Aluminosilicate [83]
(+)-isobornyl acetate	Borneol	Hydrolysis of bornylacetate to borneol [58]	

Instead of using chromium compounds as stoichiometric oxidants, organic peroxides have also been applied, together with heterogeneous catalysts containing chromium [28]. However, since organic peroxides are not very cheap [84], there is an economic incentive to use molecular oxygen [32]. α-Pinene was also recently epoxidized with hydrogen peroxide as an oxidant over gold supported on mesoporous organosilicates [52].

The products from the oxidation of β-pinene are pinocarveol, pinocarvone and myrtenol, whereas oxidation of limonene leads to carvone and carveol [78]. These products have various applications, for example pinocarvone and pinocamphene have been applied as insect pheromones [33], pinocarveol inhibits bacterial growth [85], and they are used in flavors and fragrances [34]. Myrtenol is utilized as a fragrance [76], whereas myrtanal has antiseptic effects [76]. Perillyl alcohol was also produced via β-pinene oxidation with benzyl oxide, followed by hydrolysis of the product [86].

Various monoterpenes, such as α-pinene [28], β-pinene, and limonene have been oxidized over heterogeneous catalysts. Recently terpene oxidation was demonstrated with molecular oxygen under mild conditions with Cr immobilized on mesoporous MCM-41 [86]. The products from α-pinene oxidation are pinocarvone [85], verbenone and verbenol [32], whereas in β-pinene oxidation over Cr-MCM-41 four different products are formed in about equal yields, that is, pinocarveol, pinocarvone, myrtenal and myrtenol [28, 32], Limonene oxidation over the same catalyst produced carveol and carvone. Verbenone is applied for the production of taxol, which is used for cancer treatment [87]. Limonene was also oxidized with toxic SeO_2 or CrO_3 to perillyl aldehyde, which can be easily reduced to perillyl alcohol [35].

11.3.3
Hydrogenation of Monoterpenes

α-Pinene has been hydrogenated in a continuous reactor over Ru [88], Pd/C [89] catalyst, preferentially to *cis*-pinane (Figures 11.17 and 11.18), which is further

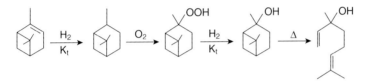

Figure 11.17 Reaction route for synthesis of linalool via hydrogenation of α-pinene to cis-pinane followed by peroxy oxidation to pinalol, which after heat treatment forms linalool [89].

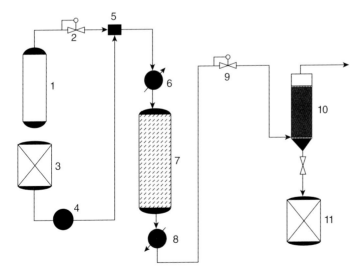

Figure 11.18 Continuous process for hydrogenation of α-pinene [88], notation: 1. reservoir, 2. and 9. control valve, 3. tank, 4. pump, 5. mixer, 6 and 8 heat exchangers, 7. reactor, 10. separator, 11 collector tank.

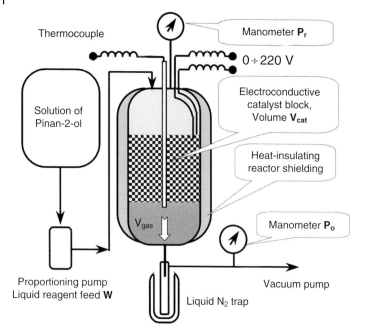

Figure 11.19 Schematic view of flow reactor for pinane-2-ol thermal isomerization into linalool [19].

oxidized to pinane-2-hydroperoxide by molecular oxygen, followed by hydrogenation over Pd/C to pinane-2-ol [89]. Thereafter, pinane-2-ol is transformed by pyrolysis over carbon catalyst block to linalool, which is used in perfumes [19, 88].

To improve the selectivity to linalool during pinane-2-ol isomerization the reaction was performed in a specially constructed heat-insulated flow reactor having no external heater (Figure 11.19) [19]. The electroconductive catalyst block, equipped with a thermocouple and capable of being heated uniformly by electric current, was located inside the reactor. The catalyst block was manufactured from stainless steel capillary with a nickel–chromium wire covered with refractory insulation inserted inside the capillary. The capillary could be warmed by passing electric current through this wire. The external surface of the catalyst block was covered with pyrolytic carbon by passing butadiene flow through the bed at 600 °C for 30 min.

11.3.4
Epoxidation of Monoterpenes

Epoxides of monoterpenes are intermediates for the synthesis of flavors and fragrances and therapeutically active agents [90]. Epoxides are either isomerized or transformed to the corresponding unsaturated aldehydes and alcohols via aldol-condensation. Epoxidation of monoterpenes is conventionally performed using

Table 11.5 Esterification and etherification of terpenes.

Compound	Product	Conventional method	Catalytic method
α-pinene	α-terpenyl acetate	Phosphoric acid [79]	Heteropolyacids [91]
(+)-limonene	α-terpinyl acetate	$Fe_2(SO_4)_3$	Aluminosilicate [83]
Camphene	(+)-isobornyl acetate	Condensation with acetic acid with sulfuric acid catalyst [92]	Bornyl acetate continuous process with Amberlyst 15 [58]
Camphene	Alkoxycamphane	Alkoxylation with sulfuric acid [93]	Alkoxylation with cation exchange resin [94]

chromic acid or peracids as homogeneous catalysts (Table 11.5). For example, α-pinene was epoxidized with CrO_3 giving α-pinene oxide [27]. A more environmentally friendly method is to use a peracid as an oxidizing agent in the production of α-pinene oxide [51], although earlier autooxidation of terpenes was also reported [78]. The drawback of using peracid as an oxidant is the formation of a corresponding acid. If an acid acceptor, such as alkaline carbonate is also added, above 90% conversion of α-pinene could be achieved, although at the same time alkali acetate is formed [51].

Epoxidation of β-pinene is conventionally carried out analogously to α-pinene epoxidation using peracid as an oxidant and giving β-pinene oxide as the main product. (R)-Limonene epoxidation followed by isomerization of the formed terpinolene oxide to karahanaenone has been demonstrated over a clay catalyst (Figure 11.20). Thereafter, the ketone was reduced over lithium aluminum hydride to karahanaenol, which has a fruity odor and can be used in fragrances [95, 96].

Epoxidation of terpenes, such as limonene, has been studied over several heterogeneous catalysts, such as niobium silicates [97] and Ti supported on mesoporous silica [53]. Hydrogen peroxide or *tert*-butylhydroxide peroxide are used as oxidants [53], with the former being a more eco-friendly and cheaper one [84].

11.3.5
Hydration of Monoterpenes

Unsaturated terpenic alcohols are easily synthesized via hydration of terpenes or their epoxides (Table 11.6). Typically, mineral acids, such as phosphoric acid, are used as catalysts in this reaction [79]. α-Pinene is transformed via terpine hydrate

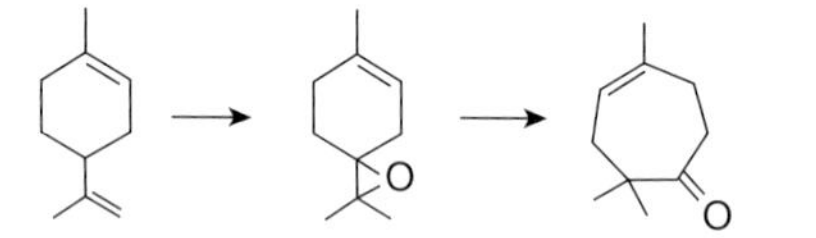

limonene terpinolene oxide karahanaenone

Figure 11.20 Synthesis of karahanaenone from terpinolene oxide [95].

Table 11.6 Micellaneous reactions.

Compound	Product	Conventional method	Catalytic method
 β-pinene	 Cis-pinane		Ru-catalyst, flow reactor [88]
 β-pinene	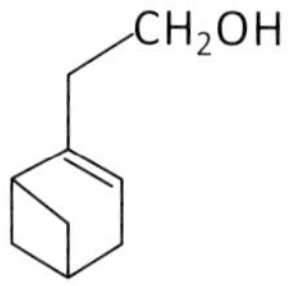 Nopol	Prins reaction with β-pinene and folmaldehyde with $ZnCl_2$ [98]	Mesoporous iron phosphate [99], Sn-SBA-15 [100] Sn-MCM-41 [101]
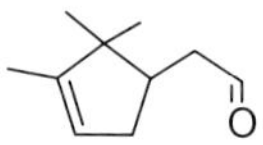 Campholenic aldehyde	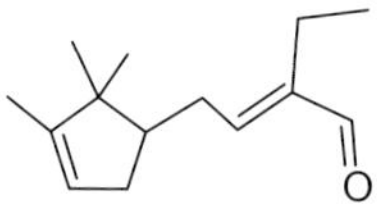 Levosandal	Sodium ethoxide [102]	Al/Mg mixed oxides [103]
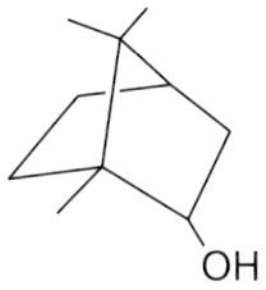 Borneol	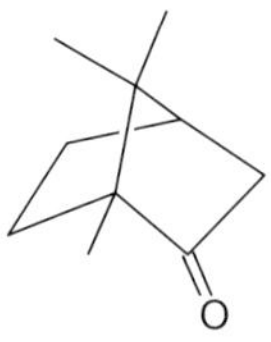 Camphor	Dehydrogenation of borneol over basic copper carbonate	[104]

to α-terpineol [80, 105], which is used in cosmetics, pesticide and flavor preparations [30]. Terpineol is listed as a top-30 commonly used flavor compound [81]. α-Terpineol is also a commercial product with a lilac odor [80].

A solid aluminosilicate is an environmentally benign catalyst for hydroxylation of limonene under an inert atmosphere in a water-containing solvent [83]. *trans*-Sobrerol is the desired product formed in the hydrolysis of α-pinene oxide (Table 11.6), which was formed in epoxidation using chromic acid as a catalyst [27, 82]. *trans*-Sobrerol is used as a drug for respiratory diseases [106, 107].

11.3.6
Esterification and Etherification of Monoterpenes

Esterification and etherification of monoterpenes are important reactions in the production of intermediates, for example, for camphor synthesis. Furthermore, monoterpene esters have found applications in fragrances.

Extremely hazardous chromium trioxide has been used as a catalyst to esterify α-pinene with acetic acid in the presence of lead tetraacetate. After alkali hydrolysis the ester was hydrolyzed to verbenyl acetate [108]. α-Terpinylacetate has been prepared using phosphoric acid as a catalyst in the esterification of α-pinene [79] or alternatively by using iron sulfate as a homogeneous catalyst together with limonene and acetic acid [109]. Solid aluminosilicates are also promising catalysts for the esterification of limonene with acetic acid to terpenyl acetate [110]. In addition, carene acetylation was demonstrated using $ZnCl_2$ [79].

In order to replace the above-mentioned methods used in terpene esterification, solid acid catalysts have also been used in, for example, acetoxylation of α-pinene. One solid acid catalyst applied for this reaction was a Keggin-type heteropolyacid, giving higher activities than achieved with Amberlyst 15 or sulfuric acid [91]. The selectivities for α-terpenylacetate were also very high, above 90% using heteropolyacid as a catalyst.

Camphene, a product from α-pinene isomerization, can be esterified with acetic acid using sulfuric acid [111] or sulfonated styrenedivinylbenzene polymers [92] as a catalyst to produce isobornylacetate, which is a perfume [80]. Bornylacetate has also been produced in a continuous process starting from camphene and using Amberlyst 15 as a catalyst [58]. In this upflow reactor both acetic acid and acetic anhydride were fed into the reactor together with camphene in order to minimize the amount of free water formed during the esterification. It was also stated that high conversion and selectivity to the desired ester were achieved when using substoichiometric amounts of camphene [58]. Catalyst deactivation, however, occurred since the feed rate decreased within 7 days by 7% and, in order to maintain high production capacity, the reaction temperature was gradually increased from the initial level of 28 to 48 °C. Furthermore, some catalyst reactivation could be observed when feeding nitrogen through the catalyst bed. Camphor is prepared from bornylacetate via hydrolysis followed by formation of bornylalcohol, which can in turn be dehydrogenated over copper carbonate to camphor (see Figure 11.4). Camphor was also prepared from borneol via catalytic dehydrogenation, when

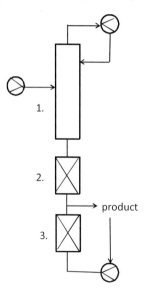

Figure 11.21 Continuous dehydrogenation of isoborneol to camphor over Cu–Ni catalyst. Notation: 1. distillation column, 2. catalytic reactor, 3. second catalytic reactor. Adapted from [104].

alcohol was led through a copper tube at 350 °C together with water vapor, followed by camphor crystallization after distillation [112]. A continuous process for dehydrogenation of isoborneol to camphor which comprises reaction and separation was reported in [104]. Dehydrogenation is carried out over Cu–Ni catalyst in toluene as a solvent using a distillation column, which has an upper rectification and lower zone (Figure 11.21). The latter zone is split into two sections with different catalysts amounts (1 : 0.15–0.35 ratio). The product is withdrawn downstream of the upper catalytic section, which has higher bed length. The reactants are fed in the distillation column, which allows one to accomplish the process at 200–210 °C with lower capital costs and better product quality [104]. Alternatively, alkyxocamphane can be synthesized via etherification of camphene using either sulfuric acid [93] or cation exchange resin as a catalyst [94]. The formed ether is thereafter oxidized, either by NO_2 or sulfuric acid, and camphor is precipitated by addition of sodium hydroxide.

11.3.7
Aldol Condensation of Monoterpene Derivatives

Campholenic aldehyde can be further transformed via base-catalyzed aldol-condensation to different conjugated aldehydes [79], which are further reduced to the corresponding alcohols, having sandalwood odor. Conventionally, homogeneous alkali, such as KOH, was used in the aldol condensation, followed by reduction of aldehyde to the corresponding alcohol with aluminum isopropoxide. For example, sodium ethoxide was used as a catalyst to prepare levosandal (Table 11.6) [102]. As, from the process intensification and environmental point of views, solid catalysts are preferred, this process was demonstrated also with solid alkali catalyst, such as mixed Al/Mg mixed oxides, which were loaded in the first reactor. The whole process was performed in a cascade reactor, in which the formed levosandal was reduced via a Meerwein–Ponndorf–Verley (MPV) reaction, using 2-propanol as a hydrogen source together with aluminum isopropoxide as a catalyst [103].

The reaction between β-pinene and formaldehyde gives as a product an unsaturated alcohol, called Nopol, which is used as a fragrance and pesticide [98, 113] (Table 11.6). Nopol has been traditionally synthesized using homogeneous catalysts, such as $ZnCl_2$ or acetic acid [100, 101]. Recently solid acid catalysts have shown very promising results in Nopol synthesis, especially mesoporous Sn-SBA-15 [100] and Sn-MCM-41 [101, 114].

11.4
Conclusions

The synthesis of fine chemicals has special features in industrial synthesis, such as typically high E-factors, thus producing a lot of waste per mass of product. Moreover, the yields and selectivities should be high. On the other hand the product capacities are relatively low, which makes the batch operation more feasible, especially taking into account the possibility to use multipurpose reactors in a campaign fashion. Conventionally, fine chemicals have been produced via scaling up the traditional organic chemistry synthesis routes. These methods are, however, far from being environmentally benign processes. There is an urgent need to develop more environmentally friendly processes also for the fine chemical industry.

The trend is to use heterogeneous solid acids and bases as catalysts, or even bifunctional catalysts which can combine for example isomerization and epoxidation steps in one-pot. The benefits of using heterogeneous catalysts are obvious as they facilitate easy catalyst separation and reuse. If heterogeneous catalysts are deactivated, there is a possibility to regenerate them, as was demonstrated especially for zeolites in fine chemicals applications.

In order to replace batch reactors with continuous ones, several types of reactors have been applied, for example fixed bed operation and spinning disc reactors. There are some specific advantages when using continuous reactors. For example, a narrow residence time can be obtained in the spinning disc reactor, which is beneficial for the synthesis of campholenic aldehyde via isomerization of α-pinene oxide, since this reaction proceeds via a complex reaction route. Moreover, continuous hydrogenation of α-pinene was demonstrated within a short residence time giving the same production capacity as the batch reactor within 6 h.

Few examples are found in the open literature confirming that the conventional fine chemicals industry approach is shifting towards more efficient processes applying heterogeneous catalysts and continuous reactor operations. The most important processes for utilizing extractives as raw materials are the production of sitostanol, myrtanal, campholenic aldehyde, linalool, and terpenol, which have applications as food additives, cosmetics and intermediates for the synthesis of pharmaceuticals. Due to the increasing production of ethanol from lignocellulose the amounts of produced CTO and turpentine are also expected to increase and thus the production of value-added chemicals derived from these fractions is also expected to grow.

11.5
Acknowledgment

This work is part of the activities at the Åbo Akademi Process Chemistry Centre within the Finnish Centre of Excellence Programme (2000–2011) by the Academy of Finland.

References

1 Mäki-Arvela, P., Holmbom, B., Salmi, T., and Murzin, D.Y. (2007) *Catal. Rev.*, **49** (3), 197–340.
2 Fengel, D., and Wegener, G. (1984) *Wood Chemistry, Ultrastructure, Reactions*, De Gruyter, New York.
3 Harada, T., and Yumoto, T. (1976) US 4076700.
4 Meyer, A. (1911) DE 272562.
5 Nagai, K. (1961) *Bull. Chem. Soc. Japan*, **34**, 1825–1827.
6 Liu, J., Zhao, J., Miao, C., Yue, Y., and Hua, W. (2011) *Chin. J. Chem.*, **29**, 1095–1100.
7 Rachwalik, R., Olejniczak, Z., Jiao, J., Huang, J., Hunger, M., and Sulikowski, B. (2007) *J. Catal.*, **252**, 161–170.
8 Solkina, Y.S., Reshetnikov, S., Estrada, M., Simakov, A.V., Murzin, D.Y., and Simakova, I.L. (2011) *Chem. Eng. J.*, **176–177**, 42–48.
9 Chimal-Valencia, O., Robau-Sanchez, A., Collins-Martinez, V., and Aguilar-Elguezabal, A. (2004) *Bioresour. Technol.*, **93**, 119–123.
10 Kaminska, J., Schwegier, M.A., Hoefnagel, A.J., and van Bekkum, H. (1992) *Recl. Trav. Chim. Pays-Bas*, **111**, 432–437.
11 Vivevic, M., Jachuck, R.J.J., Scott, K., Clark, J.K., and Wilson, K. (2004) *Green Chem.*, **6**, 533–537.
12 Hu, J., Peng, C., Huang, W., Zhang, J., and Fang, Y. (2010) CN 101885671.
13 Ravasio, N., Zaccheria, F., Guidotti, M., and Psaro, R. (2004) *Top. Catal.*, **27**, 157–168.
14 Derfer, J. (1966) US patent 3 278 623.
15 Deliy, I.V., and Simakova, I.L. (2008) *React. Kinet. Catal. Lett.*, **95**, 161–174.
16 Deliy, I.V., and Simakova, I.L. (2008) *Russ. Chem. Bull.*, **57**, 2021–2028.
17 de la Torre, O., Renz, M., and Corma, A. (2010) *Appl. Catal. A: Gen.*, **380**, 165–171.
18 Corma, A., Iborra, S., and Velty, A. (2007) *Chem. Rev.*, **107**, 2411–2502.
19 Semikolenov, V.A., Ilyna, I.I., and Simakova, I.L. (2002) *J. Mol. Catal. A: Chem*, **182–183**, 383–393.
20 European Commission (13.3.2003) Opinion of the Scientific Committee on Food on Applications for Approval of a Variety of Plant-Sterol-Enriched Foods, SCF/CS/NF/DOS/15 ADD 2 Final, http://europa.eu.int/comm./food/fs/sc/scf/index_en-html (accessed 13.12.2011).
21 Masten, S. (2002) Turpentine, Review of Toxicological Literature, National Institute of Environmental Health

Sciences, Research Triangle Park, North Carolina, N01-ES-65402.
22 Thompson, A., Cooper, J., and Ingram, L.L., Jr. (2006) *Forest Prod. J.*, **56**, 46–48.
23 Drew, J., and Pylant, G.D. (1966) *Tappi*, **49**, 430–438.
24 Kimland, B., and Norin, T. (1972) *Sven. Papperstidning*, **75**, 403–409.
25 Osadchii, S.A., and Tolstikov, G.A. (1997) *Khim. Interes. Ust. Razv.*, **5**, 79–93.
26 Vegezzi, S. (1980) US 4190675.
27 Rothenberg, G., Yatziv, Y., and Sasson, Y. (1998) *Tetrahedron*, **54**, 593–598.
28 Lempers, H.E.B., and Sheldon, R.A. (1996) *Appl. Catal. A: Gen.*, **143**, 137–143.
29 Semikolenov, V.A., Simakova, I.L., Maksimchuk, N.V., and Ancel, J.E. (2003) Pat. Russia 2235712 (Publ. 2004).
30 Ancel, J.E., Maksimchuk, N.V., Simakova, I.L., and Semikolenov, V.A. (2004) *Appl. Catal. A: Gen.*, **272**, 109–114.
31 Simakova, I.L., and Semikolenov, V.A. (2003) *Chem. Sust. Dev.*, **11**, 271–275.
32 Robles-Dutenhefner, P.A., Brandao, B.B.N.S., De Sousa, L.F., and Gusevskaya, E.V. (2011) *Appl. Catal. A: Gen.*, **399**, 172–178.
33 Mitsubishi Chemical Industries Co, Ltd (1981) Jpn Kokai Tokkyo Koho, 8130905, *Chem. Abs.* **95**, 37124f.
34 Arctander, A. (1969) *Perfume and Flavor Chemicals*, vol. 2, Monteclair, New York, pp. 2624–2622.
35 Chastain, D.E., Mody, N., and Majetich, G. (1998) US5994598.
36 Centi, G., and van Santen, R.A. (2007) *Catalysis for Renewables*, Wiley-VCH Verlag GmbH, Weinheim.
37 Sheldon, R.A. (1997) *Chem. Ind.*, **1**, 12–15.
38 Anastas, P.T., and Warner, J.C. (1998) *Green Chemistry: Theory and Practice*, Oxford University Press, New York.
39 Bruggink, A., Schoevaart, R., and Kieboom, T. (2003) *Org. Proc. Res. Dev.*, **7**, 622–640.
40 Swift, K.A.D. (2004) *Top. Catal*, **27** (1–4), 143–155.
41 Grundy, S.M., and Mok, H.Y.I. (1976) Effects of low-dose phyto-sterols on cholesterol absorption in man, in *Lipoprotein Metobolism* (ed. H. Greten), Springer-Verlag, Berlin, Heidelberg, New York, 112–118.
42 Pollak, O.J. (1953) *Circulation*, **7**, 702–706.
43 Miettinen, T., Vanhanen, H., and Wester, I. (1996) US 5502045.
44 Erikson, B.A. (1971) DE2035069.
45 Heinemann, T., Leiss, O., and von Bergmann, K. (1986) *Atherosclerosis*, **61**, 219–223.
46 Lindroos, M., Mäki-Arvela, P., Kumar, N., Salmi, T., and Murzin, D.Y. (2002) *Catalysis of Organic Reactions*, Marcel Dekker Inc., New York, pp. 587–594.
47 Mäki-Arvela, P., Martin, G., Simakova, I., Tokarev, A., Wärnå, J., Hemming, J., Holmbom, B., Salmi, T., and Murzin, D.Y. (2009) *Chem. Eng. J.*, **154**, 45–51.
48 Wärnå, J., Geant, M.F., Salmi, T., Hamunen, A., Orte, J., Hartonen, R., and Murzin, D.Y. (2006) *Ind. Eng. Chem. Res.*, **45**, 7067–7076.
49 Haario, H. (2001) *MODEST User's Manual*, Profmath Oy, Helsinki, Finland.
50 Moulijn, J.A., Makkee, M., and van Diepen, A. (2001) *Chemical Process Technology*, John Wiley & Sons, Inc., New York.
51 Häberlein, H., and Scheidl, F. (1978) DE-Os 2 835 940.
52 Yu, N., Ding, Y., Lo, A.-Y., Huang, S.-J., Wu, P.-H., Liu, C., Yin, D., Fu, Z., Yin, D., Lei, Z., and Liu, S.-B. (2011) *Micropor. Mesopor. Mater.*, **143**, 426–434.
53 Guidotti, M., Batonneau-Gener, I., Gianotti, E., Marchese, L., Mignard, S., Psaro, R., Sgobba, M., and Ravasio, N. (2008) *Micropor. Mesopor. Mater.*, **111**, 39–47.
54 Simakova, I.L., Solkina, Y.S., Moroz, B.L., Simakova, O.A., Reshetnikov, S.I., Prosvirin, I.P., Bukhtiyarov, V.I., Parmon, V.N., and Murzin, D.Y. (2010) *Appl. Catal. A: Gen.*, **385**, 136–143.
55 Schwend, E., and Schmidt, K. (1931) DE 578 569.
56 Gscheidmeier, M., Häberlein, H., Häberlein, H., Häberlein, J., and Häberlein, M. (1998) US 5826202.
57 Erman, W.E. (1985) *Chemistry of Monoterpenes: An Encyclopedic Handbook*, Marcel Dekker, New York.
58 Gscheidmeier, M., Gutmann, R., Wiesmuller, J., and Riedel, A. (1997) US patent 5596127.

59 Schering-Kahlbaum (1931) DE 578 569.
60 Etzel, G. (1949) US 2551795.
61 Gscheidmeier, M., Häberlein, H., Häberlein, H.H., Häberlein, J.T., and Häberlein, M.C. (1966) US 758145.
62 Froment, G.F., and Bischoff, K.B. (1979) *Chemical Reactor Analysis and Design*, John Wiley & Sons, Inc., New York.
63 Comelli, N.A., Ponzi, E.N., and Ponzi, M.I. (2006) *Chem. Eng. J.*, **117**, 93–99.
64 Grzona, L., Masini, O., Comelli, N., Ponzi, E., and Ponzi, M. (2005) *React. Kinet. Catal. Lett.*, **84**, 199–204.
65 Settine, R.L. (1970) *J.Org. Chem.*, **35**, 4266–4267.
66 Rudakov, S. (1955) *Zh. Obshch. Khim.*, **25** (3), 627–631.
67 Ohnishi, R., and Tanabe, K. (1974) *Chem. Lett.*, 207–210.
68 Brown, C.A. (1978) *Synthesis*, 754–755.
69 Gorzawski, H., and Hoeldrich, W.F. (1999) *J. Mol. Catal. A: Chem.*, **144**, 181–187.
70 Harwood, L.M., and Julia, M. (1980) *Synthesis*, 456–457.
71 Andrianome, M., and Delmond, B. (1985) *J. Chem. Soc., Chem. Commun.*, 1203–1204.
72 Midland, M.M., Petre, J.E., Zderic, S.A., and Kazubski, A. (1982) *J. Am. Chem. Soc.*, **104**, 528–531.
73 Brown, H.C., Zaidlewcz, M., and Bhatt, K.S. (1989) *J. Org. Chem.*, **54**, 1764–1766.
74 Suh, Y.-W., Kim, N.-K., Ahn, W.-S., and Rhee, H.-K. (2003) *J. Mol. Catal. A: Chem.*, **198**, 309–316.
75 Liebens, A., Mahaim, C., and Hölderich, W.F. (1997) *Stud. Surf. Sci. Catal.*, **108**, 587–594.
76 Corma, A., Renz, M., and Susarte, M. (2009) *Top. Catal.*, **52**, 1182–1189.
77 Chastain, D.E., Sanders, W.E., Jr., and Sanders, C.C. (1992) US 5110832.
78 Langner, S., and Rollinson, P. (1953) US 761686.
79 Bledsoe, J.O. (1997) Terpenoids, in *Kirk-Othmer Encyclopedia of Chemical Technology*, vol. 23, 4th edn (eds J.I. Korschwitz and M. Howe-Grant), J. Wiley & Sons, pp. 833–882.
80 Brown, W.B. (1936) *J. Soc. Chem. Ind.*, **55**, 321.
81 Bauer, K., Garbe, D., and Surburg, H. (2001) *Common Fragrance and Flavor Materials. Preparation, Properties and Uses*, 4th edn, Wiley-VCH Verlag GmbH, Weinheim, p. 304.
82 Corvi-Mora, C. (1987) US 4639469.
83 Hiroki, I., and Yasunao, K. (2010) JP 2010010532 (A).
84 Hoelderich, W.F., and Kollmer, F. (2000) *Pure Appl. Chem.*, **72**, 1273–1287.
85 Hideaki, M., and Izuru, Y. (1988) JP 63030409.
86 Willis, C.R., and Walling, C.T. (1976) CA 981695.
87 Wender, P.A., and Mucciaro, T.P. (1992) *J. Am. Chem. Soc.*, **114**, 5878–5879.
88 Laitinen, A., Aaltonen, O., and Kaunisto, J. (2002) WO2002072508.
89 Semikolenov, V.A., Ilyna, I.I., and Simakova, I.L. (2001) *Appl. Catal. A: Gen.*, **211**, 91–107.
90 Orloff, G., Winter, B., and Fehr, C. (1991) in *Perfumes Art Sciences and Technology* (eds P. Muller and D. Lamparsky), Elsevier, New York, 287–332.
91 Robles-Dutenhefner, P.A., Silva, K.A., Siddiqui, M.R.H., Kozhevnikov, I.V., and Gusevskaya, E.V. (2001) *J. Mol. Catal. A: Chem.*, **175**, 33–42.
92 Klose, W., and Bochow, K. (1967) DD69586.
93 Kitajina, M., and Noguchi, M. (1956) *Res. Jpn. Assoc. Camphor Ind. Eng.*, **21**, 188.
94 Kane, B.J., and Albert R.M., Jr. (1968) US 3 383 422.
95 Monteiro, J.L., and Veloso, C.O. (2004) *Top. Catal.*, **27** (1–4), 169–180.
96 Roy, A. (1999) *J. Agric. Food Chem.*, **47**, 5209–5210.
97 Feliczak-Guzik, A., and Nowak, I. (2009) *Catal. Today*, **142**, 288–292.
98 Bain, J.P. (1946) *J. Am. Chem. Soc.*, **68**, 638–641.
99 Pillai, U.R., and Sahle-Demessie, E. (2004) *Chem. Commun.*, **7**, 826–827.
100 Ramaswamy, V., Shah, P., Lazar, K., and Ramaswamy, A.V. (2008) *Catal. Surv. Asia*, **12**, 283–309.
101 de Villa, P.A.L., Alarcón, E., and de Correa Montes, C. (2002) *Chem. Commun.*, 2654–2655.
102 Klein, E., and Brunke, E.J. (1982) US4318831.
103 Corma Canos, A., Iborra Chornet, S., and Velty, A. (2010) US0312018 A1.

104 Gendelman, B.A., Patlasov, V.P., Klabukova, J.N., Seratinov, L.A., and Gotlib, V.A. (1989) SU 1696421 (A1).
105 Fahlbusch, K.G., Hammerschmidt, F.J., Panten, J., Pickenhagen, W., Schatlowski, D., Bauer, K., Garbe, D., and Surburg, H. (2003) Flavors and fragrances, in *Ullmann's Encyclopedia of Industrial Chemistry*, vol. 14, 6th edn (ed. Bohnet et al.), John Wiley & Sons, Inc., New York, p. 103.
106 Ventura, P., Schiavi, M., Serafini, S., and Selva, A. (1985) *Xenobiotica*, **15**, 317–325.
107 Allegra, L., Bossi, R., and Braga, P.C. (1981) *Respiration*, **42**, 105–109.
108 Chikao, N., and Hisao, T. (1981) JP 56030940.
109 Gainsford, G.J., Hosie, C.F., and Weston, R.J. (2001) *Appl. Catal. A: Gen.*, **209**, 269–277.
110 Ito, H., and Kuriyama, Y. (2010) JP 2010070532 (A).
111 Matsubara, Y., and Takei, K. (1966) JP 6827107.
112 Etzel, G. (1933) DE 578569.
113 Kirk, R.E., and Othmer, D.F. (1969) *Encyclopedia of Chemical Technology*, vol. 19, John Wiley & Sons, Inc., New York, p. 803.
114 Alarcón, E.A., Correa, I., Montes, C., and Villa, A.L. (2010) *Micropor. Mesopor. Mater.*, **136**, 59–67.

12
Environmental Assessment of Novel Catalytic Processes Based on Renewable Raw Materials – Case Study for Furanics

Martin K. Patel, Aloysius J.J.E. Eerhart, and Deger Saygin

12.1
Introduction

Global production of all bulk materials is likely to increase substantially in the coming decades in view of population increase, rising wealth, and the urgent need to overcome extreme poverty for more than 20% of the global population. The most important bulk materials in terms of volume – and likewise energy use – are iron and steel, non-ferrous metals, non-metallic minerals (especially cement), plastics and other chemicals, and pulp and paper. Production will especially grow in developing countries, which are expected to increase their share of the global industrial production (in physical terms) from 65% in 2006 to approximately 75% in 2030 and 80% in 2050 [1]. Among all bulk materials, plastics have experienced the largest growth, with a compound annual growth rate of 9% between 1950 and 2010 as opposed to, for example, iron and steel with a compound annual growth rate of 3% [2, 3]. In 2010, 265 million tonnes of plastics were produced worldwide. This translates to a worldwide average per capita consumption of somewhat less than 40 kg per capita (global population: 6.84 billion people), while in high-income countries the threshold of 100 kg per capita was passed in the 2000–2010 timeframe. If these consumption levels are reached across the globe, with a worldwide population stabilizing at 9 billion people (between 2050 and 2100), plastics production will have to triple to around 900 million tonnes per year. These are unprecedented production volumes, to which other products of the chemical industry, for example, fertilizers, detergents, lubricants and fuel additives must be added.

However, this growth comes at a cost. Already today, worldwide, the chemical and petrochemical sector consumes approximately 37 exajoules (EJ) of final energy per year, representing more than 30% of the global industrial energy use (including feedstocks [4]. As for all other energy intensive sectors, the chemical and petrochemical sector is faced with concerns about the declining resources of fossil fuels, increasing demand for fossil fuels by emerging economies, supply security issues in view of fossil fuel dependence on politically unstable regions, and climate change. For the production of chemicals in general, and plastics in particular,

Catalytic Process Development for Renewable Materials, First Edition. Edited by Pieter Imhof and Jan Cornelis van der Waal.
© 2013 Wiley-VCH Verlag GmbH & Co. KGaA. Published 2013 by Wiley-VCH Verlag GmbH & Co. KGaA.

examples of further important sustainability issues are the scarcity and the toxicity of (some) catalysts and additives, and littering caused by the final products (especially plastics marine litter) which are creating additional pressure on our habitats and health.

Given these challenges there is urgent need to move towards more sustainable solutions. *Novel catalytic processes based on renewable raw materials* offer solutions for at least some of the problems just discussed due to their potential to reduce non-renewable resource use and to curb greenhouse gases (GHG).

Biobased materials and green chemistry have been studied and discussed for approximately 20 years [5] but for most of this period of time, societal mainstream considered them to be too costly and, hence, illusionary. In contrast, with the increased economic importance of these products and the related technologies, more and more critical questions about the environmental and social trade-offs (next to their economics) are now being asked. With increased implementation, the strengths and weaknesses of these "green" technologies become clearer. Given the increased understanding of the need for environmentally, economically, and socially sustainable solutions, the expectations are becoming more demanding, and improved assessment methodologies and approaches are being developed and applied in a variety of contexts. It is obvious that, in the years to come, sustainability claims will be challenged more rigorously.

The most widely accepted method for the evaluation of environmental sustainability is life cycle assessment (LCA) which is increasingly being applied in all stages of a product life cycle, from R&D to existing plants. It is a tool which has gained enormous importance in the past 10 years and is nowadays acknowledged by companies, researchers, policy makers and NGOs. A remarkable body of environmental assessments following the LCA methodology was prepared in the recent past on polymers from renewable raw materials, now providing a rather good overview of how these new materials score, at least with regard to non-renewable energy use (NREU) and GHG emissions [6]. Very few studies go beyond these impact categories by covering, for example, water use, particulate matter emissions, or toxicity. Moreover, there are virtually no LCA studies which explicitly study catalysis and its contribution to energy savings and emission reduction for the production of biobased and petrochemical materials (one of the very few exceptions is the environmental assessment of the production of enzymes for use as biocatalyst by [7]).

Against this background it is the objective of this chapter to provide insight on how an environmental LCA is conducted for novel catalytic processes based on renewable raw materials and to what extent energy can be saved and emissions reduced. To this end, we provide in Section 12.2 a top-down estimate of energy savings by catalytic processes. We then present the foundations of an environmental LCA in Section 12.3, and we discuss a furan-based polymer as a case study in Section 12.4. We end with a discussion and conclusions in Section 12.5.

12.2
Energy Savings by Catalytic Processes

Catalysis allows the reduction of energy use and the associated emissions in various ways. First, a catalyst (new or improved) reduces the activation energy, thereby decreasing the process heat requirements (see Figure 12.1). Second, it increases the yield of the main product, thereby reducing the amount of waste and/or the need for internal recycling and the related energy use. Third, it influences the selectivities of the byproducts which is relevant for two reasons: (i) to increase the quantity of desired byproducts (hence providing a credit, thereby indirectly reducing the energy use related to the main product) and (ii) to reduce the effort related to downstream processing (separation and purification of the desired products). Fourth and finally, improved catalysts retain a high level of activity for a longer period of time, and/or they ensure a higher throughput, thereby reducing the effort associated with catalyst replacement and regeneration (energy and materials, downtime, labor time).

In conclusion, the amount of energy to be saved by a (new) catalyst differs from case to case, calling for in-depth studies, at least for the most important processes. Since this is beyond the scope of this chapter we conduct a crude analysis which we discuss in the following.

Based on a bottom-up model of the chemical and petrochemical sector's 58 most energy intensive products (covered by a total of 81 production processes), we estimate total process energy use at approximately 9 EJ of final energy (equivalent to 60% of the 15 EJ of process energy consumed by the entire sector). About 3 EJ of this modeled consumption is related to steam cracking, which is the sector's single largest user of energy (steam cracking is a non-catalytic process and is nowadays

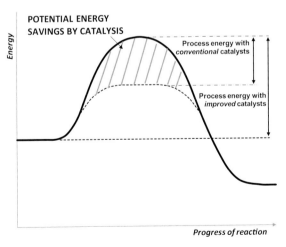

Figure 12.1 Comparison of the energy profile of a catalytic production process with conventional and improved catalysts. Adapted from Fogler [8].

essential for the chemical industry because it is the source of the currently most important bulk chemicals, namely ethylene and propylene, next to supplying butadiene and aromatics). Approximately 2 EJ is related to the production of another 27 basic chemicals, intermediates (e.g., toluene diisocyanate (TDI)), inorganics and polymers, which are currently also produced via non-catalytic processes. The remaining 4 EJ (=9 EJ – 3 EJ – 2 EJ) is required to produce 30 chemicals in catalytic production processes (see black filled bars in Figure 12.2). Due to lack of available data, we are not able to model the remaining 6 EJ (=15 EJ – 9 EJ) of the sector's process energy use. However, since by far the most chemical products are reported to be produced via catalytic routes [9], it is likely that essentially all of this energy consumption is related to catalytic processes.

Tonkovich [10] and Tonkovich and Gerber [11] analyzed the potential savings for the top-50 commodity chemicals in the US chemical industry. We apply their methodology to estimate to what extent today's energy use for the 30 catalytically produced products referred to above (4 EJ) can be further reduced. We estimate the energy efficiency potentials by improved catalysis for each of these products by assuming that the gap between the current yields and a thermodynamically optimal catalytic process (i.e., 100%) is closed by some 60 to 80% through successful R&D in the next decades. As a result of the improvements in yields, both process energy is saved and less feedstock is required. We express the feedstock savings in energy terms by multiplying the avoided feedstock use by its energy content and by a correction factor.[1] In this way, we calculate that worldwide between 0.8 and 1.1 EJ a^{-1} of final process and feedstock energy could be saved through improved catalysis at 2007 production levels. Compared to the total final process energy use of the 30 products analyzed, this is equivalent to a potential saving of 20–25%. Scaling these potentials to include all catalytic processes, we estimate total final energy savings of 2.0–2.7 EJ a^{-1} or 5–7% of the sector's final energy demand. Savings of process energy and feedstock energy contribute equally to the total potentials.

In addition, catalysis may contribute to energy savings and emissions reductions in several other processes which currently do not make use of catalysis. The largest potential exists for steam cracker products; for these, Ren *et al.* [12] provide the energy saving potentials for transition to catalytic processes (for the main hydrocarbon feedstocks, namely naphtha, ethane and propane, which together account for 86% of today's feedstock consumption for steam cracking). Based on this information, we estimate that catalysis could save between 0.5 and 1.6 EJ a^{-1} of process energy. This is equivalent to 15–50% savings compared to the current process energy use of the steam cracking process (3 EJ).

To conclude, improved catalysis could potentially save a total of 2.5 to 4.3 EJ a^{-1} of process and feedstock energy, which is equivalent to 9% of the chemical and petrochemical sector's total energy use, including feedstocks. In addition to the

---

1) We multiply the achievable total feedstock savings by an estimated correction factor of one-third because unconverted feedstock is typically recycled internally and because feedstock converted to undesired products is often burnt to generate process heat.

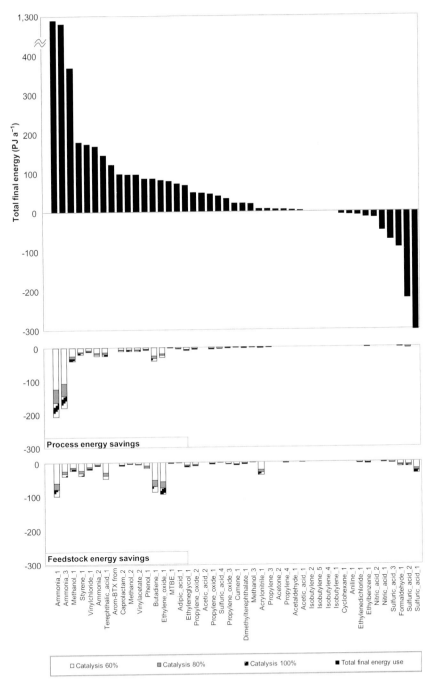

Figure 12.2 Energy saving potential by application of catalysis in the global chemical and petrochemical sector, 2007. Black bars indicate the total final energy use modeled for each chemical by process routes. Process energy and feedstock energy savings achievable by catalysis are shown separately. Savings indicated with the white bars show the potentials if the gap between the current yields and the thermodynamic limit is closed by 60%. The additional energy savings achievable if the gap is closed by 80% and if the thermodynamic limit is reached are shown by gray filled bars and hatched bars, respectively.

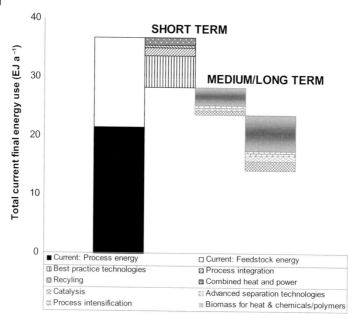

Figure 12.3 Potentials of technologies to reduce the non-renewable final energy use of the global chemical and petrochemical sector in the short and medium/long term. No production growth is assumed between today and the future. Sources: IEA [1]; Saygin et al. [4, 17].

processes covered so far, catalysis can contribute to energy savings by realizing shorter reaction pathways (with less reaction steps). A recent example is propane ammoxidation to acrylonitrile which avoids propylene as an intermediate. Using this technology, a joint venture of Thailand's PTT and Japan's Asahi Kasei started up a plant with a capacity of $200\,000\,t\,a^{-1}$ in Thailand in 2011 [13]. In the past decades, direct conversion of methane to chemicals (e.g., methanol) via catalytic routes also attracted attention (e.g., [14, 15]) and promising lab scale results were reported (e.g., [16]). Given these potentials, catalysis is expected to play an important role in reducing the chemical and petrochemical sector's energy use, and the related CO_2 emissions, in the coming decades; as shown in Figure 12.3 other important measures are the implementation of best practice technologies and the transition to biobased feedstocks. Together, these measures can reduce the total non-renewable energy use of the chemical and petrochemical sector by more than 60% (see Figure 12.3).

12.3
LCA Methodology

Life cycle assessment (LCA) is a standardized method for the environmental evaluation of products, technologies and services. The principles of LCA have been laid down in two International Standard Organization (ISO) standards, that is, ISO 14040 [18] and ISO 14044 [19]. An LCA is prepared in several steps which will be addressed in the example of a concrete case study in Section 12.4. The steps are:

1) **Goal and Scope definition:** In this step, the purpose of the analysis is specified, the so-called functional unit is defined, the system boundaries are set for the product system to be analyzed. and the environmental impacts to be evaluated are chosen.

2) **Inventory Analysis:** The material and energy flows of all processes are compiled and all emissions are calculated; the outcome is the so-called inventory table in which all relevant inputs and outputs of the product system are reported.

3) **Impact Assessment:** The potential environmental impacts associated with the inputs and outputs specified in the inventory table are established; this typically implies determining so-called mid-point indicators, such as the contribution to global warming (expressed in CO_2 equivalents), acidification, eutrophication, toxicity, and so on.

4) **Interpretation:** The results of the inventory and the assessment of the impacts are discussed in relation to the objectives of the study.

ISO does not prescribe how to deal with a number of critical issues that typically occur when conducting an LCA study. According to Reap *et al.* [20] these critical issues include the choice of the functional unit, setting of system boundaries, cut-off rules and the allocation methodology. We briefly discuss these aspects in the following.

The *functional unit* describes in quantitative terms the primary function(s) fulfilled by the system. When conducting an LCA study for a material, the chosen functional unit is often 1 kg (or tonne) of that material. The functional unit is the measure of the performance of the product system and it provides a reference to which the inputs and outputs can be related. Regarding *system boundaries*, there is no freedom of choice upstream in the process chain because all activities up and until the extraction of natural resources (mining, agriculture, forestry etc.) must be included (referred to as "cradle"). It is, however, possible to choose where the system ends downstream, for example, whether the process chain ends with an intermediate (e.g., an unprocessed bulk material) at the factory gate, a final product at the factory gate, or whether the use phase by the consumer and/or post-consumer waste management (the "grave") is also included. We apply *cut-off* for capital equipment, that is, we disregard the environmental impacts related to the manufacture of the production equipment. Cut-off is also applied related to the physical activities of the workforce, their food intake, work travel and such like.

Allocation is a problem that occurs in the inventory analysis if a process (or subprocess) yields more than one product. In this case the inputs to the process and the environmental impacts caused by the process need to be distributed (allocated) across the various products. ISO (2006) recommends avoiding allocation wherever possible through the expansion of the system boundary or by dividing the unit process into two or more sub-processes, and by separately assessing these sub-processes. System expansion, which implies the inclusion of all related product systems, indeed avoids allocation (in the sense of partitioning), while subdivision of the unit process only shifts the problem of partitioning to smaller subsystems (which can be expected to improve accuracy compared to partitioning of the larger system). For partitioning, the most common methods are allocation based on physical attributes (often mass), monetary values (prices) or energy content.

While the aspects discussed above are relevant for all LCA studies, very specific issues are encountered when performing an LCA of bio-based materials (see Pawelzik et al. [21]).

12.4
Case Study: Energy Analysis and GHG Balance of Polyethylene Furandicarboxylate (PEF) as a Potential Replacement for Polyethylene Terephthalate (PET)

This case study focuses on polyethylene furandicarboxylate (PEF), which is being developed as a biobased alternative to the petrochemical plastic polyethylene terephthalate (PET). The largest application of PET is the bottle market, which amounts to about 15 Mt or approximately 6% of the global plastics market, and represents roughly 0.2% of the global primary energy consumption [22]. The main component of PET is purified terephthalic acid (PTA) [23], which could be replaced by biobased 2,5-furandicarboxylic acid (FDCA). The diol used is ethylene glycol (EG), which is identical for PET and PEF.

The objective of the case study presented here, which was published in more detail by Eerhart et al. [22], is to determine the mass and energy balances, the non-renewable energy use (NREU, comprising fossil and nuclear energy) and GHG emissions of PEF. Due to the early stage of development and the lack of measured data from an industrial facility, no further environmental impact categories were considered. The functional unit is set at 1 kg of PEF. The chosen system covers all steps from cradle to grave, however, excluding the use phase (assumed to be stationary and hence not causing additional impacts). For post-consumer waste treatment of PEF (grave), disposal in a municipal solid waste incinerator (MSWI) without energy recovery is assumed. Cut-off was avoided, with the exception of the activities specified above in Section 12.3 (these are typically disregarded in LCA studies for bulk materials). Allocation across the intermediates hydroxymethylfurfural (HMF), HMF ethers, levulinic acid (LA) and LA esters was performed by partitioning based on mass; economic allocation was conducted as sensitivity analysis (for details see [22]).

Figure 12.4 shows the entire process chain for the production of PEF starting from agriculture, which is the most important source of feedstocks for this product. The production of PEF consists of the following sub-processes:

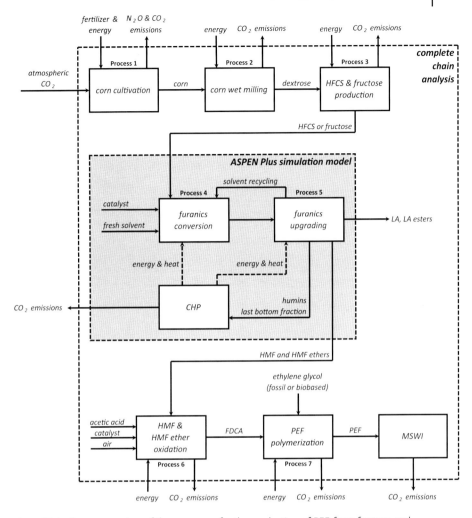

Figure 12.4 System overview of the processes for the production of PEF from fructose and HFCS.

1) cultivation of corn (maize).
2) corn wet milling (CWM) to produce corn starch from corn.
3) conversion of corn starch into fructose and high fructose corn syrup (HFCS).
4) further conversion of fructose and HFCS into furanics.
5) recovery and upgrading of furan compounds into HMF and HMF ethers.
6) oxidation of HMF and HMF ethers into FDCA.
7) polymerization of FDCA and ethylene glycol (EG) into PEF.

The core of the process, that is, step 4, was modeled with the chemical process simulation model ASPEN Plus. In more detail, step 4 consists of three process areas, that is, the actual furan conversion, furan intermediates recovery and upgrading (i.e., downstream processing), and a combined heat-and-power (CHP) plant. The recovery and upgrading section consists of eight distillation towers in total. In the CHP plant the low value by-products, the so-called humins, are burned for onsite supply of utilities. The results of the ASPEN model were transferred to the PEF chain analysis; the entire process, the modeling approach and intermediate results are explained in detail by Eerhart et al. [22].

In line with the LCA methodology, all energy and material inputs, and the associated environmental impacts were taken into account. This includes, for example, nitrogen fertilizer production and application, leading to nitrous oxide (N_2O) emissions on the land (N_2O is a strong greenhouse gas).

As shown in Figure 12.5, PEF allows the curbing of non-renewable energy use and GHG emissions by at least 50% compared to petrochemical PET (PET data originate from [24]). Figure 12.5 shows variants of PEF in order to consider that the co-monomer ethylene glycol (EG) can either be produced from petrochemical feedstocks (which is nowadays the dominating pathway) or from biobased resources; three different options of sourcing EG are distinguished, that is, production from maize, sugarcane today and sugarcane by 2020. The results in Figure 12.5 refer to the best performing case from the ASPEN model.

Not only in comparison with PET but also when comparing PEF with other biobased plastics (not presented here; for details see [22], PEF scores very well for the two environmental impact categories considered. However, the following caveats should be considered:

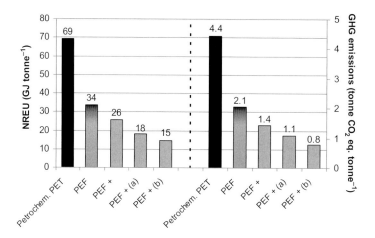

Figure 12.5 Non-renewable energy use and greenhouse gas emissions of PEF (in comparison with PET) for the system cradle-to-grave (with end-of-life treatment by municipal solid waste incineration without energy recovery). Petrochemical components are presented in black, while gray refers to a biobased component. Abbreviations: PEF = use of petrochemical EG; PEF+ = use of EG from maize; PEF + (a) = use of EG from sugarcane today; PEF + (b) = use of EG from sugarcane 2020.

- The inventory analysis builds on ASPEN Plus modeling for the core of the furanics process, which so far has not yet been operated at industrial scale; actual industrial application needs to be demonstrated and the environmental performance needs to be proven.

- Humins are expected to be an excellent fuel for the CHP unit (since humins are free of water) but to the authors' knowledge, no commercial application currently exists.

- The NREU and CO_2 emissions of fructose and HFCS had to be estimated since no process data from industry processes were available.

- Compared to PET polymerization, PEF polymerization has some process advantages (lower temperature and pressure, as well as reduced acetic acid oxidation to CO_2) which were not taken into account.

- For allocation, other approaches could be chosen (economic allocation was performed by Eerhart et al. next to mass allocation and was found not to change the results significantly).

The use of food resources (maize or sugar cane in this analysis) for non-food purposes is one of the most critically viewed features. Major efforts are being made to valorize lignocellulosic biomass for the production of biofuels which – if successful – could also be applied to biobased materials. Since, so far, the commercial viability remains to be proven, the assessment of PEF relies on conventional feedstocks (e.g., corn). This may, however, cause additional GHG emissions due to the increased pressure on land resources, among them tropical rainforests. Apart from direct land use change, indirect land use change (ILUC) is increasingly being considered. ILUC is land use change that occurs outside the biobased product's feedstock production area but that is induced elsewhere by increasing the production quantity of that feedstock. Several different approaches have been developed to quantify ILUC-induced GHG emissions; however, the main disadvantage is that they are subject to large uncertainties which have led to a controversial discussion about the impacts of land use change. Since 2008, more research into LUC factors has been done and estimates for the ILUC factor are now between 0.15 and 0.30 g per MJ ethanol. Using the latter, most disadvantageous case, GHG emission reduction by PEF compared to PET decreases from 54% (default for maize-based ethylene glycol according to Figure 12.5) to 38%. It is important to note here that the research into ILUC factors is still under development and an increased understanding of the many mechanisms involved with ILUC will decrease the uncertainty regarding the ILUC values.

To conclude, even though the PEF production process from fructose and HFCS is still under development, the analyses presented above show that the process is likely to offer reductions around 50% for NREU and GHG emissions if land change effects are disregarded. If the effect of ILUC is included, the reductions for GHG emissions drop from 54% (case 1; no ILUC) to 38% (values for the best performing case from the ASPEN model, see Eerhart et al. [22]).

12.5
Discussion and Conclusions

Given the increasing importance of sustainability aspects, it is essential to obtain a thorough understanding of the environmental performance of products and services and, in particular, of novel materials which are connected to environmental claims. New plastics and chemicals, produced from renewable raw materials by application of novel catalytic processes, fall into this category. Life cycle assessment is a powerful tool which allows evaluation of the environmental sustainability of processes, products and services. It can be applied in the early stages of R&D in order to support decision making regarding feedstocks, auxiliaries, process design and other features; it is also regularly applied for existing processes, for example, to establish a benchmark or to identify hotspots. So far, the focus has been primarily on a few environmental impact categories, with non-renewable energy use and GHG emissions being the most prominent. Other impact categories such as water use, acidification, and eutrophication are increasingly considered. The full-fledged application of LCA to a wide range of environmental and health impacts is primarily hampered by lack of data and, to some extent, also by the lack of reliable and broadly accepted assessment methodologies (e.g., for toxicity).

A meta-analysis of 44 LCA studies conducted by Weiss *et al.* [25] indicated that biobased materials save, relative to conventional materials, a substantial amount of non-renewable energy and GHG emissions, which coincides with the finding of the case study on PEF presented in this chapter. However, Weiss *et al.* also show that biobased materials may increase eutrophication and stratospheric ozone depletion; their findings are inconclusive with regard to acidification and photochemical ozone formation. The variability in the results found by Weiss *et al.* [25] highlights the difficulties in drawing general conclusions.

To conclude, the body of work on environmental assessments and our own analyses on catalytic conversions lead to the following important conclusions: (i) catalysis has contributed to substantial energy use and it has the potential to continue to do so; (ii) substantial progress has been made towards the replacement of conventional fossil fuel-based chemicals and plastics by novel alternatives based on renewable raw materials; and (iii) further work is needed, especially in order to reduce the impacts other than energy use and GHG emissions. In this context increased attention will also have to be paid to the toxicity of catalysts and additives where more sustainable solutions need to be sought by substitution and/or by process improvement.

References

1 IEA (2009) Energy Technology Transitions for Industry. Strategies for the Next Industrial Revolution, OECD/IEA, Paris, France.

2 Simon, C.-J., and Schnieders, F. (2011) Business Data and Charts 2009/2010. Status April 2011. Plastics Europe Market Research Group

(PEMRG), Plastics Europe, Brussels, Belgium.

3 WSA (2012) Crude steel production statistics. World Steel Association, Brussels, Belgium.

4 Saygin, D., Patel, M.K., Worrell, E., Tam, C., and Gielen, D.J. (2011) Potential of best practice technology to improve energy efficiency in the global chemical and petrochemical sector. *Energy*, 36 (9), 5779–5790.

5 Bozell, J.J., and Landucci, R. (eds) (1993) Alternative Feedstocks Program–Technical and Economic Assessment–Thermal/Chemical and Bioprocessing Components. Prepared by five National Labs for the U.S. Department of Energy Office of Industrial Technologies.

6 Chen, G.-Q., and Patel, M.K. (2012) Plastics derived from biological sources: present and future: a technical and environmental review. *Chem. Rev.*, 112, 2082–2099.

7 Nielsen, P.H., Oxenbøll, K.M., and Wenzel, H. (2007) Cradle-to-gate environmental assessment of enzyme products produced industrially in Denmark by Novozymes A/S. *Int. J. Life Cycle Assess.*, 12 (6), 432–438.

8 Fogler, H.S. (2002) *Elements of Chemical Reaction Engineering*, 3rd edn, Prentice Hall PTR, Prentice Hall Inc., Upper Saddler River, NJ, USA.

9 SAND (1997) Catalyst technology roadmap report. Sandia Report, SAND97-1424, June 1997. Sandia National Laboratories, Albuquerque, NM, USA, http://www.osti.gov/bridge/servlets/purl/544046-UGHBzl/webviewable/544046.pdf (accessed January 2013).

10 Tonkovich, A.L.Y. (1994) Impact of Catalysis on the Production of the Top 50 U.S. Commodity Chemicals, PNL-9432, March 1994, Pacific Northwest Laboratory, Richland, WA, USA.

11 Tonkovich, A.L.Y., and Gerber, M.A. (1995) The Top 50 Commodity Chemicals: Impact of Catalytic Process Limitations on Energy, Environment and Economics, PNL-10684, August 1995, Pacific Northwest Laboratory, Richland, WA, USA.

12 Ren, T., Patel, M., and Blok, K. (2006) Olefins from conventional and heavy feedstocks: energy use in steam cracking and alternative processes. *Energy*, 31, 425–451.

13 ICIS (2012) Chemical profile: acrylonitrile. ICIS Chemical Business. Surry, United Kingdom. http://www.icis.com/Articles/2012/02/27/9535512/chemical-profile-acrylonitrile.html (accessed 24 January 2013).

14 Holmen, A. (2009) Direct conversion of methane to fuels and chemicals. *Catal. Today*, 142, 2–8.

15 Ren, T., Patel, M.K., and Blok, K. (2008) Steam cracking and methane to olefins: energy use, CO_2 emissions and production costs. *Energy*, 33, 817–833.

16 Palkovits, R., Antonietti, M., Kuhn, P., Thomas, A., and Schüth, F. (2009) Solid catalysts for the selective low-temperature oxidation of methane to methanol. *Angew. Chem. Int. Ed.*, 48, 6909–6912.

17 Saygin, D., Gielen, D.J., Draeck, M., Worrell, E., and Patel, M.K. (2012) Assessment of the technical and economic potentials of biomass use for the production of steam, chemicals and polymers. Submitted paper, under review.

18 International Organization for Standardization (ISO) (2006) ISO14040. Environmental Management–Life Cycle Assessment–Principles and Framework.

19 International Organization for Standardization (ISO) (2006) ISO14044. Environmental Management–Life Cycle Assessment–Requirements and Guidelines.

20 Reap, J., Roman, F., Duncan, S., and Bras, B. (2008) A survey of unresolved problems in life cycle assessment, part 1: goal and scope and inventory analysis. *Int. J. Life Cycle Assess.*, 13, 290–300.

21 Pawelzik, P., Carus, M., Hotchkiss, J., Narayan, R., Wellisch, M., Selke, S., Weiss, M., Wicke, B., and Patel, M.K. Critical aspects in the life cycle assessment (LCA) of bio-based materials – reviewing methodologies and deriving recommendations. Submitted paper, under review.

22 Eerhart, A.J.J.E., Faaij, A.P.C., and Patel, M.K. (2012) Replacing fossil based PET with biobased PEF; process analysis,

energy and GHG balance. *Energy Environ. Sci.*, 5, 6407–6422.

23 Chauvel, A., and Lefebvre, G. (1989) *Petrochemical Processes – Technical and Economic Characteristics. Two Volumes*, Editions Technip, Paris, France.

24 PlasticsEurope (May 2011). Ecoprofile of Polyethylene Terephthalate (PET) (Bottle Grade), http://www.plasticseurope.org/plastics-sustainability/eco-profiles/browse-by-list.aspx (accessed January 2013).

25 Weiss, M., Haufe, J., Carus, M., Brandao, M., Bringezu, S., Hermann, B., and Patel, M.K. (2012) A review of the environmental impacts of biobased materials. *J. Ind. Ecol.*, 16 (S1), S169–S181.

13
Carbon Dioxide: A Valuable Source of Carbon for Chemicals, Fuels and Materials

Michele Aresta and Angela Dibenedetto

13.1
Introduction

Nature has always made C-products from CO_2, capturing it from the atmosphere. Because of the low concentration of atmospheric CO_2 (385 ppm today vs 275 ppm in the preindustrial era) several organisms use concentration apparatuses to fix the entire CO_2 molecule into carboxylates or to reduce it into energy-rich forms. The products derived from CO_2 have generated in a million years the fossils that provide 85% of the energy our society is using today, and will use for the next 40 years or so.

The natural carbon cycle is unable to control the anthropogenic emission of CO_2: the consequence is the increase in the atmospheric concentration of CO_2 that is causing worries about the possible change of climate and the insurgence of extreme events that will be out of the human control. Although the role of CO_2 in such changes is not neatly demonstrated, nevertheless the parallel trend of: (i) the consumption of fossil fuels, (ii) the growth of the concentration of CO_2 in the atmosphere, and (iii) the increase in the planet temperature seems to point to a strict relation and to the need to adopt preventive measures that may reduce the accumulation of CO_2 in the atmosphere or its emission.

Besides a shift in fuels, use of perennial and renewable energies, efficiency strategies, in the conversion of fossil fuels into energy and in the use of the energy obtained, other containment measures are under evaluation, such as the capture and storage of CO_2, a technology that has the potential of disposing of large volumes of CO_2, but has high costs (energetic and economic), and still suffers uncertainty about the permanence of disposed CO_2.

In carbon capture and storage (CCS) a large portion of the total cost is due to the capture of CO_2. Once CO_2 has been captured it can be either stored or used. The utilization of CO_2 is a much more challenging approach to avoiding CO_2 than the disposal in natural fields and it would bring to our society many more benefits than the disposal. In fact, the use of CO_2 to make chemicals and fuels will avoid the extraction of fossil carbon, while the disposal will demand the extraction of more fossil energy.

The conversion of CO_2 launches a great challenge to chemists and biotechnologists. A success in mimicking the photosynthetic paths by developing a "man-made photosynthesis" will allow our society to be able to save fossil carbon for the next generations, while the implementation of innovative synthetic technologies will support the development of a sustainable industry.

13.2
The Conditions for Industrial Use of CO_2

Anthropogenic carbon dioxide (ca. $36\,Gty^{-1}$) is generated in the utilization of chemicals (ca. 8%) and fuels (ca. 92%). These two classes of products are characterized by a different molecular structure and energy content. Fuels are, in general, hydrocarbons, with a simple molecular structure, but require a large energy input for their synthesis from CO_2 as a unique carbon source. Products of the chemical industry can be formed by either fixing CO_2 onto fossil fuel-derived substrates, or by using CO_2 as a unique source of carbon. The energy input is clearly quite different in the three cases mentioned above.

The utilization of CO_2 as a sole source of carbon for making products of the chemical industry or of the energy industry must respond to three key issues: (i) environmental, (ii) energetic, (iii) economic.

13.2.1
Environmental Issue

As mentioned above, the key issue is to reduce the emission of CO_2 with respect to the actual standard. Therefore, any use of CO_2 must be accompanied by the reduction of its emission with respect to the synthetic technologies on stream. The implementation of such a condition is not a guarantee of a better environmental performance. In fact, if the CO_2 emission is reduced and other categories (Table 13.1) are affected to a greater extent by the new technology, then the global effect may not be in the direction of a lower environmental burden. However, the benefit of the implementation of a new technology must be verified by using a

Table 13.1 Impact categories affected by the implementation of industrial technologies.

Carcinogens (non-carcinogens)	Respiratory inorganics
Ionizing radiations	Ozone layer depletion
Respiratory organics	Aquatic ecotoxicity
Terrestrial acidification	Terrestrial nutrification
Land occupation	Aquatic acidification
Climate change	Aquatic euthrophication
Non-renewable energy	Mineral extraction

methodology that assures a complete analysis of the impacts: this is the life cycle assessment (LCA).

In order to "avoid" CO_2, in general. the new CO_2-based synthetic methodology must be less energy-and material-intensive than the existing one, must have a higher C-fraction utilization (CFU), and have an energy consumption ratio (ECR) (energy accumulated in the product/energy used for the synthesis) close to 1 or higher than 1. If these conditions hold, and CO_2 is used instead of toxic compounds (as a phosgene substitute, for example), then the new process will be more "environmentally friendly" than the old one. It may be worth emphasizing that "utilization" of CO_2 is not a synonym for "avoiding" CO_2: in fact, the amount of CO_2 used is given by the stoichiometry of the reaction, while the quantity of "avoided" CO_2 can be calculated only by applying LCA. The avoided CO_2 can be much more than the used CO_2. An example is given in Scheme 13.1 where the "used" and "avoided" CO_2 for the synthesis of ethene carbonate are reported.

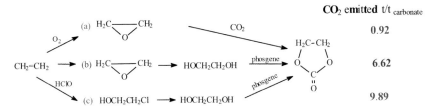

Scheme 13.1 Emitted CO_2 for the synthesis of 1 t of ethane carbonate. Moving from the use of phosgene to the carboxylation of the epoxide circa 9 tCO_2/t carbonate is avoided. Per each t of carbonate, 0.5 t CO_2 are used.

An important point is that, unless CO_2 is fixed into organic polymers (polycarbonates, polyurethanes) or inorganic carbonates, then when the product made from CO_2 is used it will release CO_2. The overall effect will be, thus, that of a delayed emission into the atmosphere. As a matter of fact, the process of formation is much faster than the fixation of CO_2 (compare the combustion of methane with the reduction of CO_2 to methane). In order to reduce the CO_2 emission, we need to avoid its formation.

However, the utilization of CO_2 in the synthesis of chemicals, fuels, and materials is a great opportunity, whose benefits must be quantified by LCA and demonstrated case by case.

13.2.2
Energy Issues

Carbon dioxide lies at the bottom of the potential energy well of C1 molecules. Figure 13.1 shows that any reaction that reduces the C/O ratio or increases the H/C ratio will require energy. Nevertheless, the production of carboxylates (carbonates, carbamates, acids, esters, etc) will lower the energy of the system.

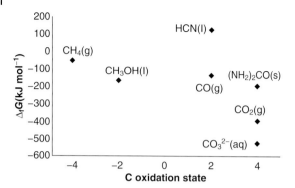

Figure 13.1 Gibbs free energy of formation versus the oxidation state of carbon in C1 molecules.

Therefore the reactions that involve CO_2 can be separated into two classes:

1) Reactions in which the entire CO_2 moiety is incorporated into an organic substrate (R–COOH, R–COOR, $(RO)_2CO$, cyclic carbonates, polycarbonates, RR'N-COOR", polyurethanes), or into inorganic carbonates, and so on.

2) Reactions that need energy for the conversion of CO_2 (reduction to other C1 molecules, production of hydrocarbons and fuels in general).

In a frame in which the main energy source is fossil carbon, Class 2 reactions have a low possibility of development. But if perennial (solar, wind, water, geothermal) energy is used, then such reactions may have a large exploitation and large volumes of CO_2 will be used with substantial carbon-recycling.

In the conversion of CO_2, thermodynamics and kinetics play a key role. Reactions involving CO_2, if they also have a negative enthalpy change, are accompanied by a negative ΔS, due to the conversion of gaseous CO_2 into liquid or solid products. As a consequence, the ΔG may be positive and the equilibrium concentration of the products can be very low, of the order of 1% or less. In order to improve the conversion yield of the substrate, high pressure can be used or, even better, supercritical CO_2 (sc-CO_2). In the former case, the solvent and the reaction conditions (pressure and temperature) may play a key role because they drive the solubility of CO_2 and the amount of it available in solution for the reaction to occur. In the latter case CO_2 acts as solvent and reagent: the best condition is that all reagents are in a single phase.

Beside such thermodynamic barriers, kinetics also play a fundamental role. In fact most reactions in which CO_2 is involved require breaking of E–O bonds (E=metal or carbon, or another element), that may be particularly rich in energy (often over 60 kcal mol^{-1}). Therefore, the chemistry of conversion of CO_2 is not an easy task and requires the development of new catalysts and finding the most suitable reaction conditions.

In some cases, such barriers can be circumvented by using an activated form of CO_2 or by coupling reactions that may improve the thermodynamics and kinetics, as will be discussed below. Energy issues are very important to master as they may make a new process based on CO_2 acceptable or not.

13.2.3
Economic Issues

From the economic point of view the utilization of CO_2 in synthetic chemistry is a viable process. In fact, any product made from CO_2 will have an added value. This is a positive point that may foster the acceptance of the capture costs and the integration of the utilization into the general scheme of CO_2 capture, storage and utilization, CCSU. The value of chemicals derived from CO_2 is very variable, depending on whether they are fuels or specialty chemicals. Depending on the nature of the chemical, is its market. In a vision of a new management of carbon, an integrated approach must be designed that merges the capture, disposal and use of CO_2, the latter for its multiple aspects.

13.3
Carbon Dioxide Conversion

In the following sections, the various possibilities of CO_2 conversion will be considered, listing them according to a tentative "time to market". Short- and medium-term options will be discussed. The long-term option is "man-made photosynthesis".

Some applications of CO_2 are already implemented or may be exploited in a short (5–7 years) time. In the following subsections the conditions for full exploitation will be discussed.

13.3.1
Carbonates

Carbonates are characterized by the "–OC(O)O–" moiety, either present in an ionic lattice or covalently bound to organic groups (Figure 13.2).

As said above, the addition of a new oxygen atom (originating from a metal oxide, M=O, or from an alkoxo-group, RO-), to the CO_2 moiety is an exothermic process, (Table 13.2) which can have positive ΔG.

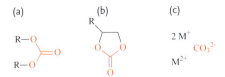

Figure 13.2 Carbonates: organic (a, b) and inorganic (c).

13 Carbon Dioxide: A Valuable Source of Carbon for Chemicals, Fuels and Materials

Table 13.2 Enthalpy and Gibbs free energy of carboxylation reactions.

Reaction	$\Delta_r H°$/kcal mol^{-1}	$\Delta_r G°$/kcal mol^{-1}
$CaO + CO_2 \rightarrow CaCO_3$	−42.58	−31.2
$2\ MeOH + CO_2 \rightarrow (MeO)_2CO$	−4.00	+6.1
$2\ EtOH + CO_2 \rightarrow (EtO)_2CO$	−3.80	Na
$2\ PhOH + CO_2 \rightarrow (PhO)_2CO$	+12.09	Na

13.3.1.1 Organic Molecular Compounds

Industrially, organic carbonates are made by reacting phosgene, $COCl_2$ with alcohols Eq. (13.1a) in the presence of a base such as NaOH.

$$2ROH + COCl_2 + 2B \rightarrow (RO)_2CO + 2B\ HCl \quad (13.1a)$$

$$2\ ROH + CO + \frac{1}{2}O_2 \rightarrow (RO)_2CO + H_2O \quad (13.1b)$$

Phosgene is very reactive and, thus, the reaction does not require any catalyst. The drawback of such a reaction is the high toxicity of phosgene, the use of chlorinated solvents, and the production of chlorinated solid waste. Scheme 13.2 shows possible solutions to the problem.

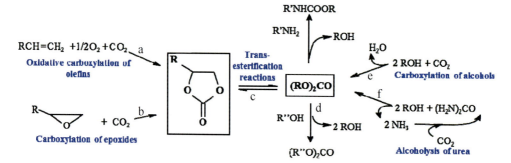

Scheme 13.2 Network of alternative reactions to the use of phosgene.

The oxidative carbonylation of alcohols Eq. (13.1b), on stream for quite a few years, has never reached a large scale application because of some drawbacks due to the corrosion of plants and other risks. Besides the above reactions, organic carbonates can be obtained by alkylation of inorganic carbonates with alkyl halides-RX Eq. (13.2a–b), at a temperature of circa 400 K under phase-transfer conditions, in ionic liquids [1]. Nevertheless, such reactions do not meet the environmental targets of these days that tend to eliminate the use of organic halides.

$$2\ RX + MCO_3 \rightarrow (RO)_2CO + MX_2 \quad (13.2a)$$

$$2\ RX + M_2CO_3 \rightarrow (RO)_2CO + 2MX \quad (13.2b)$$

Scheme 13.2 shows other routes to acyclic carbonates which do not make use of halogenated compounds, such as the direct carboxylation of alcohols (Scheme 13.2e) or the use of urea or the trans-esterification reaction.

The use of urea is an interesting reaction as urea can be considered an activated form of CO_2, (see Figure 13.1), moreover the ammonia produced can be recovered and either re-used in the conversion of CO_2 into urea again, or in other reactions requiring ammonia.

13.3.1.2 Synthesis of Acyclic Carbonates via Carboxylation of Alcohols

Bifunctional catalysts are needed for the synthesis of linear carbonates, a reaction that has thermodynamic and kinetic limitations. A basic and acid activation of the alcohol is necessary for the synthesis of an organic carbonate. Eqs (13.3)–(13.7)

$$ROH + B \rightarrow RO^- \; BH^+ \tag{13.3}$$

$$RO^- + CO_2 \rightarrow ROC(O)O^- \tag{13.4}$$

$$ROH + A \rightarrow R^+ + AOH^- \tag{13.5}$$

$$RC(O)O^- + R^+ \rightarrow ROC(O)OR \tag{13.6}$$

$$2ROH + CO_2 \rightleftarrows (RO)_2CO + H_2O \tag{13.7}$$

At the reaction temperature (often above 400 K) the equilibrium position of the overall reaction (13.7) is shifted to the left and the equilibrium concentration of the carbonate may be as low as less than 1%, conditions not suitable for an industrial process. In order to shift the equilibrium to the right various attempts have been made, such as water elimination, high CO_2 pressure, and various alcohol/CO_2 molar ratios.

Aldols [2] and ketals [3] or orthofomates [4] or else cyanides [5] have been used as water traps, but the resulting compounds are soluble in the reaction medium which will increase the cost of product recovery. Moreover, the organic water-trap has to be used in equimolar amount with the carbonate formed, that does not meet industrial applications. Conversely, carbodiimides (RN=C=NR) are quite interesting agents as their hydrated form (RHN–C(O)–NHR, urea) can be insoluble in the reaction medium and can be easily separated. DCC [6] –CyN=C=NCy is not a simple water trap, but an *organic catalyst* able to quantitatively convert alcohols into the relevant carbonates under very mild reaction conditions (330 K and 0.2 MPa CO_2, while metal systems require temperatures higher than 410 K and high pressures). DFT (density functional theory) calculations have demonstrated that, in the presence of DCC, the formation of DMC has a $\Delta G°$ equal to $-146\,kJ\,mol^{-1}$ (see Scheme 13.3), because of the simultaneous formation of dicyclohexylurea-DCU. The dehydration of DCU [7] closes the cycle (Scheme 13.3) so that the synthesis of organic carbonates can take place using DCC alone in two steps, a strategy useful for the synthesis of small-scale carbonates or, in the case of the alcohol, brings moieties sensitive to the temperature.

Recent reviews on the synthesis of linear carbonates based on homogeneous catalysts or organic catalysts are available in ref. [8a–d]. Here, we discuss the most

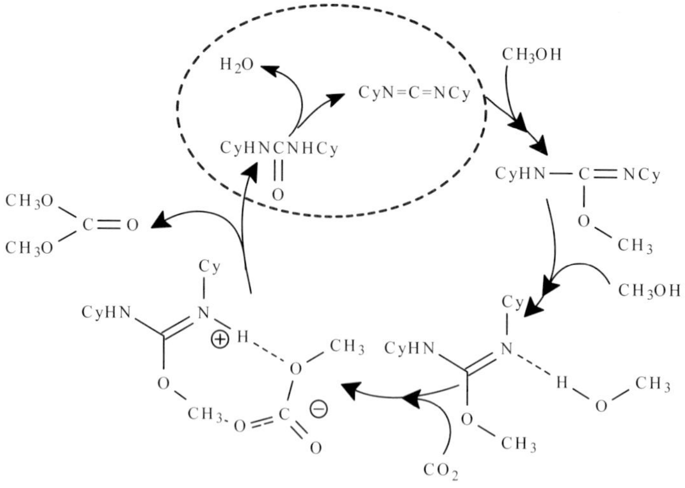

Scheme 13.3 The synthesis of organic linear carbonates catalyzed by DCC.

recent results relevant to attempts to improve the conversion yield of alcohols. It is worth emphasizing that much attention has been paid in recent years to the synthesis of dimethylcarbonate from methanol and CO_2, but recently the synthesis of diethylcarbonate (DEC) [9, 10] has become a target. Such interest is due to the fact that DMC and DEC have quite similar properties and uses. The preparation from bio-ethanol gives a "bio-derived" label to DEC and features such as "green" and "sustainable" to the processes in which the latter is used and the products derived from it.

For the synthesis of DMC and DEC, homogeneous, heterogeneous and heterogenized catalysts have been used, or else organic catalysts [6]. The reaction mechanism has been elucidated for a few of them [11].

The mechanism depends on the catalyst used. Two molecules of methanol can be activated through a base mechanism to give the "E(OMe)$_2$" moiety that undergoes CO_2 insertion to produce "E(OMe)OC(O)OMe". The latter undergoes an intramolecular Me-transfer from E-OMe to EOC(O)OMe, producing DMC and a species containing an "E=O" moiety that may lead to oligomers that, as in the case of Sn-catalysts, need to be converted into a monomer for the catalysis to continue [12], or else it may result in an inactive species (Scheme 13.4a).

Alternatively, and this is the case for the Nb-alcoxo catalyst (Scheme 13.4b), two molecules of methanol are activated through a *base-plus-acid* mechanism and DMC and water are formed out of the co-ordination sphere of the metal, regenerating the catalyst. The latter must be recovered once the equilibrium is reached in order to avoid water converting it into an inactive oxo-species E=O. (Scheme 13.4c) What drives the reaction is the energy of the TS for the alkylation of the E[OC(O)OMe] moiety: either an intramolecular or an intermolecular alkyl-moiety transfer may be initiated. The organic catalyst CyN=C=NCy (DCC) follows the same reaction path (a) in Scheme 13.4 and produces urea (CyHNCONHCy) as an inactive form of DCC.

(a) $EX_2 + 2B + 2MeOH \rightarrow E(OMe)_2 + 2 HX$ (E= Metal system)

$E(OMe)_2 + CO_2 \rightarrow E(OMe)(OCOOMe) \rightarrow 1/n\ (E-O)n\ (or\ E=O) + (MeO)_2CO$

(b) $E(OMe)(OCOOMe) + 2MeOH \rightarrow E(OMe)_2 + H_2O + (MeO)_2CO$

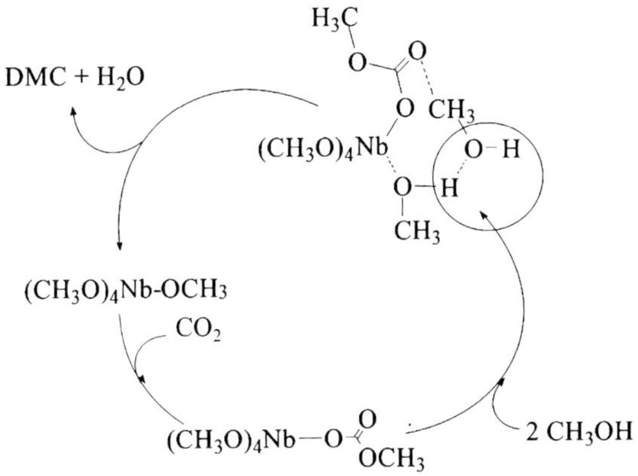

(c) $Nb(OMe)_5 + H_2O \rightarrow (CH_3O)_3Nb=O + 2\ MeOH$

Scheme 13.4 Different mechanisms in the synthesis of carbonates.

Heterogeneous catalysts (based mainly on W and Ce oxides) show the same mechanism of activation of methanol and insertion of CO_2 to afford E(OMe)(OCOOMe): for Al- or Nb-loaded CeO_2 the closure of the cycle has been shown to proceed according to Scheme 13.4b [13].

The main limitations for such reactions are: (i) the deactivation of the catalysts due to the formation of water and (ii) the low equilibrium concentration of the carbonates. Mixed oxides [14–16] are more resistant than single oxides and operational times of weeks have been demonstrated [13]. The elimination of water using membranes may help in setting a continuous process [13], but the low conversion yield (a few units percent) does not make such a process interesting for industrial application.

Attempts have been made to shift the equilibrium position by combining the carboxylation of alcohols with a second reaction that may at the end capture water. This is the case for the conversion of DCC into DCU discussed above. For example, butene oxide is reported to act as a water trap, producing butenediol. The mechanism is not very clear as it is known that oxiranes easily react with CO_2 to produce cyclic carbonates, and the latter react with alcohols to produce acyclic carbonates and diols (see below). Such reactions are not always beneficial as often the conversion yield is increased but the selectivity is lost [17]. The use of alkoxy-ionic-liquids [18] (AIL) seems to improve the conversion of methanol to 12% with 90% selectivity at 400 K and 7.8 MPa. Heteropolymetalates [19] also catalyze the formation of DMC from methanol and CO_2, but methyl formate and dimethoxymethane are easily formed, reducing the selectivity.

13.3.1.3 Synthesis of Carbonates via Transesterification or Alcoholysis of Urea

The direct carboxylation of alcohols with CO_2 is, thus, limited by a low conversion. Finding alternative routes characterized by a conversion yield of the alcohol of a few tens percent would push greatly towards the exploitation of the new routes. The transesterification of cyclic carbonates and the reaction of alcohols with urea represent two approaches that may contribute to implementing new sustainable synthetic routes. The transesterification of a cyclic carbonate, such as ethene carbonate Eq. (13.8a) with methanol produces DMC and ethene glycol ($HOCH_2CH_2OH$). The issue here is that the co-product (glycol) must be recovered and converted back to ethene carbonate: the recycling is essential in order to avoid wasting the ethene glycol. As the conversion of the glycol into the cyclic carbonate occurs via reaction with urea Eq. (13.8b), the question is: why not react methanol directly with urea? This topic will be discussed later in this section.

$$\text{ethene carbonate} + 2\,CH_3OH \longrightarrow (CH_3O)_2CO + HOCH_2CH_2OH \tag{13.8a}$$

$$HOCH_2CH_2OH + (H_2N)_2CO \longrightarrow 2NH_3 + \text{ethene carbonate} \tag{13.8b}$$

By using the *reactive distillation* ethene carbonate has been reacted with ethanol in the presence of sodium ethoxide [20a] affording DEC with a 91% conversion yield. Zn–Y oxides have also been used as catalysts in the transesterification of ethene carbonate with methanol [20b]. Ethene carbonate is today produced by reaction of ethene epoxide with CO_2.

The alcoholysis of urea has great potential and is investigated with much attention all around the world. The process goes in two steps [21], that is, formation of H_2NCOOR Eq. (13.9) that is then converted into DMC Eq. (13.10), using either a single ZnO catalyst [22] or two different catalysts [21].

$$(H_2N)_2CO + ROH \rightarrow ROC(O)NH_2 + NH_3 \tag{13.9}$$

$$H_2NCOOR + ROH \rightarrow (RO)_2CO + NH_3 \tag{13.10}$$

The reaction of urea with alcohols [22–25] or polyols [26] is under study and it is clear now that the second stage is the rate-determining step. The route based on urea has the advantage of having an interesting alcohol conversion (>25%) that may rise to 90% with polyols [26], coupled to the easy availability of urea that is produced at a scale of over $130\,Mt\,a^{-1}$ with an yearly foreseen increase of 7%. The reaction of polyols with urea gives a higher conversion as the formation of cyclic carbonates is thermodynamically favored with respect to linear ones. Moreover, when polyols are used, as they are high-boiling alcohols, working under vacuum shifts the equilibrium to the right as NH_3 is eliminated.

13.3.1.4 Synthesis of Cyclic Carbonates and Polymers

The synthesis of cyclic carbonates via carboxylation of epoxides Eq. (13.11) has been known since 1943 [27–29].

$$\text{epoxide} + CO_2 \longrightarrow \text{cyclic carbonate} \qquad (13.11)$$

The reaction may afford polymers [30] as well, depending on the catalyst and the reaction conditions [31–41].

Several catalysts have been used, including organic catalysts [42], metal halides [43], transition metal complexes [44], Lewis acids [33], metal phthalocyanines [45] and heterogeneous catalysts [46–52]. Optically active epoxides have been converted into the relevant carbonates using metal oxides with total retention of configuration [52]. Racemic epoxides have been carboxylated with an ee = 22% using Nb(IV) complexes with optically active (N, O, P as donor atoms) ligands, the low ee being due to de-anchoring of the ligand from the metal center [52]. Ionic liquids (IL) are good media for the carboxylation of oxiranes which is promoted by microwaves [53]. This subject has been largely reviewed in recent years [54–58].

Sc-CO_2 as a solvent [59], also in combination with ILs [60, 61], has often been used.

The oxidative carboxylation of olefins is an approach to the synthesis of cyclic carbonates from cheap and easily available reagents, such as olefins, CO_2 and O_2, Eq. (13.12), that avoids the production of epoxides.

$$\text{olefin} + \tfrac{1}{2} O_2 + CO_2 \longrightarrow \text{cyclic carbonate} \qquad (13.12)$$

Such reaction is illustrated by only a limited number of examples [62–68] and requires research in order to avoid the loss of olefin due to the addition of dioxygen across the C=C double bond forming aldehydes or the relevant acids [69]. Using homogeneous Rh-complexes and labeled $^{16(18)}O\text{–}^{18(16)}O$ peroxo groups it has been possible to show that the metal-bound O-atom of the peroxo group is transferred to an oxophile [70].

The interest of such a reaction lies in the fact that dioxygen is used instead of organic hydroperoxides or hydrogen peroxide [71]. A two-step reaction in which a metal oxide in a high oxidation state transfers lattice oxygen to an olefin and is then re-oxidized using O_2 has been used with success for avoiding the double-bond splitting [72].

The carboxylation of epoxides may produce polycarbonates. Al-porphyrin complexes [73] or Zn-compounds [74] have been used for a long time [75, 76] and have made the copolymerization of CO_2 and epoxides quite popular, and recently fully reviewed [77, 78]. The amount of polycarbonates used around the world has grown with a quasi-exponential trend in the last few years reaching, over 4.5 Mt y^{-1} in

2011. Polycarbonates are used in fields such as: automotive, construction, medical care, packaging and electronics industries and also to make optical discs. The use in construction, optical discs and electronics covers over 90% of the total use.

Recent advancements in this specific field have allowed the production of polymers [79–82] with a TOF of $15\,000\,h^{-1}$, working with a quite high substrate/catalyst ratio (50 000). Several new approaches [83–89] demonstrate the interest in this field and the potential of such technology.

A point of interest in this field is the use of epoxides or olefins characterized by different molecular structures which may produce polymers with different structural properties. A key issue is the production of epoxides using a clean route. The use of hydrogen peroxide represents the cleanest industrial route for the conversion of olefins into epoxides: due to the limited quantity of hydrogen peroxide produced today ($500\,kt\,y^{-1}$) it would be better to use dioxygen, and this makes the oxidative carboxylation Eq. (13.12) a key technology if properly developed.

By transesterification it is possible to produce carbonates and polymers. Zn-metal-organic-frameworks (MOF) [90] based on 1,4-benzenedicarboxylate produce polycarbonate-diol from diphenylcarbonate and 1,6-hexanediol.

Enzymes (lipases) [91] are also used as catalysts in the reaction of dialkylcarbonates and diols or polyols and a diester [92]. Diethylcarbonate, 1,8-octanediol and tris(hydroxymethyl)ethane have been polymerized in the presence of a lipase-B extracted from *Candida antartica* to produce a polycarbonate polyol [93].

13.3.2
Carbamates and Polyurethanes

13.3.2.1 Synthesis of Molecular Carbamates

Carbamates are common species in Nature and are involved either in CO_2 activation in photosynthetic processes (Calvin Cycle) or in physiological removal of N-compounds. Amines (monoethanolamine, mainly, or poly-amines) [94, 95] are used as a means of capturing CO_2 from gas mixtures (flue gases). The reaction of carbon dioxide with amines is also used in synthetic chemistry [96].

The simplest species, obtained by reaction of ammonia with CO_2, is carbamic acid, H_2NCOOH, a labile compound that easily decomposes back to NH_3 and CO_2, but can be stable as a solid in the zwitterionic form $^+NH_3COO^-$ at low temperature [97–99]. Only recently, secondary amines, with suitable structural features [100, 101] (such as dibenzylamine or a Co-aminophosphane complex) have afforded, at 300 K, isolable carbamic acids, characterized by XRD, as H-bonded dimeric solids, Eq. (13.13). Also, some amino-silanes have been found to produce dimeric carbamic acids $[(RO)_3SiCH_2CH_2NHCH_2CH_2NHCOOH]_2$ at 273 K [102].The reaction of amines with CO_2 easily affords solid ammonium carbamates, Eq. (13.14).

$$RR'NH + CO_2 \rightarrow \frac{1}{2}[RR'NCO_2H]_2 \qquad (13.13)$$

$$2RR'NH + CO_2 \rightarrow (RR'NH_2)^{+-}OOCNRR' \qquad (13.14)$$

The formation of the carbamate "RHNCOO⁻" moiety is also promoted by metal salts, Eq. (13.15), via the formal insertion of CO_2 in the M–N bond of amides resulting from the reaction of the metal center with the amine employed [103].

$$RR'NH + L + MBPh_4 + CO_2 \rightarrow M^{+-}(O_2CNRR') + [HL]BPh_4 \qquad (13.15)$$

R' = H, alkyl, L = RR'NH; R = aryl, R' = H, L = NR''3 (R'' = alkyl); M = Li, Na, K

The carbamate moiety "RR'NCO$_2^-$" formed in this way can be transferred *to an electrophilic center* forming either *ionic or covalent bonds*. Of particular importance is the transfer to an organic substrate that can produce organic carbamates, or even isocyanates and ureas. The latter approach represents an attractive and innovative alternative to the problematic phosgene-based technologies. The synthesis of organic carbamates from amines and CO_2 has been extensively investigated over the past years, and recently reviewed [96]. Such products find application in pharmacology [104], as agrochemicals, or in synthetic chemistry as protecting groups [105, 106], or precursors, not only for ureas and isocyanates but also for polymers [107, 108]. An issue in such syntheses is that the carbamate moiety "RR'NCO$_2^-$" can react with an electrophile, not only at the O-anionic, Eq. (13.16), end but also at the N-atom, causing CO_2 elimination, Eq. (13.17) [8, 96].

Despite the great number of metal-carbamates isolated, only a few examples of the transfer of the carbamic group from the metal center to an alkyl halide or to a sulfide have been documented [103, 109, 110]. An extensive review has recently been published by Chaturvedi [111].

The general reaction between CO_2, amines and alkyl halides to produce the relevant carbamates is represented by Eqs. (13.16) and (13.17).

$$\underset{R^2}{\overset{R^1}{\diagdown}}NH + CO_2 + R^3X \xrightarrow{-HX} \underset{R^2}{\overset{R^1}{\diagdown}}N-\overset{O}{\underset{OR^3}{\diagup\!\!\!\diagdown}} \qquad (13.16)$$

$$R^1R^2N\text{-COOM} + R^3X \longrightarrow R^1R^2R^3N + MX + CO_2 \qquad (13.17)$$

The electrophilicity of the cation in ammonium or metal carbamates is a crucial factor for mastering the transfer reaction toward the selective alkylation at oxygen for the production of organic carbamates [112]. Complexing agents such as crown ethers and cryptands have been successfully used to this end [113]. Sterically hindered strong organic bases, such as amidines or pentaalkylguanidines (CyTMG [114], phosphazenes [114–116], and DBU [1,8-diazabicyclo[5.4.0]undec-7-ene]) have also been used successfully for this purpose. Tetraethylammonium superoxide [117], basic resins [118], Triton B (benzyl-trimethyl-ammonium hydroxide) [119], and inorganic bases such as K_2CO_3 in the presence of catalytic amounts of $(Bu_4N)I$ [120] also produce good results. Polyethylene glycol [121] (PEG) is a suitable solvent, and phase-transfer-catalysts (PTC) convert amines, CO_2 and alkyl halides into organic carbamates in the presence of K_2CO_3 [121].

Alcohols can substitute alkyl halides, avoiding the use of halides and making the whole process more ecofriendly. Unfortunately, this reaction requires quite drastic conditions in order to overcome the thermodynamic and kinetic limitations.

Interestingly, aziridines in sc-CO_2 produce cyclic oxazolidin-2-ones [96], precursors of polyurethanes. Epoxides produce carbamates by reaction with CO_2 and amines, Eq. (13.18).

$$R^1R^2NH + CO_2 + \text{epoxide}(R^3, R^4) \longrightarrow R^1R^2N-C(=O)-O-CH(R^3)-CH(R^4)-OH \quad (13.18)$$

After the first example reported by Yoshida and Inoue, who used $Ti(NMe_2)_4$ as promoter of the reaction of CO_2 and 1,2-epoxycyclohexane [122], many other metal amides have been employed, such as $TiCp(NMe_2)_3$, $W(NMe_2)_6$ [123], $EtZn(NPh_2)$, $Et_2Al(NPh_2)$ [124], and also a variety of different epoxides, such as propylene and styrene oxide [125], or chloromethyloxirane [126], producing cyclic carbamates as the final product. α-Haloacylophenones behave as oxiranes precursors [127], and oxietanes produce mono-carbamates of 1,3-propanediols [128].

Synthesis of carbamates from amines, CO_2 and alkenes, Eq. (13.19), can be found in the literature, namely the case of norbornene, dicyclopentadiene, 1,5-cyclopentadiene, reacted with $PdCl_2$ [129].

$$R^1R^2NH + CO_2 + \text{alkene}(R^3, R^4) \longrightarrow R^1R^2N-C(=O)-O-CH(R^3)-CH(R^4) \quad (13.19)$$

Pd(0)-phosphine complexes catalyze the formation of allylic carbamates and, more recently, $Pd(PPh_3)_4$ has been used to prepare vinyloxazolidinones. In both cases DBU has been used to achieve good yields [130, 131].

1-Alkynes have been used for the preparation of vinyl carbamates, Eq. (13.20), with $Ru_3(CO)_{12}$ as catalyst [132], better if in sc-CO_2 than in organic solvents [133]. The mechanism has been proposed as involving Ru-vinylidene intermediates [134]. A few examples of other catalysts are documented and $ReBr(CO)_5$ was reported to catalyze the synthesis of N,N-diethyl-carbamates in very good yields [135].

$$R^1R^2NH + CO_2 + \equiv\!\!-R^3 \longrightarrow R^1R^2N-C(=O)-O-CH=CH-R^3 \quad (13.20)$$

Propargyl alcohols, primary amines and CO_2 produce cyclic carbamates, also in sc-CO_2, in the presence of Cu(I) halides [136]. The guanidine-catalyzed synthesis of carbamates and carbonate in sc-CO_2 has been documented [137]. Propargylamines and CO_2 produce 5-methylene-2-oxazolidinones, with high yields in the presence of strong organic bases under mild conditions [138], or in presence of solid bases (alumina, hydrotalcites) or supported organic bases in sc-CO_2 [139].

The electrochemical reduction of CO_2 in conventional solvents [140] or ionic liquids, in the presence of alkyl halides and aromatic or aliphatic amines produces carbamates [140–142].

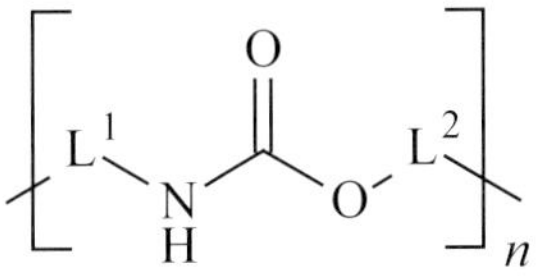

$$\left[L^1 \underset{H}{\overset{}{N}} \underset{O}{\overset{O}{\|}} O^{L^2} \right]_n$$

Figure 13.3 Polyurethane.

13.3.2.2 Indirect Synthesis of Carbamates

An interesting and clean way to carbamates is the reaction of amine with organic carbonates, Eq. (13.21).

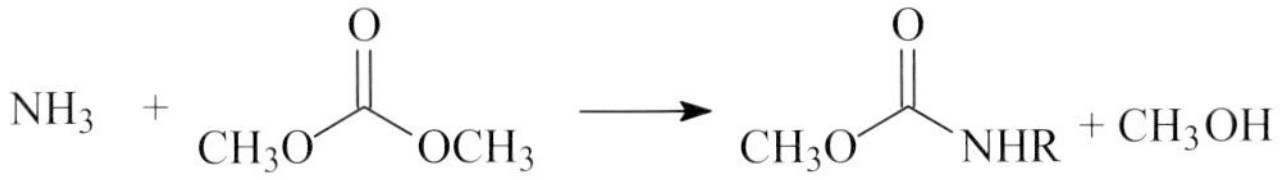

$$NH_3 + \underset{CH_3O}{\overset{O}{\|}} OCH_3 \longrightarrow \underset{CH_3O}{\overset{O}{\|}} NHR + CH_3OH \quad (13.21)$$

The organic carbonate (DMC) used for the synthesis of carbamate can be produced by methanol carboxylation, as discussed above (see Scheme 13.2e). Another route to carbamates is the reaction of carbonates with urea [143].

The urethane moiety is the fundamental constituent part of polyurethanes, known since 1937 : Perlon (Figure 13.3) was made by Bayer by reacting 1,6-diisocyanato-hexane with 1,4-dihydroxy-butane [144].

Such technology is still valid today for the synthesis of polymers with different properties, either linear or branched, in a two- or three-dimensional network. The structure and property of a polyurethane depend on the ratio of the reagents employed and are influenced by several parameters, such as temperature, catalyst, additives, and reactor feeding and volumes [145].

As anticipated, the fundamental reaction in the synthesis of polyurethanes is the condensation between an isocyanate and an alcohol, Eq. (13.22a).

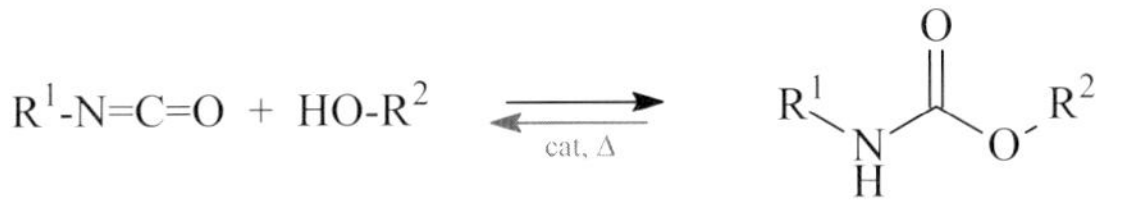

$$R^1\text{-}N\text{=}C\text{=}O + HO\text{-}R^2 \underset{cat, \Delta}{\rightleftharpoons} R^1\underset{H}{\overset{}{N}}\underset{O}{\overset{O}{\|}}R^2 \quad (13.22a)$$

$$O\text{-}C\text{-}N\text{-}L^1\text{-}N\text{=}C\text{=}O + HO\ L^2\ OH \longrightarrow \left[L^1 \underset{H}{\overset{}{N}} \underset{O}{\overset{O}{\|}} O^{L^2} \right]_n \quad (13.22b)$$

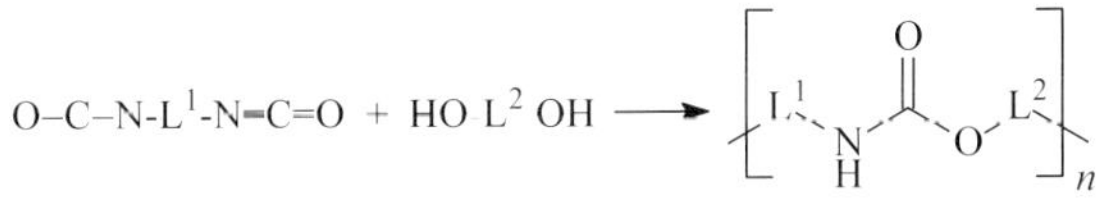

The synthesis of polyurethanes can be carried out by condensation of diisocyanates and polyalcohols, Eq. (13.22b). Reaction (22a) is reversible and its reverse can be used for the synthesis of isocyanates from carbamates of primary amines. This will bring a very interesting process for the synthesis of isocyanates from CO_2, as the routes currently used are quite complex and waste producing.

Phosgene is the main source, and the reductive carbonylation of nitro-aromatic compounds or the oxidative carbonylation of amines represent not very much exploited alternatives [146, 147]. The condensation process is not straightforward as side reactions may occur, involving mainly the isocyanate functionality that

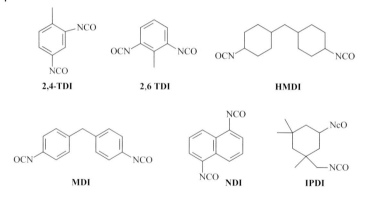

Figure 13.4 Aromatic and cyclic diisocyanates used for polyurethanes synthesis.

reacts with electrophiles and nucleophiles [148, 149]. Isocyanates easily associate into dimers and trimers. They can also polymerize in alkaline conditions and low temperatures not used in the polyurethane synthesis. Trimers have hyperbranched structures with improved thermal stability [145].

The most commonly used diisocyanates are the aliphatic 1,6-diisocyanato-hexane, used by Bayer, and the aromatic toluene-diisocyanate (TDI) (a mixture of 2,4- and 2,6- isomers), 4,4'-methylene-bis (phenyl-isocyanate) (MDI), and 1,5-naphthylene-diisocyanate (NDI). In the last decade, cyclic isocyanates such as isophorone diisocyanate (IPDI) and 4,4'-methylene-bis (cyclohexyl-isocyanate) (HMDI) have been successfully used for the production of polymers less susceptible to photo-degradation (Figure 13.4) [145].

A direct use of CO_2 for polyurethane synthesis is represented by its copolymerization with N-analogs of epoxides, that is, aziridines or azetidines [77] that afford both cyclic urethanes and polymers, Eq. (13.23) [150–152]. The reaction conditions [153] play a key role for controlling the alternate insertion of CO_2 and the co-monomer, with respect to the simple polymerization of the co-monomer.

$$\underset{R}{\overset{H}{\underset{}{\triangle}}}\!\!\!\!\!\!\!\!N + CO_2 \longrightarrow \left[\begin{array}{c} R \\ \diagdown \\ N \\ H \end{array}\right]_n \left[\begin{array}{c} O \\ \| \\ N-O \\ | \\ R \end{array}\right]_m \quad (13.23)$$

Polyurethanes are used for the production of elastomers as protective coatings (automobile seats), or for construction, or else as plastic equipment, and textile fibers [145]. The modulation of the mechanical and elastic properties of polyurethanes can be mastered by selecting the chemical structure of the single constituents, that is, isocyanates and polyols. Polyethers give more flexible polymers than polyesters [154]. The above-mentioned symmetrical diisocyanates, MDI and HMDI, are used to produce materials with high mechanical strength [155]. In general polymers can be subjected to post-synthetic modifications in order to regulate their mechanical properties.

Polyurethanes can be produced by using "biogenic blocks" and this is now an area of great interest [145, 156, 157]. They are largely used in the biomedical field

[158] as bio-compatible materials for cardiac valves, regenerative membranes or tissues [145, 159, 160]. Newly synthesized polyurethane ionomers can be easily dispersed in water and used as "green" varnishes for different substrates [161]; when combined with other polymers they produce water-resistant adhesives or polymeric coatings [145, 162]. Other materials such as very hard polymer-ceramic composites, films used for capacitors, and materials with interesting opto-electronic properties have recently been prepared [145, 163].

13.4 Energy Products from CO_2

The conversion of CO_2 into energy-rich products that may be used as fuels requires either hydrogen or energy in the form of heat or electrons. This requirement has limited so far the interest in such conversion, considering that 90% of hydrogen is now produced from fossil carbon, according to Eqs. (13.24)–(13.26) that are all strongly endoergonic.

Water gas reaction

$$C + H_2O_{(vap)} \rightarrow CO + H_2; \Delta H^\circ_{298k} = 131 \text{ kJ mol}^{-1} \tag{13.24}$$

Wet-reforming of methane

$$CH_4 + H_2O_{(vap)} \rightarrow CO + 3H_2; \Delta H^\circ_{298k} = 206 \text{ kJ mol}^{-1} \tag{13.25}$$

Dry reforming of methane

$$CH_4 + CO_2 \rightarrow 2CO + 2H_2; \Delta H^\circ_{298k} = 247 \text{ kJ mol}^{-1} \tag{13.26}$$

Usually the production of dihydrogen is further incremented by reacting CO with $H_2O_{(v)}$ in the so-called water-gas-shift reaction, Eq. (13.27), that is also endoergonic.

Water-gas-shift reaction

$$CO + H_2O(v) \rightarrow CO_2 + H_2; \Delta H^\circ_{298k} = -41 \text{ kJ mol}^{-1} \tag{13.27}$$

The latter produces CO_2 and raises the issue that such routes are not at all the correct strategy for reducing CO_2 emission or for converting CO_2. If CO_2 has to be converted into fuels, a new approach is needed that does not make use of fossil-C for the production of H_2 and this is the use of water (water-splitting) coupled to a non-fossil source of energy. Water splitting, Eq. (13.28), is strongly endoergonic ($\Delta G_0 = 237.2 \text{ kJ mol}^{-1}$).

$$H_2O \xrightarrow{\Delta} H_2 + \frac{1}{2}O_2 \tag{13.28}$$

However, if hydrogen is available, then the conversion of CO_2 can occur via known chemical routes Eqs. (13.29)–(13.33), e.g., conversion into formic acid, methanol, even methane or hydrocarbons or olefins.

$$CO_2 + H_2 \rightleftarrows HCOOH \tag{13.29}$$

$$CO_2 + 2H_2 \rightarrow H_2CO + H_2O \tag{13.30}$$

$$CO_2 + 3H_2 \rightarrow CH_3OH + H_2O \tag{13.31}$$

$$CO_2 + 4H_2 \rightarrow CH_4 + 2H_2O \tag{13.32}$$

$$(n+2)CO_2 + [3(n+2)+1]H_2 \rightarrow CH_3(CH_2)_n CH_3 + 2(n+2)H_2O \tag{13.33}$$

Such an approach would allow continuation of the use of already known fuels with the advantage of maintaining existing infrastructures for mobility and other applications. Merging the production of hydrogen with CO_2 reduction would eliminate the problem of storage, shipping and utilization of H_2. Although, in future, the use of "artificial photosynthesis" may lead to a large conversion of CO_2 into useful products [164], in the short term only a few selected options can be considered as of potential exploitation: (i) the use of excess electric energy for H_2O electrolysis or for the direct reduction of CO_2 in water, (ii) the application of PV (photovoltaics) for the same purposes as in (i), (iii) scission of water using concentrators of solar power (CSP), (iv) the exploitation of thermodynamic cycles [165].

The use of CSP could also make possible the direct dissociation of CO_2 into CO and O_2 [166]. The former can give energy by reaction with dioxygen. A cycle can be set based on the oxidation of CO to CO_2 –reduction of CO_2 to CO (Scheme 13.5).

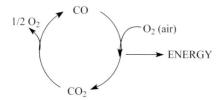

Scheme 13.5 Cycle of reduction of CO_2 oxidation of CO.

Among CO_2-derived energy products, methanol plays a central role as it can find a direct utilization as fuel (in cars or in fuel cells), or as a versatile bulk chemical (production of acetic acid (CH_3COOH), ethene and derived polymers such as LDPE or HDPE, ethers, hydrocarbons). Methanol is today produced from CO/CO_2 mixtures and can be produced from CO_2 alone: the synthetic process is well mastered at the plant level. The same is true for the conversion of methanol into other chemicals or fuels [167].

Therefore, if hydrogen is available it will find a use in recycling CO_2-producing chemicals and fuels, thus reducing the need to extract fossil carbon. Products derived from the conversion of CO_2 can also be considered as a strategy to *store excess electric energy* (e.g., energy produced in a period when lower power is demanded by the grid): the fuels could be used in cars (substituting fuels produced from fossil carbon) or for generating electric energy during the peak hours. Such practice is already implemented in some power stations which convert electric energy into potential energy (by raising up-hill masses of water) which is then again converted into electric energy (falling water will generate electricity during peak hours).

Interestingly, if the volume energy density of fuels (15–36 GJ m^{-3}) and batteries (0.1–1.5 GJ m^{-3}) are compared, the former are a better storage option than the latter. However, the conversion of electric energy into fuels by reduction of CO_2 also gives an answer to an old issue: the storage of electric energy.

Wind power could be used for running an electrolyzer so that wind towers would be in operation also when the grid is saturated and cannot accept further electric energy.

The electrolysis of water is run with an efficiency equal to 70–80%, that coupled to the selectivity of conversion of H_2 and CO_2 into methanol (close to 100%) would make such a strategy quite efficient for energy storage in chemicals, despite the loss of energy for the compression of H_2 (ca. 3.6 kWh kg^{-1}) [168] to the pressure required for the synthesis of methanol (3.0 to 5.0 MPa).

The direct electro-reduction of CO_2 in water is another approach which needs improvements in relation to the life of the electrodes, that can be improved by using electrocatalysts, the overpotential, the current density, the solvent.

13.5 Production of Inorganic Carbonates

The carbonation of natural silicate minerals, Eqs. (13.34) and (13.35), or industrial slag (with an estimated potential of sequestration of 180 Mt y^{-1} of CO_2) or mine tailings rich in Group 2 metal oxides, such as magnesium oxide (MgO) and calcium oxide (CaO), is an interesting option for long-term storage of CO_2 via the production of useful materials. Serpentine and olivine contain high concentrations of MgO, while pyroxenes and amphiboles are a potential source of both CaO and MgO [169–171].

$$Mg_3Si_2O_5(OH)_4 \text{ (s)} + 3CO_2 \text{ (g)} \rightarrow 3MgCO_3 \text{ (s)} + 2SiO_2 \text{ (s)} + 2H_2O \tag{13.34}$$

$$CaSiO_3 \text{ (s)} + CO_2 \text{ (g)} \rightarrow CaCO_3 \text{ (s)} + SiO_2 \text{ (s)} \tag{13.35}$$

The carbonation of natural rocky materials has as a serious drawback the amount of material to be mined per ton of CO_2: 2.5–3 t of magnesium silicate mineral need to be mined per t_{CO2}, equivalent to ~8 t of mineral per ton of coal used. Although the carbonation of such basic oxides is exothermic, the need to work in water to accelerate the carbonation rate makes the entire process quite energy intensive, in addition to the cost of chemicals and the necessary recycling of water.

The use of slag would be much more interesting. Slag by-products from iron- and steelmaking may produce valuable precipitated calcium carbonate (PCC). Such carbonates have special applications and have an economic value. In addition to the above industrial sectors, the pulp and paper industry may also be providers of waste or by-products rich in basic oxides that can be converted to added value materials whose use will depend on their purity which, in turn, depends on the source of Group 2 cations. In fact, slag or mining retailings can be accompanied by a variety of other elements, difficult to separate from Ca or Mg, that will depreciate the carbonates [169]. Special PCC has a price of € 100 per ton [169]: such purity requires an extraction of Ca and Mg cations from the slag and their selective precipitation, that has a cost depending on the source.

The key issues in such an approach are: (i) the development of low energy and low environmental impact extraction or activation of the reactive component, MgO or CaO, from silicate mineral or slag, (ii) speeding-up the carbonation kinetics of basic oxide produced in whatever way, (iii) obtaining products of good purity that may find a market.

13.6
Enhanced Fixation of CO_2 into Aquatic Biomass

The fixation of CO_2 into aquatic biomass (microalgae or macroalgae) under non-natural conditions (concentration of CO_2 in the gaseous phase up to 150 times higher than natural conditions) makes the production of such biomass quite attractive for the production of fuels, chemicals and materials. Biodiesel, biogas, bioethanol or bio-hydrogen can be produced from aquatic biomass. Macro-algae, in general, are produced at lower costs than micro-algae. Species of micro-algae rich in lipids (30–70% dry weight) are suitable for bio-oil production. Table 13.3 shows, as a comparison, the amount (L) of oil per hectare per year of different types of biomass [172, 173]. Macro-algae, in general, present a lower content of lipids than micro-algae and a larger variability [174], depending on the cultivation technique and on the period of the year in which they are collected [175, 176].

The most economic utilization of the aquatic biomass is to simultaneously produce chemicals and fuels [177–182]. In general, chemicals extracted from algae have a quite high added value and some algae are (or have been made) specialized in the production of costly chemicals. Often, almost 50–60% of the biomass can be used besides energy product extraction, and can give pharmaceuticals and nutraceuticals, pigments, proteins, and new materials.

Different species of micro-algae are produced and used for pharmaceuticals and nutraceuticals purposes. *Chlorella*, produced by several companies (the largest producer is Taiwan Chlorella Manufacturing&Co which produces 400 t of dry algal biomass per year) [183] is known to have therapeutic effects on gastric ulcers, wounds, and constipation, together with preventive action against both atherosclerosis and hyper-cholesterol, and antitumor activity.

An important active substance is β-1,3-glucan which is believed to be an active immune-stimulator, free radical scavenger and a reducer of blood lipids. Unfortunately the validation of many of these claims is still awaited [183–187].

Table 13.3 Yield of oil from various types of biomass.

Biomass	Corn	Safflower	Sunflower	Coconut	Palm	Microalgae
Yield ($l\,ha^{-1}\,y^{-1}$)	170	785	965	2840	6000	47 250 to 142 000

Spirulina (*Arthrospira*) is used in human nutrition because of its high protein content, excellent nutrient value [183], essential fatty acid content (linolenic acid is not synthesized by humans) [182] and "nutraceuticals", the food supplements with claimed nutritional and medicinal benefits. The largest production plant is Earthrise Farms in California, USA covering over 444 000 m^2, producing algal tablets and powder sold in over 20 countries and owned by DIC in Japan [188]. Cyanotech, in Kona, Hawaii, produces a powder under the name Spirulina Pacifica [189]. The market for dried *Spirulina* was estimated to be US$ 40 million in 2005 [190].

Algal pigments, such as chlorophylls a and b (chlorins), chlorophyll c (porphyrins), phycobilipigments (open tetrapyrrols), and carotenoids (polyisoprenoids with terminal cyclohexane rings; carotenes and xanthophylls) are produced on a large scale (Figure 13.5).

The most commercially valuable groups of pigments are carotenoids (β-, ε-, α carotenes, xanthophylls) and phycobilins with several available products derived mainly from green microalgae (e.g. *Dunaliella*) and Cyanobacteria (e.g. *Spirulina*) [191, 192]. Lutein, zeaxantin and canthaxantin are also commercially used, for example, in poultry feed, and astaxanthin as a colorant in aquaculture [193] and also in human health and nutrition [194, 195]. Astaxanthin can be used in fish farming and as a dietary supplement or anti-oxidant. It can be produced by *Haematococcus*, a freshwater alga that normally grows in puddles, birdbaths, and other shallow fresh water depressions. *Haematococcus* can contain up to 3% astaxanthin,

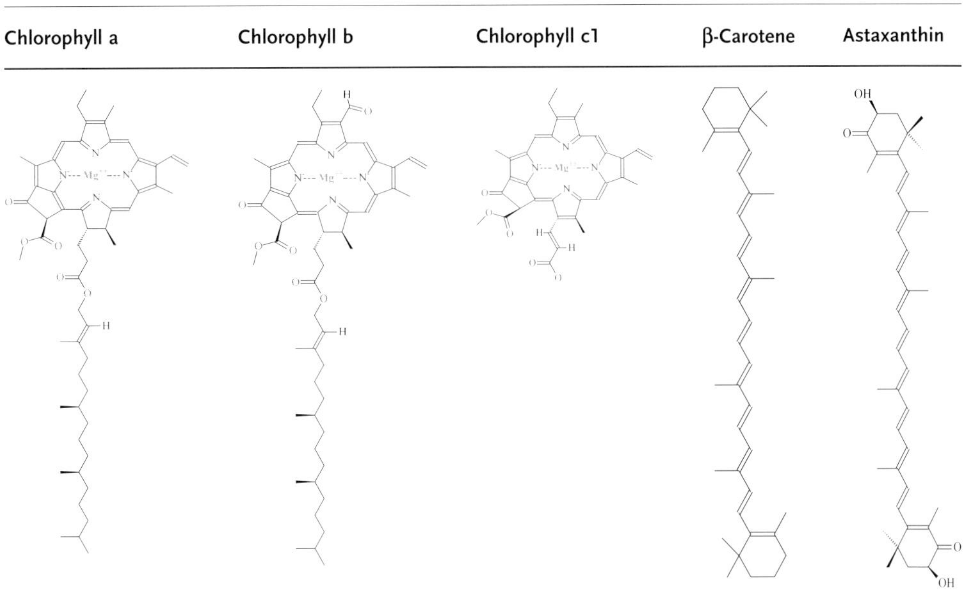

Figure 13.5 Structures of some pigments.

but it requires a two-stage culture process which is not suited to open pond cultivation.

The first stage of the process is designed to optimize algal biomass (green thin-walled flagellated stage with optimum growth at a temperature 22–25 °C), and the second stage (thick-walled resting stage) under intense light and nutrient poor conditions during which astaxanthin is produced [183]. Also, if natural astaxanthin is characterized by high price, it has a quite large application in carp, chicken and red sea bream diets due to enhanced natural pigment deposition, regulatory requirements and consumer demand for natural products [183]. Commercial production is being carried out in Hawaii, India and Israel, where Algatech sell a crushed *Haematococcus* biomass on the pharmaceutical market [183, 196, 197].

Phycobiliproteins (phycoerythrin, different phycocyanins and allophycocyanins) from red algae and Cyanobacteria have a long tradition of use as dyes in food, cosmetics and as fluorescent markers in biomedical research [198–200].

Proteins are present in algae in a diverse range of forms and cellular locations, for example, as a component of the cell wall, as enzymes, and bound to pigments and carbohydrates. Green algal cell walls contain hydroxyproline-rich glycoproteins, one of the few protein groups to contain high levels of hydroxyproline [201]. Algal proteins are important as a source of essential amino acids, as well as for the bioactive potentials of specific phycobiliproteins and lectins, with great bioactive potential such as antibiotic, mitogenic, cytotoxic, antiinflammatory, antiadhesion, anti-HIV, reduction to pain sensitivity, human platelet aggregation inhibition and anti-cancer, antimicrobial, anti-viral, and anti-hypertensive activities [202].

Very interesting is the utilization of synthetic biology and genomics to enhance the productivity and increase the utility of algae to produce advanced green plastics and chemicals from biopolymers. The productivity of aquatic biomass can be improved through genetic engineering of the (micro)-organisms. An alternative is to generate new plants through the inclusion of bacterial genes in the mother plant so that the new form may produce new materials that are extracted and processed. Such materials would have the advantage of being derived from renewable resources, fully biodegradable or compostable and produced through eco-friendly synthesis. They would be characterized by high processability with good mechanical properties and low cost. The engineering of such new materials would take advantage of a possible functionalization *in vivo*, introducing in the molecular structure functional groups that would be useful for an interaction with other fibers or for making materials with defined properties, such as thermal stability or brittleness. The production of copolymers using only biomass-derived monomers or a mix of fossil-fuel-derived and biomass-derived monomers is also a quite interesting approach. Several biomass-derived polymers such as collagen, gelatine, alginates, casein, elastin, and zein, have been used for some time. They are used for the production of medical textiles.

The market for bioplastics is currently a tiny niche in the global plastics market but is expected to double during 2012 as rising oil prices (that should make the

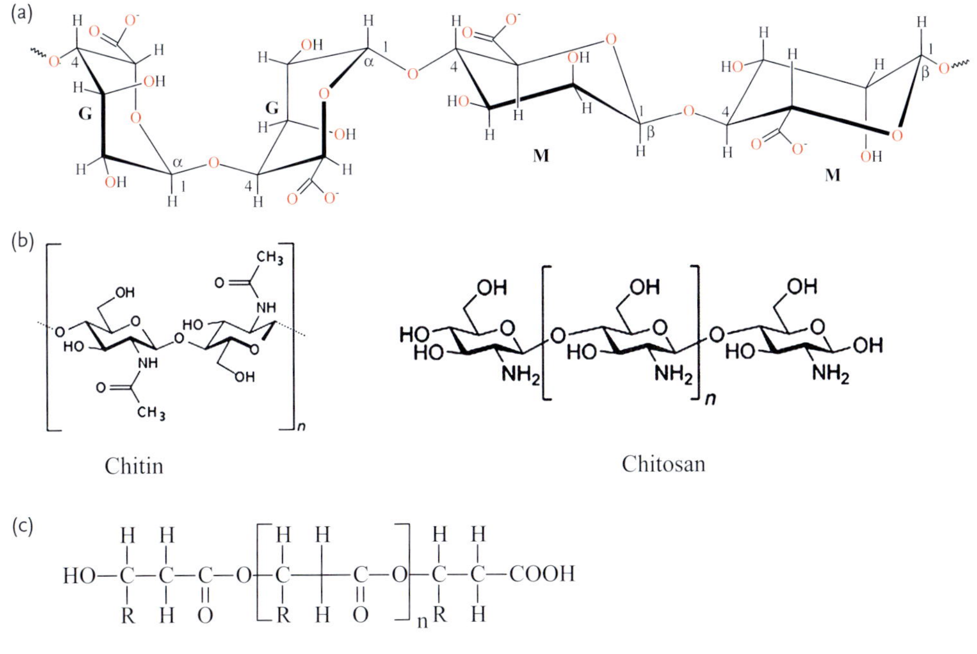

Figure 13.6 Chemical structure of (a) alginate, (b) chitin and chitosan, (c) polyalkanoates.

cost of bio-derived plastics competitive) and environmental regulations crimp petroleum-based products. The various plastics that can be made from algae feedstock include: (-) hybrid plastics obtained by adding denatured algae biomass to petroleum based plastics like polyurethane and polyethylene as fillers (filamentous green algae are suitable for this use); (ii) cellulose-based plastics, derived from cellulose (algal strains that contain cellulose are suitable for feedstocks for cellulose-based plastic); (iii) poly-lactic acid (PLA) where lactic acid, produced by bacterial fermentation of algal biomass, is polymerized to produce polylactic acid; (iv) bio-polyethylene where ethanol used to produce ethylene can be derived from bacterial digestion of algal biomass, or directly from algae.

Polymers derived from aquatic biomass are:

1) *alginate* (Figure 13.6a) [203], used for its ability, when dissolved in water, to increase the viscosity of aqueous solutions, to form gels (gels form when a calcium salt is added to a solution of sodium alginate in water) and to form films of sodium or calcium alginate and fibers of calcium alginates.

2) *chitin and chitosan* (Figure 13.6b), used for the fabrication of anti-bacterial, anti-fungal, anti-viral materials. They are non-toxic and non-allergic and originate fibers that have good breathability, smoothness, and absorbence.

3) *polyalkanoates*, (Figure 13.6c) made either from a single monomer (polyhydroxybutyrate–PHB) or from mixed alkanoates, such as poly(3HB-co-3MP),

a polymer with a thioester linkage from 3-hydroxobutyrate and 3-mercaptopropionate.

4) *tri-block co-polymers,* made for example from caprolactam co-polymerized with a hydrophilic natural oligoagarose and a polypeptide (polylysine).

5) *algaenan,* a non-hydrolyzable, insoluble biopolymer [204] isolated from a variety of unicellular algae, resulting from the reticulation of low molecular weight aldehydes and unsaturated hydrocarbons (long chain, up to 40-C). Reticulation occurs via acetal and ester links that build the primary structure of the polymer. Such material is produced by *Botriococcus braunii, Chlorella* and other microalgae.

In this way a large number of polymers having different properties can be built. Interestingly, as mentioned above, polymers made from monomers derived from oil and biomass can be synthesized, the biogenic part being used to modify some properties of oil-derived polymers. This will help to reduce the dependence on fossil-carbon and to produce more biodegradable and compostable materials.

13.7
Conclusion and Future Outlook

The utilization of CO_2 as source of carbon for the synthesis of chemicals, materials and fuels is a strategy that is today of great interest, as demonstrated by the National and International programs that support research in this area. The utilization in the production of chemicals may find an industrial application in the medium term, while the production of fuels requires a longer term and the full development of the production of hydrogen from water.

The production of chemicals from CO_2 may not require a large energy input, as several reactions are exoergonic, if the co-reagent (olefin, amine, epoxide, etc.) is an energy-rich species.

Particular interest lies in the synthesis of polymers from CO_2. Polycarbonates and polyurethanes are chemicals with large markets (4–6 Mty^{-1}) that may incorporate amounts of CO_2 higher than 1 Mty^{-1}, with the double benefit of reducing the use of fossil carbon and sequestrating CO_2 for a long time.

The synthesis of fine chemicals from CO_2 is an interesting approach as it would be possible to develop quite attractive synthetic routes with low waste production. This is the case when the CO_2 moiety is incorporated into an organic substrate producing carboxylate functionalities. The latter are not generated with high selectivity using existing technologies and often require high energy inputs, while producing large volumes of waste.

A key reaction that may be coupled to CO_2 utilization is the C–H activation that would allow simple production of carboxylic acids.

The reduction of CO_2 to methanol requires large volumes of non-fossil-carbon derived hydrogen. When this is available, new synthetic routes will open for a number of methanol-derived chemicals.

In perspective, the market of CO_2 for the synthesis of chemicals will grow from an actual $130\,Mt\,a^{-1}$ to circa $200\,Mt\,a^{-1}$ in the short term and $>300\,Mt\,a^{-1}$ in the medium term, reaching more interesting figures of the order $>700–800\,Mt\,a^{-1}$ when large volumes of H_2-derived from water become available.

References

1 Jorapur, Y.R., and Chi, D.Y. (2005) *J. Org. Chem.*, **70**, 10774.
2 Isaacs, N.S., O'Sullivan, B., and Verhaelen, C. (1999) *Tetrahedron*, **55**, 11949.
3 (a) Sakakura, T., Saito, Y., Choi, J.C., Masuda, T., Sako, T., and Oriyama, T. (1999) *J. Org. Chem.*, **64**, 4506; (b) Sakakura, T., Saito, Y., Choi, J.C., and Sako, T. (2000) *Polyhedron*, **19**, 573.
4 Sakakura, T., Saito, Y., Okano, M., Choi, J.C., and Sako, T.J. (1998) *Org. Chem.*, **63**, 7095.
5 Honda, M., Kuno, S., Begum, N., Fujimoto, K., Suzuki, K., Nakagawa, Y., and Tomishige, K. (2010) *Appl. Catal. A: Gen.*, **384**, 165.
6 Aresta, M., Dibenedetto, A., Fracchiolla, E., Giannoccaro, P., Pastore, C., Papai, I., and Schubert, G. (2005) *J. Org. Chem.*, **70** (16), 6177.
7 Aresta, M., Dibenedetto, A., and Stufano, P. (2009) Italian Patent, MI2009A001221; (b) Aresta, M., Dibenedetto, A., Stufano, P., Aresta, M., Maggi, S., Papai, I., Rokob, T.A., and Gabriele, B. (2010) *Dalton Trans.*, **39**, 6985.
8 Dibenedetto, A., Aresta, M., and Stufano, P. (in press) in *Comprehensive Inorganic Chemistry II* (ed. J. Reedijk), Elsevier, Amsterdam (b) Ballivet-Tkatchenko, D., and Dibenedetto, A. (2010) in *Carbon Dioxide as Chemical Feedstock* (ed. M. Aresta), Wiley-VCH Verlag GmbH, Weinheim, p. 169; (c) Leini, E., Maki-Arvela, P., Eta, V., Murzin, D. Yu., Salmi, T., and Mikkola, J.P. (2010) *Appl. Catal. A: Gen.*, **383**, 1; (d) Keiler, N., Rebmann, G., and Keller, V. (2010) *J. Mol. Catal. A: Chem.*, **317** (1–2), 1.
9 Yamazaki, N., Nakahama, S., and Higashi, F. (1979) *Ind. Eng. Chem. Prod. Res. Dev.*, **18**, 249.
10 Kizlink, J., and Pastucha, I. (1994) *Collect. Czech. Chem. Commun.*, **59**, 2116.
11 Ballivet-Tkatchenko, D., dos Santos, J.H.Z., Philippot, K., and Vasireddy, S. (2011) *C. R. Chim.*, **14** (7), 780; (b) Dibenedetto, A., Aresta, M., Angelini, A., Papai, I., Pastore, C., and Aresta, B.M. (2010) *J. Catal.*, **269**, 44.
12 Ballivet-Tkatchenko, D., Dauteau, O., and Steizman, S. (2000) *Organometallics*, **19** (45), 63.
13 Dibenedetto, A., Aresta, M., Angelini, A., Ethiraj, J., and Aresta, B.M. (2012) *Chem. Eur. J.*, **18** (33), 10324–10334.
14 Aresta, M., Dibenedetto, A., Pastore, C., Cuocci, C., Aresta, B., Cometa, S., and De Giglio, E. (2008) *Catal. Today*, **137**, 125.
15 Tomishige, K., and Kunimori, K. (2002) *Appl. Catal. A: Gen.*, **237**, 103.
16 Wang, W., Wang, S., Ma, X., and Gong, J. (2009) *Catal. Today*, **148**, 323.
17 Leino, E., Maki-Arvela, P., Eränen, K., Tenho, M., Murzin, D.Y., Salmi, T., and Mikkola, J.P. (2011) *Chem. Eng. J.*, **176–177**, 124.
18 Eta, V., Maki-Arvela, P., Salminen, E., Salmi, T., Murzin, D.Y., and Mikkola, J.P. (2011) *Catal. Lett.*, **141** (9), 1254.
19 Aouissi, A., Apbiett, A.W., Al-Othman, Z.A., and Al-Amro, A. (2010) *Transit. Met. Chem.*, **35**, 927.
20 Qiu, P., Wang, L., Jiang, X., and Yang, B. (2012) *Energy Fuels*, **26**, 1254–1258; (b) Wang, L., Wang, Y., Liu, S., Lu, L., Ma, X., and Deng, Y. (2011) *Catal. Commun.*, **16** (1), 45.
21 Aresta, M., Dibenedetto, A., Devita, C., Bourova, O.A., and Chupakhin, O.N. (2004) *Stud. Surf. Sci. Catal.*, **153**, 213.
22 Zhang, J., Wang, F., Wei, W., Xiao, F., and Sun, Y. (2010) *Korean J. Chem. Eng.*, **27** (6), 1744.

23 Wang, H., Wang, M., Zhao, W., Wei, W., and Sun, Y. (2010) *React. Kinet. Mech. Catal.*, **99**, 381.

24 Wanga, D., Zhang, X., Gao, Y., Xiao, F., Wei, W., and Sun, Y. (2010) *Catal. Commun.*, **11**, 430.

25 Wang, D., Zhang, X., Zhao, W., Peng, W., Zhao, N., Xiao, F., Wei, W., and Sun, Y. (2010) *J. Phys. Chem. Solids*, **71**, 427.

26 Aresta, M., Dibenedetto, A., Nocito, F., and Ferragina, C. (2009) *J. Catal.*, **268**, 106; (b) Dibenedetto, A., Angelini, A., Aresta, M., Ethiraj, J., Fragale, C., and Nocito, F. (2011) *Tetrahedron*, **67**, 1308.

27 Pacheco, M.A., and Marshall, C.L. (1997) *Energy Fuels*, **11**, 2.

28 Ainsworth, S.J. (1992) *Chem. Eng. News*, **70**, 9.

29 Weissermel, K., and Arpe, H.J. (1997) *Industrial Organic Chemistry*, 3rd edn, Wiley-VCH Verlag GmbH, Weinheim, p. 162.

30 Coates, G.W., and Moore, D.R. (2004) *Angew. Chem. Int. Ed.*, **43**, 6618; (b) Darensbourg, D.J. (2007) *Chem. Rev.*, **107**, 2388; (c) Darensbourg, D.J., Mackiewicz, R.M., Phelps, A.L., and Billodeaux, D.R. (2004) *Acc. Chem. Res.*, **37**, 836.

31 Limura, N., Takagi, M., Iwane, H.J., and Ookago, H. (1995) Kokai Tokkyo Koho, Japanese Patent, 07267944.

32 Inoe, K., and Oobkuko, H. (1995) Kokai Tokkyo Koho, Japanese Patent, 07206846.

33 Inoe, K., and Oobkubo, H. (1995) Kokai Tokkyo Koho, Japanese Patent, 07206847.

34 Inoe, K., and Oobkubo, H. (1995) Kokai Tokkyo Koho, Japanese Patent, 07206848.

35 Inaba, M., Hasegawa, K., and Nagaoka, H. (1997) Kokai Tokkyo Koho, Japanese Patent, 09067365.

36 Ichikawa, S., and Iwane, H. (1997) Kokai Tokkyo Koho, Japanese Patent, 09235252.

37 Tojo, M., and Fukuoka, S. (1991) Kokai Tokkyo Koho, Japanese Patent, 03120270.

38 Bobyleva, L.I., Kryukov, S.I., Bobylev, B.N., Liakumovich, A.G., Surovstev, A.A., Karpov, O.P., Akhmedyanova, R.A., and Koneva, S.A. (1992) Russian Patent, 1781218.

39 Mais, F.J., Buysch, H.J., Mendoza-Frohn, C., and Klausener, A. (1993) European Patent, 543249.

40 Kuran, W., and Listos, T. (1994) *Macromol. Chem. Phys.*, **195**, 977.

41 Sakai, T., Kihara, N., and Endo, T. (1995) *Macromolecules*, **28**, 4701.

42 Sakai, T., Tsutsumi, Y., and Ema, T. (2008) *Green Chem.*, **10**, 337.

43 Sone, M., Sako, T., and Kamisawa, C. (1999) Kokai Tokkyo Koho, Japanese Patent, 11335372.

44 Darensbourg, D.J., and Holtcamp, M.W. (1996) *Coord. Chem. Rev.*, **153**, 155.

45 Marquis, E.T., and Sanderson, J.R. (1994) US Patent, 5 283 365.

46 Li, Y., Zhao, X.Q., and Wang, Y.J. (2005) *Appl. Catal. A: Gen.*, **279**, 205.

47 Zhang, X., Zhao, N., Wei, W., and Sun, Y. (2005) Chemical fixation of carbon dioxide to propylene carbonate over organic base modified solids. Presented at the International Conference on Carbon Dioxide Utilisation – ICCDU VIII, Oslo.

48 Yano, T., Matsui, H., Koike, T., Ishiguro, H., Fujihara, H., Yoshihara, M., and Maeshima, T. (1997) *Chem. Commun.*, 1129.

49 Yamaguchi, K., Ebitani, K., Yoshida, T., Yoshida, H., and Kaneda, K. (1999) *J. Am. Chem. Soc.*, **121**, 4526.

50 Aresta, M., and Dibenedetto, A. (2001) 221st ACS National Meeting, San Diego, Abstract 220.

51 Aresta, M., Dibenedetto, A., Gianfrate, L., and Pastore, C. (2003) *J. Mol. Catal. A: Gen.*, **245**, 204.

52 Aresta, M., Dibenedetto, A., Gianfrate, L., and Pastore, C. (2003) *Appl. Catal. A: Gen.*, **255**, 5.

53 Machac, J.R., Jr., Marquis, E.T., and Woodrum, S.A. (2000) US Patent, 654438.

54 Ballivet-Tkatchenko, D., and Dibenedetto, A. (2010) Synthesis of linear and cyclic carbonates, in *Carbon Dioxide as Chemical Feedstock* (ed. M. Aresta), Wiley-VCH Verlag GmbH, Weinheim, p. 169.

55 Song, J., Zhang, B., Jiang, T., Yang, G., and Han, B. (2011) *Front. Chem. China*,

6, 21; (b) Ulusoy, M., Kilic, A., Durgun, M., Tasci, Z., and Cetinkaya, B. (2011) *J. Organomet. Chem.*, **696**, 1372; (c) Liang, S., Liu, H., Jiang, T., Song, J., Yang, G., and Han, B. (2011) *Chem. Commun.*, **47**, 2131; (d) Buchard, A., Kember, M.R., Sandeman, K.G., and Williams, C.K. (2011) *Chem. Commun.*, **47**, 212; (e) Shibata, A., Mitani, I., Imakuni, A., and Baba, A. (2011) *Tetrahedron Lett.*, **52**, 721; (f) Dengler, J.E., Lehenmeier, M.W., Klaus, S., Anderson, C.E., Herdtweck, E., and Riege, B. (2011) *Eur. J. Inorg. Chem.*, 336; (g) Kilic, A., Ulusoy, M., Durgun, M., Tasci, Z., Yilmaz, I., and Cetinkaya, B. (2010) *Appl. Organomet. Chem.*, **24**, 446; (h) Ulusoy, M., Sahin, O., Kilic, A., and Buyukgungor, O. (2010) *Catal. Lett.*, **141**, 717.

56 Wang, J.L., Miao, C.X., Dou, X.Y., Gao, J., and He, L.N. (2011) *Curr. Org. Chem.*, **15** (5), 621.

57 Dai, W., Luo, S., Yin, S., and Au, C. (2010) *Front. Chem. Eng. China*, **4** (2), 163.

58 Dai, W.L., Luo, S.L., Yin, S.F., and Au, C.T. (2009) *Appl. Catal. A: Gen.*, **366** (1), 2.

59 Kawanami, H., and Ikushima, Y. (2000) *Chem. Commun.*, 2089.

60 Jairton, D., Roberto, F.D.S., and Paulo, A.Z.S. (2002) *Chem. Rev.*, **102**, 3667.

61 Kawanami, H., Sasaki, A., Matsui, K., and Ikushima, Y. (2003) *Chem. Commun.*, 896.

62 Jacobson, S.E. (1984) European Patent Application, 117147.

63 Aresta, M., Quaranta, E., and Ciccarese, A. (1987) *J. Mol. Catal.*, **355**, 41.

64 Aresta, M., and Dibenedetto, A. (2002) *J. Mol. Catal.*, **399**, 182.

65 Aresta, M., Ciccarese, A., and Quaranta, E. (1985) *C1 Mol. Chem.*, **1**, 267.

66 Aresta, M., Quaranta, E., and Tommasi, I. (1994) *New J. Chem.*, **18**, 133.

67 Aresta, M., Dibenedetto, A., and Tommasi, I. (2001) *Eur. J. Inorg. Chem.*, 1801.

68 Aresta, M., Dibenedetto, A., and Tommasi, I. (2000) *Appl. Organomet. Chem.*, **14**, 799.

69 San, J., Liang, L., Sun, J., Jiang, Y., Lin, K., Xu, X., and Wang, R. (2011) *Catal. Surv. Asia*, **15**, 49.

70 Aresta, M., Tommasi, I., Dibenedetto, A., Fouassier, M., and Mascetti, J. (2002) *Inorg. Chem.*, **330** (1), 63.

71 Clerici, M.G., Belussi, G., and Romano, U. (1991) *J. Catal.*, **129**, 159.

72 Dibenedetto, A., Aresta, M., Fragale, C., Distaso, M., Pastore, C., Venezia, A.M., Liu, C., and Zhang, M. (2008) *Catal. Today*, **137**, 44.

73 Sugimoto, H., and Inoue, Y (2004) *J. Polym. Sci. A: Polym. Chem.*, **42**, 5561; (b) Sugimoto, H., Ohtsuka, H., and Inoue, S. (2004) *Stud. Surf. Sci. Catal.*, **153**, 243.

74 Super, M., Berluche, E., Costello, C., and Beckman, E. (1997) *Macromolecules*, **30**, 368.

75 Inoue, S. (1987) in *Carbon Dioxide as a Source of Carbon: Chemical and Biochemical Uses*, NATO ASI Series, Ser. C, vol. 206 (eds M. Aresta and G. Forti), Reidel, Dordrecht, p. 331.

76 Rokicki, A., Kuran, W., and Macromol, J. (1981) *Sci. Rev. Macromol. Chem.*, **135**, C21.

77 Darensbourg, D.J., Andreatta, J.R., and Moncada, A.I. (2010) in *Carbon Dioxide as Chemical Feedstock* (ed. M. Aresta), Wiley-VCH Verlag GmbH, Weinheim, p. 213.

78 Kember, M.R., Buchard, A., and Williams, C.K. (2011) *Chem. Commun.*, **47**, 141.

79 Li, H., and Niu, Y. (2011) *Polym. J.*, **43**, 121.

80 Gosh, A., Ramidi, P., Pulla, S., Sullivan, S.Z., Collom, S.L., Gartia, Y., Munshi, P., Biris, A.S., Noll, B.C., and Berry, B.C. (2010) *Catal. Lett.*, **137**, 1.

81 Ren, W.M., Zhang, X., Liu, Y., Li, J.F., Wang, H., and Lee, X.B. (2010) *Macromolecules*, **43**, 1396.

82 Kim, B.E., Varghese, J.K., Han, Y.G., and Lee, B.Y. (2010) *Bull. Korean Chem. Soc.*, **31**, 829.

83 Lee, K., Ha, J.J.Y., Cao, C., Park, D.W., Ha, C.S., and Kim, I. (2009) *Catal. Today*, **148**, 389.

84 Darensbourg, D.J., and Wilson, S.J. (2011) *J. Am. Chem. Soc.*, **133** (46), 18610.

85 Darensbourg, D.J., and Moncada, A.I. (2010) *Macromolecules*, **43** (14), 5996.

86. Suriano, F., Coulembier, O., Hedrick, J.L., and Dubois, P. (2011) *Polym. Chem.*, **2**, 528.
87. Islam, M.M., Seo, D.W., Jang, H.H., Lim, Y.D., Shin, K.M., and Kim, W.G. (2011) *Macromol. Res.*, **19** (12), 1278.
88. Colonna, M., Berti, C., Binassi, E., Fiorini, M., Sullalti, S., Acquasanta, F., Karanam, S., and Brunelle, D.J. (2011) *React. Funct. Polym.*, **71** (10), 1001.
89. Fov, E., Farrell, J.B., and Higginbotham, C.L. (2009) *J. Appl. Polym. Sci.*, **111** (1), 217.
90. Wang, L., Xiao, B., Wang, G.Y., and Wu, J.Q. (2011) *Sci. China Chem.*, **54** (9), 1468.
91. Yang, Y., Yu, Y., Zhang, Y., Liu, C., Shi, W., and Li, Q. (2011) *Process Biochem.*, **46** (10), 1900.
92. Gross, R.A., and Jiang, Z.-Z. (2011) US Patent, 7 951 899.
93. Liu, C., Jiang, Z., Decatur, J., Xie, W., and Gross, R.A. (2011) *Macromolecules*, **44** (6), 1471.
94. Aresta, M. (2003) *Carbon Dioxide Recovery and Utilisation*, Kluwer, Dordrecht.
95. Vaidya, P.D., and Kenig, E.Y. (2007) *Chem. Eng. Technol.*, **30**, 1467.
96. Quaranta, E., and Aresta, M. (2010) in *Carbon Dioxide as Chemical Feedstock* (ed. M. Aresta), Wiley-VCH Verlag GmbH, Weinheim, p. 121.
97. Terlouw, J.K., and Schwarz, H. (1987) *Angew. Chem. Int. Ed. Engl.*, **26**, 805.
98. Kaminskaia, N.V., and Kostic, N.M. (1997) *Inorg. Chem.*, **36**, 5917.
99. Remko, M., Liedl, K.R., and Rode, B.M. (1993) *J. Chem. Soc., Faraday Trans.*, **89**, 2375.
100. Aresta, M., Ballivet-Tkatchenko, D., Belli Dell'Amico, D., Bonnet, M.C., Boschi, D., Calderazzo, F., Faure, R., Labella, L., and Marchetti, F. (2000) *Chem. Commun.*, 1099.
101. Jamroz, M.H., Dobrowolski, J.C., Rode, J.E., and Borowiak, M. (2002) *J. Mol. Struct.*, **618**, 101.
102. Dibenedetto, A., Aresta, M., Fragale, C., and Narracci, M. (2002) *Green Chem.*, **4**, 439.
103. Aresta, M., Dibenedetto, A., and Quaranta, E. (1995) *Journal of the Chemical Society, Dalton Transactions*, 3359; (b) Belli Dell'Amico, D., Calderazzo, F., Labella, F., Marchetti, F., and Pampaloni, G. (2003) *Chem. Rev.*, **103**, 3857.
104. Ray, S., Pathak, S.R., and Chaturvedi, D. (2005) *Drugs Future*, **30**, 161.
105. Gupta, R.C. (ed.) (2006) *Toxicology of Organophosphate and Carbamate Compounds*, Elsevier Academic Press, Burlington, MA.
106. Green, T.W., and Wuts, P.G.M. (2007) *Protective Groups in Organic Synthesis*, John Wiley & Sons, Inc., New York.
107. Wu, C., Cheng, H., Liu, R., Wang, Q., Hao, Y., Yu, Y., and Zhao, F. (2010) *Green Chem.*, **12**, 1811.
108. Li, J., Guo, X.G., Wang, L.G., Ma, Y.Z., Zhang, Q.H., Shi, F., and Deng, Y.Q. (2010) *Sci. China Chem.*, **53**, 1534; (b) Peterson, S.L., Stucka, S.M., and Dinsmore, C.J. (2010) *Org. Lett.*, **12**, 1340.
109. Tsuda, T., Washida, H., Watanabe, K., Miva, M., and Saegusa, T. (1978) *J. Chem. Soc., Chem. Commun.*, 815.
110. Yoshida, Y., Ishii, S., Watanabe, M., and Yamashita, T. (1989) *Bull. Chem. Soc. Jpn.*, **62**, 1534.
111. Chaturvedi, D. (2012) *Tetrahedron*, **68**, 15.
112. Aresta, M., and Quaranta, E. (1992) *Tetrahedron*, **21**, 1515.
113. Aresta, M., and Quaranta, E. (1993) Italian Patent, 1237207.
114. McGhee, W.D., Riley, D., Kevin, C., Pan, Y., and Parnas, B. (1995) *J. Org. Chem.*, **60**, 2820.
115. Shi, M., and Shen, Y.M. (2001) *Helv. Chim. Acta*, **84**, 3357.
116. McGhee, W.D., and Talley, J.J. (1994) US Patent, 5 302 717.
117. Singh, K.N. (2007) *Synth. Commun.*, **37**, 2651.
118. Chaturvedi, D., Mishra, N., and Mishra, V. (2006) *Chin. Chem. Lett.*, **17**, 1309.
119. Chaturvedi, D., and Ray, S. (2006) *Monatsh. Chem.*, **137**, 459.
120. Chaturvedi, D., Kumar, A., and Ray, S. (2002) *Synth. Commun.*, **32**, 2651.
121. Kong, D.L., He, L.N., and Wang, J.Q. (2011) *Synth. Commun.*, **41**, 3259.
122. Yoshida, Y., and Inoue, S. (1978) *Bull. Chem. Soc. Jpn*, **51**, 559.

123 Yoshida, Y., and Inoue, S. (1980) *Polym. J.*, **12**, 763.
124 Yoshida, Y., Ishii, S., Kawato, A., Yamashita, T., Iano, M., and Inoue, S. (1988) *Bull. Chem. Soc. Jpn*, **61**, 2913.
125 Yoshida, Y., and Inoue, S. (1978) *Chem. Lett.*, 139; (b) Yoshida, Y., and Inoue, S. (1979) *J. Chem. Soc., Perkin Trans. 1*, 3146.
126 Asano, T., Saito, N., Ito, S., Hatakeda, K., and Toda, T. (1978) *Chem. Lett.*, 311.
127 Toda, T. (1977) *Chem. Lett.*, 957.
128 Ishii, S., Zhou, M., Yoshida, Y., and Noguchi, H. (1999) *Synth. Commun.*, **29**, 3207.
129 McGhee, W.D., and Riley, D.P. (1992) *Organometallics*, **11**, 900.
130 McGhee, W.D., Riley, D.P., Christ, M.E., and Christ, K.M. (1993) *Organometallics*, **12**, 1429.
131 Yoshida, M., Ohsawa, Y., Sugimoto, K., Tokuyama, H., and Masataka, I. (2007) *Tetrahedron Lett.*, **48**, 8678.
132 Sasaki, D., and Dixneuf, P.H. (1986) *J. Chem. Soc., Chem. Commun.*, 790.
133 Rohr, M., Geyer, C., Wandeler, R., Schneider, M.S., Murohy, E.F., and Baiker, A. (2001) *Green Chem.*, **3**, 123.
134 Mahe, R., Sasaki, Y., Bruneau, C., and Dixneuf, P.H. (1989) *J. Org. Chem.*, **54**, 1518.
135 Jiang, J.L., and Hua, R. (2006) *Tetrahedron Lett.*, **47**, 953.
136 Jiang, H., Zhao, J., and Wang, A. (2008) *Synthesis*, 763.
137 Della Ca, N., Gabriele, B., Ruffolo, G., Veltri, L., Zanetta, T., and Costa, M. (2011) *Adv. Synth. Catal.*, **353**, 133.
138 Costa, M., Chiusoli, G.P., Taffurelli, D., and Dalmonego, G. (1998) *J. Chem. Soc., Perkin Trans. 1*, 1541.
139 Maggi, R., Bertolotti, C., Orlandini, E., Oro, C., Sartori, G., and Selva, M. (2007) *Tetrahedron Lett.*, **48**, 2131.
140 Feroci, M., Orsini, M., Rossi, L., Sotgiu, G., and Inesi, A. (2007) *J. Org. Chem.*, **72**, 200.
141 Feroci, M., Casadei, M.A., Orsini, M., Palombi, L., and Inesi, A. (2003) *J. Org. Chem.*, **68**, 1548.
142 Ikeda, S., Takagi, T., and Ito, K. (1987) *Bull. Chem. Soc. Jpn*, **60**, 2517.
143 Guo, X., Shang, J., Li, J., Wang, L., Ma, Y., Shi, F., and Deng, Y. (2011) *Synth. Commun.*, **41**, 1102.
144 Bayer, O., Siefken, W., Rinke, H., Orthner, L., and Schild, H. (1937) German Patent, DRP 728981.
145 Krol, P. (2007) *Prog. Mater. Sci.*, **52**, 915.
146 Paul, F. (2000) *Coord. Chem. Rev.*, **203**, 269.
147 Giannoccaro, P., Dibenedetto, A., Gargano, M., Quaranta, E., and Aresta, M. (2008) *Organometallics*, **27**, 967; (b) Giannoccaro, P., Cornacchia, D., D'Oronzo, S., Mesto, E., Quaranta, E., and Aresta, M. (2006) *Organometallics*, **25**, 2872.
148 Saunders, J.H., and Frisch, K.C. (1983) in *Polyurethanes, Chemistry and Technology, Part I. Chemistry* (ed. F.L. Malabar), Interscience Publishers, New York, p. 106–107.
149 Kaji, A., Arimatsu, Y., and Murano, M. (1992) *J. Appl. Polym. Sci. A: Polym. Chem.*, **30**, 287.
150 Soga, K., Hosoda, S., Nakamura, H., and Ikeda, S. (1976) *J. Chem. Soc., Chem. Commun.*, **16**, 617.
151 Soga, K., Chiang, W.Y., and Ikeda, S. (1974) *J. Polym. Sci. Polym. Chem. Ed.*, **12**, 121.
152 Inoue, S. (1976) *Chemtech*, **6**, 588.
153 Ihata, O., Kayaki, Y., and Ikariya, T. (2004) *Angew. Chem. Int. Ed.*, **43**, 717.
154 Liaw, D.J. (1997) *J. Appl. Polym. Sci.*, **66**, 1251.
155 Krol, P., and Wojturska, J. (2002) *Polimery*, **47**, 6.
156 Meng, Z., Yong-hong, Z., Xiao-hui, Y., and Li-hong, H. (2011) *Adv. Mater. Res.*, **974**, 250.
157 Tschan, M.J.-L., Brule, E., Haquette, P., and Thomas, C.M. (2012) *Polym. Chem.*, **3**, 836.
158 Szycher, M. (2011) in *An Introduction to Biomaterials* (ed. J.O. Hollinger), CRC Press, Boca Raton, FL, p. 281.
159 Poussard, L., Burel, F., Couvercelle, J.P., Merhi, Y., Tabrizian, M., and Bunel, C. (2004) *Biomaterials*, **25**, 3473.
160 Wang, J.H., and Yao, C.H. (2000) *J. Biomed. Mater. Res.*, **51**, 761.
161 Santerrre, J.P., and Brash, J.L. (1997) *Ind. Eng. Chem. Res.*, **36**, 1352.
162 Kro, P., Kro, B., Pikus, S., and Skrzypiec, K. (2005) *Colloids Surf. A: Physicochem. Eng. Aspects*, **259**, 35.

163 Sebastian, M.T., and Jantunen, H. (2010) *Int. J. Appl. Ceram. Technol.*, **7**, 415.
164 Aresta, M., and Dibenedetto, A. (in press) *Philos. Trans. A R. Soc.*
165 Brown, L.C., Besenbruch, G.E., Scultz, K.R., Marshall, A.C., Showalter, S.K., Pickard, P.S., and Funk, J.F. (2002) General Atomics Report, GA-A23944, p. 1.
166 Winter, C.J., Sizmann, R.L., and Vant-Hull, L.L. (eds) (1991) *Solar Power Plants, Fundamentals, Tecnology, Systems, Economics*, Springer, New York.
167 Olah, G.A., Goeppert, A., and Prakash, G.K.S. (eds) (2010) *Beyond Oil and Gas: the Methanol Economy*, Wiley-VCH Verlag GmbH, Weinheim.
168 Drnevich, R. (2003) Strategic Initiatives for Hydrogen Delivery Workshop, Tonawanda, NY, 7 May 2003.
169 Zevenhoven, R., Eloneva, S., and Teir, S. (2006) *Catal. Today*, **115**, 73.
170 Eloneva, S., Teir, S., Salminen, J., Fogelholm, C.J., and Zevenhoven, R. (2008) *Energy*, **33** (9), 1461–1467.
171 Eloneva, S., Puheloinen, E.M., Kanerva, J., Ekroos, A., Zevenhoven, R., and Fogelholm, C.J. (2010) *J. Cleaner Production*, **18** (18), 1833–1839.
172 Briggs, M. (2004) Widescale Biodiesel Production From Algae. University of New Hampshire Biodiesel Group, http://www.unh.edu/p2/biodiesel/article_alge.html (accessed 31 March 2012).
173 Riesing, T.F. (2006) Cultivating algae for liquid fuel production, http://oakhavenpc.org/cultivating_algae.htm (accessed 31 March 2012).
174 Aresta, M., Dibenedetto, A., Carone, M., Colonna, T., and Fragale, C. (2005) *Environ. Chem. Lett.*, **3** (3), 136.
175 Khotimchenko, S.V. (2003) *Bot. Mar.*, **46**, 455.
176 Al-Hasan, R.H., Hantash, F.M., and Radwan, S.S. (1991) *Appl. Microbiol. Biotechnol.*, **35**, 530.
177 Huntley, M., and Redalje, D.G. (2007) *Mitigat. Adapt. Strat. Global Change*, **12**, 573.
178 Rosenberg, J.N., Oyler, G.A., Wilkinson, L., and Betenbaugh, M.J. (2008) *Curr. Opin. Biotechnol.*, **19** (5), 430.
179 Sheehan, J., Dunahay, T., Benemann, J., and Roessler, P.A. (1998) NREL close out report.
180 John, R.P., Anisha, G.S., Nampoothiri, K.M., and Pandey, A. (2011) *Bioresour. Technol.*, **102** (1), 186.
181 Ueda, R., Hirayama, S., Sugata, K., and Nakayama, H. (1996) US Patent, 5578472.
182 Adams, J.M., Gallagher, J.A., and Donnison, I.S. (2009) *J. Appl. Phycol.*, **21**, 569.
183 Spolaore, P., Joannis-Cassan, C., Duran, E., and Isambert, A. (2006) *J. Biosci. Bioeng.*, **101**, 87.
184 American Cancer Society, Chlorella, American Cancer Society, Atlanta, GA (2007) http://www.cancer.org/docroot/ETO/content/ETO_5_ 3X_Chlorella.asp (accessed 31 March 2012).
185 U.S. Department of Health & Human Services (2001) Dietary Fads and Frauds, http://profiles.nlm.nih.gov/NN/B/C/R/R/_/nnbcrr.pdf (accessed 31 March 2012).
186 Singh, S., Bhushan, K.N., and Banerjee, U.C. (2005) *Crit. Rev. Biotechnol.*, **25** (3), 73.
187 Henrikson, R. (1989) *Earth Food Spirulina How This Remarkable Blue-Green Algae Can Transform Your Health and Our Planet*, Ronore Enterprises, Laguna Beach, Calif., p. 174, ISBN 10 0962311103.
188 Earthrise Nutritionals LLC (2004) Earthrise the Company. Earthrise Nutritionals, http://www.earthrise.com/company.asp?page=page5.html (accessed 31 March 2012).
189 Cyanotech Corporation (2007) Spirulina Pacifica. Cyanotech, http://www.cyanotech.com/spirulina.html (accessed 31 March 2012).
190 Aresta, M., Narracci, M., and Tommasi, I. (2003) *Chem. Ecol.*, **19**, 451.
191 Chaneva, G., Urnadzhieva, S., Minkova, K., and Lukavsky, J. (2007) *J. Appl. Phycol.*, **19**, 537.
192 Prasanna, R.A., Sood, A., Suresh, S., Nayak, S., and Kaushik, B.D. (2007) *Acta Bot. Hung.*, **49**, 131.
193 Pulz, O., and Gross, W. (2004) *Appl. Microbiol. Biotechnol.*, **65**, 635.
194 Hussein, G., Sankawa, U., Goto, H., Matsumoto, K., and Watanabe, H. (2006) *J. Nat. Prod.*, **69**, 443.

195 Vilchez, C., Forjan, E., Cuaresma, M., Bedmar, F., Garbayo, I., and Vega, J.M. (2011) *Mar. Drugs*, **9**, 319.

196 Algatech (2004) Astaxanthin–The Algatech Story, www.algatech.com (accessed 31 March 2012).

197 Borowitzka, M.A. (2006) Biotechnological & Environmental Applications of Microalgae, http://www.bsb.murdoch.edu.au/groups/beam/BEAM-Appl0.html (accessed 31 March 2012).

198 Prasanna, R.A., Sood, A., Jaiswal, S., Nayak, S., Gupta, V., and Chaudhary, V. (2010) *Appl. Biochem. Microbiol.*, **46**, 119.

199 Eriksen, N. (2008) *Appl. Microbiol. Biotechnol.*, **80**, 1.

200 Sekar, S., and Chandramohan, M. (2008) *J. Appl. Phycol.*, **20**, 113.

201 Gotelli, I.B., and Cleland, R. (1968) *Am. J. Bot.*, **55**, 907.

202 Harnedy, P., and FitzGerald, R.J. (2011) *J. Phycol.*, **47**, 218.

203 Smidsrod, O., and Skjak-Braek, G. (1990) *TIBTECH*, **8**, 71.

204 Tegelaar, E.W., de Leeuw, J.W., Derenne, S., and Largeau, C. (1989) *Geochim. Cosmochim. Acta*, **53**, 3103.

Index

a
acetic acid   251, 255, 256, 341
acetoacetate decarboxylase (AAD)   202
acetone   202
acetyl CoA   202, 203
acid catalysts
– see also individual acid catalysts
– aromatic chemicals   193, 205
– extractives transformation   330, 334, 343
– furfural from carbohydrates   97, 100–102, 103–107, 109, 112
– non-thermochemical biorefineries   252
– turpentine derivatives   339, 341
acid removal, ethanol dehydration   163
acyclic carbonates   368, 369–371
acyclic pathways   107, 108
adiabatic reactions   161, 162
Advanced Research Projects Agency-Energy program (ARPA-E)   39
agricultural crops
– see also sugar-based biomass
– carbohydrate source   84, 85
– corn   153, 154, 204, 205, 357
– sitosterol hydrogenation   321–325
Alcell pulping process   10, 253
alcohols
– alcoholysis of urea   372
– aldehyde/ketone functions conversion to   190
– butanols   4, 11, 12
– carbamates production   365, 377
– carboxylation   368, 369–371
– characteristics   170
– coniferyl alcohol   225, 226
– fatty alcohols   177–180
– oxidative carbonylation   368
– perillyl alcohol   318, 320, 334
– propargyl alcohols   376

aldehydes   190, 200, 205, 318, 332, 333, 340, 342
aldol condensations   342, 343
algal biomass
– algaenan   386
– alginate   385
– carbohydrate source   84, 85, 92–94
– carbon dioxide fixation   382–386
– product selection criterion   47
– types   180
alkali earth elements   283
alkanes   197, 272–274
alkenes   272, 273, 376
N-alkylpyrrolidones   21
alumina catalysts   159, 160
aluminium oxide catalysts   330, 331
aluminium-containing MCM-41   106
Amazon rain forest   152, 153
amides   170, 376
amines   170, 374–377
p-aminobenzoic acid   221, 222
ammonia fiber expansion (AFEX)   252
Amyris   4, 13, 54–78, 180
anaerobic digestion   249, 252, 253
Anderson–Schultz–Flory (ASF) distributions   273
Anellotech process   195, 198
aniline   220–222
anionic polymerization   62, 63
anthranilic acid   220–222
aphid alarm pheromones   59
application strategies   263, 264
aquatic carbohydrates   92–94
aqueous halides   101, 102
aqueous phase reforming (APR)   192, 193
aqueous sugar streams   192
arabinogalactans   87, 91
arabinoglucuronoxylans   88, 91

arabinose 98, 100, 255
arabinoxylans 88, 91, 92
Archer Daniels Midland (ADM) Company 39, 44
AroE enzyme 212
aromatic chemicals
– aromatic acid pathway 209
– bio-oil derived 197
– biological routes 199–226
– BTX biorefinery 186–199
– Diels–Alder β-farnesene adducts 73–75
– diisocyanates 378
– extraction, oils/fuels 96
– lignin 226–228
– production from bio-based feedstocks 185–230
– yields 229
aromatics from biobased feedstocks, yields 229
artemisinin 54
Arthrospira 383
artificial photosynthesis 364, 367, 374, 380
ASPEN simulation model 357–359
availability of raw materials 41–43
Avantium Chemicals 17, 22, 207
azetidines 378
aziridines 376, 378

b

base oils 77
batch reactors 128, 129
benzene–toluene–xylene (BTX) process 185–199
Billion-Ton study (US DOE/DOA) 185
bio-based plastics, *see* plastics
bio-isobutanol 12, 13, 199–201
bio-oils
– *see also* fats/oils
– aromatic chemicals production 188, 189
– Fischer–Tropsch process 289
– product selection criterion 47, 48
– production process 16, 17
– pyrolysis 195–198
Biofene 13
biofuels
– *see also* diesel biofuels; *individual fuels*
– from carbon dioxide 379–381
– life cycle analyses 33, 34
– renewable
– – current situation 9–18
– – existing biorefinery infrastructures 44
– – outcomes comparison in product selection 40
– target selection 33

Biofuels Digest polls 2
Biological and Chemical Catalysts Technologies Program (US DOE) 32
biological routes
– aromatic chemicals 199–226
– – purified terephthalic acid from *p*-xylene 199
– biorefineries using 249
– common aromatic pathway 209–221
– enzymic reaction types 49
– fats/oils as raw material 173, 174
– isobutylene production 201–203
– organosolv biorefining 245–267
– polycarbonate production 374
– valine pathway to isobutylene 200, 201
biomass feedstocks
– *see also* sugar-based biomass
– aromatic chemicals production 185–230
– – chemistry 187–192
– energy aspects 352, 353, 359
– fats/oils as raw material 174, 175
– 5-hydroxymethyl furfural synthesis 111
– lignocellulosic
– – biomass gasification 14, 15
– – current situation 10, 11
– – gasification 14, 15
– – organosolv biorefining 245–267
– – oxygen removal 14, 16
– pretreatment/gasification for FT process 288–299
– pyrolysis 195–198
– residues available 18
– statistical design of experiments 135–137
– usage by companies 4, 5
biomass-to-liquids (BTL)
– Fischer–Tropsch process 271–312
– – basics 271–278
– – biomass pretreatment/gasification 288–299
– – cobalt catalysis 278–285
– – concept 277, 278, 299–308
– – energy/carbon efficiencies 310–312
– – pilot/demo plants 308–310
– – process concepts 299–308
– – reactors 285–288
biopolymers
– *see also individual polymers and plastics*
– cyclic carbonates synthesis 373, 374
– definition 151
– polyethylene from ethanol 151–164
– polymerization reactions 164, 257
– polyurethanes synthesis 377–379
– precursor rational selection 28
– storage carbohydrates 84, 86, 87

– structural carbohydrates 86–97
– terpenes 60–68
biorefineries
– *see also* companies
– anaerobic digestion catalysts 252, 253
– benzene–toluene–xylene process 186–199
– existing infrastructures 43, 44
– organosolv process 245–267
– thermochemical treatments 248, 249
– types 247–251
biotechnology 47, 246
biphasic reactors 99
Blue Tower concept 296, 298, 300
boiling point/NIR spectra correlation 144
borneol 336, 340
bottles for beverages 156, 157, 194, 195, 356–359
Bouveault–Blanc reduction 178
Braskem
– ethanol to polyethylene process 151–164
– – commercial plants 154–157
– – development reasons 151, 152
– – legislation/certification 157, 158
– – polymerization 164
– – process description 158–163
– – Triunfo plant 155, 156, 164
brassicasterol 322
brown macroalgae 93
building blocks 19, 35, 36, 58–60, 83–113
bulk chemicals 352, 380
butanols 4, 11, 12
byproduct use product selection criterion 46

c

calcium oxide 381, 382
campesterol 322
camphene
– esterification/etherification 339, 341
– monoterpenes isomerization 325–327
– α-pinene isomerization 331
– terpenes isomerization 318
campholenic aldehyde 318, 332, 333, 340, 342
camphor 327, 340
candidate bio-based product selection 27–50
carbamates synthesis 374–376, 377–379
carbide mechanism 274
carbohydrates
– *see also* sugar-based biomass
– aquatic 92–94
– building blocks 19
– dehydration to fuels 16–18
– dehydration to furans 94–112
– DOE 2010 Report on products from 37, 38
– furan-based building blocks from 83–113, 205
– potential products 39
– sources 84–94
– storage carbohydrates 84, 86, 87
– – cellulose 84, 86, 87, 89, 253, 254
– – lignocellulosic biomass 10, 11, 14, 15, 245–267
– – starch 43, 44
– structural 86–92
– top chemical opportunities list 35
carbon
– capture and storage (CCS) 363, 367
– efficiency 230, 303, 310–312
– flux 209, 210, 212, 213
– natural cycle 363
– sources, carbon dioxide 363–387
carbon dioxide
– carbamates 374–379
– carbon source 363–387
– conversion possibilities 367–379
– cyclic carbonates/polymers synthesis 373, 374
– cyclic reduction 380
– energy products from 379–381
– Fischer–Tropsch process 301, 302, 305, 306
– fixation into aquatic biomass 382–386
– industrial use 364–367
– inorganic carbonates production 381, 382
– removal, Braskem's ethanol dehydration 163
– transesterification/alcoholysis of urea 372
carbon monoxide 15, 161, 163, 248, 299
– *see also* Fischer–Tropsch process
carbonates
– carbon dioxide conversion 367–374
– inorganic 381, 382
– organic linear 369, 370
– urea transesterification/alcoholysis 372
carbon–carbon bond cleavage 49
carbon–oxygen cleavage 49
carboxylation of epoxides 373, 374
carboxylic acids 170
trans-carveol 320
carvone 320
catalysts
– anaerobic digestion biorefineries 252, 253
– biomass-derived sugars 193
– biorefining processes 251–253

– Braskem's ethanol dehydration   159, 160
– catalytic hydrogenolysis   191
– catalytic reforming   191
– coking   147, 162, 195
– deactivation   327–330
– energy savings   351–354
– fats/oils
– – as raw material   173–177
– – transformation requirement   170–173
– fatty alcohols case study   178, 179
– furfural from carbohydrates   101–107, 112
– kinetics/pseudo-kinetics   131, 132
– loadings design   135, 136
– long term performance/deactivation   145–148
– novel, environmental assessment   349–360
– α-pinene isomerization   327
– principal component analysis   139–142
– process conceptual development   173–177
– process energy profiles   351
– product selection criteria   48, 49
– terpenes   53–78
– – polymerization   61–64
– testing equipment for conceptual process design   128–131
catechol-O-methyl transferase (COMT)   211, 217
cationic polymerization   63
cellulose   84, 87, 89, 99, 253, 254
– see also lignocellulosic biomass feedstocks
certification   157, 158
chalcones   222–226
char   289, 291
chemical structure preservation   256
Chemrec technology   295, 297, 300
Chemurgy movement   27, 53
chitin   385
chitosan   385
Chlorella   382
chlorophyll pigments   383
cholesterol-suppressing agents   321–325
Choren technology   208, 209, 295, 297, 299, 300
chorismate pathway   209, 215
chromic acid   325, 335, 339, 341
chromium catalysts   100
chromium trioxide   341
trans-cinnamic acid decarboxylase (CADC)   220
circulating-bed gasifiers   292–294
clean-up   301, 302
Clostridium spp.   202
coal, bio-coal   249

Cobalt Biofuels   11, 12
cobalt catalysts
– Fischer–Tropsch process   276–285
– – activation/preparation   278–280
– – activity   280–283
– long-term performance   146, 148
Coca-Cola beverage bottles   156, 157
codes of conduct   157, 158
coking   147, 162, 195
commercial interest/activities
– see also companies
– aromatics from biobased feedstocks   228–230
– carbon dioxide   364–367
– cobalt FT catalyst formulations   283, 284
– ethanol dehydration   154–157
– furfural production/applications   95, 96
– organosolv biorefinery products   257–259
– status quo   1–23
– terpene building blocks   57, 58
commodity aromatic chemicals   185–230
common aromatic pathway   209–221
companies
– see also individual companies
– bio-isobutanol   12, 13
– Biofuels Digest 2010-2011 top 50   2
– carbohydrate potential products   38–40
– cobalt FT catalyst formulations   284
– collaboration   7, 8
– ethanol dehydration for polymers   157
– Fischer–Tropsch process   272, 308–310
– geographic spread   4, 5
– Lignol biorefinery process development   265
– market drivers   44, 45, 262, 263
– pilot biomass gasification plants for FT process   295, 296
– platform chemicals production   3
– polyethylene from ethanol   151
complex heteroxylans   92
concentrators of solar power (CSP)   380
conceptual process design (CPD)   123–145
conditioning   301, 302
coniferyl alcohol   225, 226
continuous reactors   129, 321
conversion process types   6
copper catalysts   179
corn (maize)   153, 154, 204, 205, 357
costs
– Alternative Feedstocks Program   32
– carbon dioxide industrial use   367
– conceptual process design   123, 124, 126–128

- Fischer–Tropsch process 307, 308
- metal catalysts 177
crude oil 47–8
crude tall oil (CTO) compounds 321–325
cryogenic distillation, ethylene 163
current situation 1–23
- production volumes/energy use 349
- renewable chemicals 18–22
- renewable fuels 9–18
- renewables arena 2–9
cyclic compounds 107, 108, 373, 374, 376–378

d
deactivation of catalysts
- cobalt FT catalysts 282, 283
- function of 327–330
- long-term performance 145–148
degree of polymerization 257
degree of unsaturation 171
dehydration route
- carbohydrates to fuels 16, 17
- carbohydrates to furans 94–112
- ethanol to polyethylene 154–157
- product selection criteria 49
dehydrogenation, isoborneol 342
3-dehydroquinic acid (DHQ) 216, 217
dehydroshikimic acid (DHS) 212, 213
Delphi analysis 33
demonstration projects 266, 294–298, 308–310
3-deoxy-D-*arabino*-heptulosonic acid 7-phosphate (DAHP) 209, 210, 212, 213
1-deoxy-D-xylulose-5-phosphate synthase (DXS) 214
1,4-diamionobutane 36
Diels–Alder reactions
- β-farnesene homo Diels–Alder products 71–74
- isoprene adducts 72
- purified terephthalic acid
- – muconic acid route from 210, 213
- – production via 5-hydroxymethyl furfural 206
- terpenes 60
diesel biofuels
- companies using/producing 6–8, 10
- – Amyris 55, 56
- β-farnesene 55
- Fischer–Tropsch process 271, 276
- partial least squares analysis 143–145
diethylcarbonate (DEC) 370
differential scanning calorimetry (DSC) 65, 66, 73, 75, 76

dihydroxyacid dehydratase (DHAD) 200
diisocyanates, aromatic/cyclic 377, 378
dimethyl carbonate (DMC) 369–371, 377
dimethyl terephthalate 206
dimethylallyl pyrophosphate (DMAPP) 56
2,5-dimethylfuran (DMF) 124–128
distilled tall oil 317
drop-in replacements
- advanced bio-fuels 44
- benzene–toluene–xylene process 185–189, 195, 196
- carbohydrate-derived building blocks 19, 20
- companies currently using 7
- Green Polyethylene from ethanol 151–164
dual bed reactors 291, 292
DuPont 21
DXP reductoisomerase (DXR) 214

e
economics, *see* costs
elastomers 378
electrical energy storage 380, 381
elimination reactions 98, 99
energy
- balance in Fischer–Tropsch process 302–305
- carbon dioxide industrial use 365–367
- case study 356–359
- current situation 349
- Lignol biorefinery process efficiency 256
- profiles for catalytic processes 351
- steam cracking 351, 352
energy-rich products, *see* biofuels
Enerkem technology 295, 298
Ensyn/Envergent joint venture 16
enthalpy 273, 366–369
Entner–Douderoff glycolysis pathway 201
entrained-flow gasifiers 291–293, 297, 300–303, 305, 307, 308
environmental assessments 349–360, 364, 365
- life cycle assessment 33, 34, 350, 354, 356, 360
enzymes
- *see also* biological routes; fermentation; *individual enzymes*
- common aromatic pathway 209–221
- polycarbonates production 374
- reaction types 49
- valine pathway to isobutylene 200, 201
epichlorohydrin Solvay facility 39

epoxidation 326, 337–340
epoxides 365, 373, 374, 376
1,2-epoxy-α-terpineol 326
1,2-epoxycarveol 326
1,2-epoxylimonene 326
erythrose-4-phosphate (E4P) 209, 210
esters/esterification 170, 172, 339, 341, 342, 372
ethane carbonate 365
ethanol
– Braskem process 151–164
– companies using/producing 4–10
– current situation 10–12
– existing biorefinery infrastructures 43
ethene carbonate 372
etherification 339, 341, 342
5-ethoxymethyl furfural (EMF) 140
ethylbenzene 193, 195–199, 353
ethylene, see polyethylene
European ethanol production 153, 154
evaluation processes
– extrudate catalysts performance 131–133
– life cycle assessment 33, 34, 350, 354–356, 360, 365
– novel catalytic processes 349–360
– product selection criteria 41–48
existing biorefinery infrastructures 43, 44
exothermal reactions 273
extractives
– catalytic transformation 317–344
– – crude tall oil compounds 321–325
– – overview 317–321
– – turpentine compounds 325–343
– components 317
– Lignol biorefinery process 255
extrudate catalysts 131–133

f

farnesenes
– Amyris 4, 180
– current situation 12, 13
– homo Diels–Alder reaction products 71–74
– structure 59, 63
– synthesis routes 58–60
– terpene production 54–78
– thermal Diels–Alder reaction 73, 75, 76
fast pyrolysis 188, 189, 289
fats/oils
– see also bio-oils
– aromatic chemicals from 96, 197, 198
– catalytic transformation requirement 170–173
– fatty alcohols case study 177–180
– lubricants 68–78
– process development/design 173–177
– raw materials 169–181
– vacuum gas oil, catalyst testing 131
fatty acid methyl ester (FAME) 6, 7, 10, 180
fatty acids 169, 170, 321–5
fatty alcohols case study 177–180
feedstocks, see biomass feedstocks
fermentation
– algal biomass 385
– Amyris Biofuel 13
– artemisinin 54
– building blocks 19
– concentration 229
– ferulic acid 224, 225
– furfural 97
– Gevo's process 11, 44
– 3-hydroxypropionic acid 46
– isobutanol removal 201
– LanzaTech process 15
– Lignol biorefinery process 254–266
– muconic acid 210, 213
– novel chemicals 58
– process design 121
– productivity 230
– simultaneous saccharification and fermentation 97, 253
– succinic acid 39
– sugars to ethanol 7, 10, 12, 14, 44, 83
– syngas 43
– terpenes 54, 60, 78
Fischer–Tropsch (FT) process
– biomass pretreatment/gasification 288–299
– biomass-to-liquids 271–312
– biorefinery processes 251, 252
– catalyst stability 146
– cobalt catalysis 278–285
– DOE biomass programs 33
– energy/carbon efficiencies 310–312
– pilot/demo plants 308–310
– process concepts 299–308
– reactors 285–288
fixation of carbon dioxide 382–286
fixed-bed reactors 161, 162, 174–176, 179, 180, 291, 292
flash pyrolysis 15, 16
fluidized-bed gasifiers 291–294, 297, 298, 303, 305
fluidized-bed reaction 160
Ford Motor Company 27, 53
formic acid 99, 100, 109, 251

fossil fuels   83, 363
free radical polymerizations   63
fructose
– dehydration through cyclic/acyclic intermediates   108, 109, 111
– 2,5-dimethylfuran synthesis   124–128
– 5-hydroxymethyl furfural synthesis   110
– – conceptual process design example   124–128
fuels
– *see also* biofuels
– fossil fuels   83, 363
– jet fuels   306, 307, 311
functional units   355
functionally equivalent replacements   45
furan-based building blocks   83–113
2,5-furandicarboxylic acid (FDCA)   21, 22, 205, 206, 356, 357
furanic polyesters production   22
furans   205
furfural
– commercial production   95, 96
– derivatives   96
– distribution/properties   95
furfuraldehyde, from xylose   205
– production from carbohydrates   83–113
– – carbohydrates dehydration   17
– – heterogeneous catalysts   103–107
– – homogeneous catalysts   101, 102
– – systems   101, 102
– recovery, Lignol biorefinery process   254, 255

g

galactoglucomannans   87, 90, 91
gallic acid   216, 217
gas chromatography   69, 70
Gas Technology Institute (GTI)   296, 298
gasification of biomass   248, 251, 290–292
gel permeation chromatography (GPC)   65, 66, 69
genetic engineering   246, 384
geographic spread of companies   4, 5
Gevo Inc.   4, 11, 12
Gibbs free energy of formation   366–368
β-(1->3,1->4)-glucans   92
glucomannans   87, 90
glucose   108, 110, 253, 254
glucuronoxylans   88, 90
glycerol   46
gold catalysts   330, 331

graphene oxide   107
grasses   84, 85
green algae   84, 85, 92, 94
green catalysts   101, 103, 104
Green Polyethylene (GP)   151–164
greenhouse gas emissions (GHG)
– carbon dioxide industrial use   364–367
– corn-based ethanol   153
– diesel biofuels   306
– Fischer–Tropsch process   310, 311
– novel catalytic processes   350, 360
growth potential   2, 3, 9

h

H2Bioil process   249
Haematococcus   383, 384
halides   61, 101, 102
heat exchangers   287
hemicelluloses   89, 99
hereditary factors   321
heterogeneous catalysts   103–107, 327
heterogeneous heteropolyacids (HPAs)   103, 104
hexose   107, 108
high temperatures   102, 271, 285
history
– biomass-to-liquids Fischer–Tropsch process   271, 272
– carbohydrate dehydration to furans   94–96
– chemical process development   27
– ethylene production   154
– feedstocks   53–54
– Lignol biorefinery process development   264–266
– organosolv biorefining   245–246
– target product selection   31–38
HMF routes to PTA   207, 208
HMG–CoA synthase   202, 203
HMG–CoA reductase pathway   56, 57
homogeneous catalysis   101, 102
HP-L® lignin   253, 258
– market drivers   262, 263
– new product opportunities   260–262
– physical/chemical characteristics   260
– properties   259
hybrid fractionation processes   97
hydration reactions   102, 109, 336, 340, 341
hydrodeoxygenation (HDO)   249
hydrogen gas   322, 323, 380
– *see also* Fischer–Tropsch process
hydrogenated fatty acid streams   198
hydrogenation   140, 190, 191, 337, 338
hydrogenolysis   178, 190, 191
hydroquinone   216

3-hydroxy-isovaleric acid (3HIVA) 202
p-hydroxybenzoic acid (pHBA) 218, 219
5-hydroxymethyl furfural (HMF)
– acyclic pathways 107, 108
– aromatics via 204, 208
– carbohydrates dehydration 17, 95
– current situation 21
– cyclic intermediates 107, 108
– energy use case study 356, 357
– fructose route 124–128
– hydration reaction 102, 109
– production from carbohydrates 83–113
– purified terephthalic acid production 206–208
hydroxystyrene 219–221

i

imidazolinium-based ionic liquids 110
impact assessment, *see* environmental assessment
impregnation 278, 279
impurities
– Braskem's ethanol dehydration 160, 161, 163
– cobalt FT catalysts activity loss 283
– ethylene purification 163
– fats/oils feedstocks 175
– Fischer–Tropsch process gas clean-up 301, 302
– statistical experimental design 138
indirect carbamates synthesis 377–379
indirect land use change (ILUC) 359
industrial activities, *see* commercial interest/activities
industrial slag 381, 382
Ineos New Plant Bioenergy 15
inorganic carbonates production 381, 382
International Standard Organization (ISO) standards 355, 356
inulin 84, 86, 87
inventory analysis 355
iodine value (IV) 171
ionic liquids 110
iron-based catalysts 271, 276, 277
isoborneol 327, 342
isobornyl acetate 327
(+)-isobornyl acetate 336
isobutanol 12, 13, 199–201
isobutylene 199–203
isocyanates 40, 375, 377, 378
isomerization
– monoterpenes 325–335
– α-pinene 330, 331
– α-pinene oxide 332

– β-pinene oxide 335
– terpenes 318
isopentyl pyrophosphate (IPP) 56
isoprene 61–63, 72
isoprenoids 56, 57, 383
isothermal reactions 161
isothermal zone length 129, 130
isovaleraldehyde dehydrogenase (IDH) 200

j

jet fuels 306, 307, 311

k

karahanaenone 339, 340
Karlsruhe Institute of Technology (KTI) 295, 297, 300, 309
Keggin-type heterogeneous heteropolyacids 103, 104, 341
kerosene 276, 306, 311
ketal products 21, 22
2-ketoacid decarboxylase (KIVD) 200
ketol-acid reductoisomerase (KARI) 200, 201
ketone functions 14, 20, 21, 188–190, 193, 322, 339
kinetics 131, 132, 274, 275, 324, 366, 367, 369
Kraft pulping process 317

l

land use 152, 153, 359
Langmuir–Hinschelwood kinetic expression 274
LanzaTech Inc. 14, 15
lavender oil 334
legislation 122, 157, 158, 355, 356
levosandal 340, 343
levulinates 17, 18
levulinic acid
– current situation 20–22
– energy use case study 356
– fructose conversion to HMF/DMF 124
– 5-hydroxymethyl furfural hydration reaction 109
– product selection criterion 46
Lewis acids 63, 95, 206, 334, 373
life cycle assessment (LCA)
– biofuels 33, 34
– carbon dioxide industrial use 365
– environmental sustainability 350, 354–356
– novel catalytic processes 360
lignans 222–226, 255

lignins
- *see also* HP-L® lignin
- chemicals derived from   122
- common aromatic pathway   226–228
- companies, product potential   39
- derivatives   258, 259
- Lignol biorefinery process   254
- native   257
- organosolv biorefineries   253
- technical   257–258
- value-added product selection   36, 37, 39, 40, 46

lignocellulosic biomass feedstocks   10, 11, 14, 15, 245–267
Lignol biorefinery process   253–266
limonene   208, 209, 336, 339, 340
linalool   318, 337
linear carbonates synthesis   369, 370
lipogenesis   173
loading (catalysts)   280, 281
loading (principal component analysis)   140
Lobry De Bruijn-Alberda Van Ekenstein transformation   107
long-term catalyst performance   145–148
low-temperature Fischer–Tropsch (LTFT) process   271–312
lubricants   68–78
- *see also* fats/oils

m

macroalgae   93, 382–386
magnesium oxide   381, 382
maize corn   153, 154, 204, 205, 357
mandelic acid   219, 220
market drivers   44, 45, 262, 263, 271, 272
market-ready opportunities   261
- *see also* drop-in replacements
Markley's definition   169
mass balance   302–305, 323
mass spectrometry   70, 71
mass transfer limitations   129–131
Meerwin–Ponndorf–Verley (MPV) reaction   343
(3-mercaptopropyl)-trimethoxysilane (MPTS)   104
mesoporous molecular sieve MCM-41   104–106
metals
- *see also* cobalt catalysts
- amides   376
- chromium   100, 341
- copper catalysts   179
- costs   177

- Fischer–Tropsch process   276, 277
- iron-based catalysts   271, 276, 277
- metal-carbamates   375
- nickel   276, 277
- niobium   103, 106, 340, 370, 371, 373
- oxides, Group 2   381, 382
- palladium   176, 177
- ruthenium   146, 148, 179, 276, 277
- sodium metal catalysts   62
- tin   107
- vanadium phosphates   106
- Ziegler catalysts   61

methane   272–274, 379
methanol   3, 251, 276, 353, 354, 370–372, 379–381
mevalonate pathway   56, 57
mevalonic diphosphate decarboxylase   202
microalgae   93, 382–386
microbial strain engineering   54–57
microporous zeolites   103
mineral acid catalysts   101, 102
molecular weights, chain lengths   273
mono ethylene glycol (MEG)   154, 156, 157
monoterpenes
- derivatives   342, 343
- epoxidation   337–340
- esterification   341, 342
- etherification   341, 342
- hydration   340–341
- hydrogenation   337, 338
- isomerization   325–335
- oxidization   337
muconic acid route   210, 212–214
multitubular reactors   161
myrtanal   318–320, 334, 335
myrtenol   320, 334

n

N-methyl-2-pyrrolidinone (NMP)   35, 36
N-vinyl-2-pyrrolidinone   35, 36
naphtha   159, 191, 192, 197, 198, 271, 276, 306
near-infrared spectra (NIR)   144
new chemical process design   121–149
new property advantages/disadvantages   45
nickel catalysts   160, 175–177, 276, 277
niobium   103, 106, 340, 370, 371, 373
non-renewable energy use (NREU)   354, 358, 359
non-thermochemical biorefineries   252
Nopol   340, 343
nuclear magnetic resonance (NMR) spectra   64, 65, 71–74

o

oils, *see* fats/oils
olefins   197, 374, 379
one-pot synthesis   333, 334
optimization statistical design methods   133, 134–138
organic catalysts   369–371
organic solvents   109, 110, 112, 250, 251
organosolv biorefining   245–267
original equipment manufacturers (OEMs)   55, 56
outcomes comparison   40, 41
oxazolidin-2-ones   376
oxygen removal   14, 16
oxygenates removal   163

p

paclitaxel   246
palladium catalysts   176–177
palm oil   4, 171, 172, 174, 198
para-hydroxybenzoic acid (pHBA)   218, 219
paraffins   194, 275, 276
parallel reactor equipment   121–149
parameter estimation   323, 324
partial least squares (PLS) regression   139, 143–145
particle size of catalysts   280
partnerships between companies   7, 8
pentose phosphate pathway   209, 212
pentose sugar feedstock   97–101
perfumes   58, 60
perillyl alcohol   318, 320, 334
petrochemical industry   1, 56, 191
phenol   218, 219
phenylalanine (PAL)   220
pheromones, aphids   59
phloroglucinol   222, 223
phosgene use   365, 368, 375, 377
phosphoric acid   340, 341
photosynthesis, artificial   364, 367, 374, 380
pigments, algal   383
pilot plants   266, 294–298, 308–310
pinane-2-ol   338
pinene oxides
– isomerization/hydrolysis   332, 335
– α-pinene oxide   326
– β-pinene oxide   326
– terpenes
– – hydration   336
– – isomerization   334
pinenes
– esterification/etherification   339
– α-pinene   330, 331, 338
– β-pinene   340
– terpenes
– – hydration   336
– – isomerization   318, 329, 332
– – oxidation   319, 320
– turpentine   325
pinocarveol   320
pinocarvone   320
pinoresinol   225, 226
Pinus ponderosa   317
plant sterols hydrogenation   322
plant-based biomass   10, 11, 14, 15, 245–267
PlantBottle®   156, 157
plasma gasifiers   292, 294, 296, 298
plastics
– *see also individual plastics*
– aquatic biomass-derived   385, 386
– beverage bottles   156, 157, 194, 195, 356–359
– global production increase   349, 350
platform chemicals   1–4, 7, 14, 20, 21, 33, 46
podophyllotoxin   225, 226
polyalanoates   385, 386
polycarbonates   373, 374
polyester of ethylene glycol (PEF)   207, 208
polyethylene
– ethanol-derived process   151–164
– – Brazil   152–154
– – commercial plants   154–157
– – development reasons   151, 152
– – legislation/certification   157, 158
– – polymerization   164
– – process description   158–163
polyethylene furandicarboxylate (PEF)   356–359
polyethylene terephthalate (PET)   154, 156, 157, 185, 199, 356–359
poly(farnesene) structures   63
polyhydroxybenzenes   221, 223
poly(isoprene)   61, 62
polymerizations   164, 257, 281
– *see also* biopolymers
polyols, sugar-derived   44
polysaccharides   84, 86–92, 385
– *see also individual polysaccharides*
– lignocellulosic biomass feedstocks   10, 11, 14, 15, 245–267
polyurethanes synthesis   377–379
polyvinylchloride (PVC)   154–157
porous solid acids   103–107, 112
powder catalysts   131, 132
pretreatment of biomass   288–299

principal component analysis (PCA) 139–142
principal component analysis, yields 141, 142
pro-chiral centers 72
process development/design
– Braskem's ethanol dehydration 158–163
– fats/oils as raw material 173–177
– Lignol biorefinery process 256, 257
– parallel reactor equipment 121–149
– target product selection 27–50
production volumes 6, 7, 349
products 33, 34, 275, 276, 306, 307
– *see also* target product selection criteria
propargyl alcohols 376
propylene 352
proteins, aquatic biomass 384
purified terephthalic acid (PTA) 356
– from *p*-xylene 199
– limonene to 208, 209
– production via 5-hydroxymethyl furfural 206–208
– *p*-xylene conversion 186
pyrogallol 216, 217, 221–223
pyrolysis
– biomass fast pyrolysis products 188, 189
– biomass-derived sugars 192–195
– biorefineries using 248, 249
– biorefinery catalysts 251
– Fischer–Tropsch process 289
– raw biomass/bio-oil 195–198
pyrolysis oils, *see* bio-oils

q
Quaker Oats Company 96, 204
quinic acid 216

r
Range Fuels 2, 15, 290, 292
rapeseed oil 175
rapid thermal processing (RPT) 16
rational selection of target products 27–50
raw materials
– *see also* biomass feedstocks
– aromatic chemicals production 187, 188
– availability, product selection criteria 41–43
– fats/oils 169–181
– lignocellulosic biomass feedstocks 10, 11, 14, 15, 245–267
reaction types/mechanisms
– acyclic carbonates from alcohols carboxylation 369–371
– biomass gasification for FT process 290
– Braskem's ethanol dehydration 158, 159

– carbamates synthesis 374–379
– carbon dioxide reactions 366
– carbonates synthesis 371
– cyclic carbonates synthesis 373, 374
– energy products from carbon dioxide 379–381
– Fischer–Tropsch process 272–275
– furan formation from sugars 95, 100
– – elimination reactions 98, 99
– hydrogenation/hydrogenolysis 190, 191
– phosgene use 368
– polyurethanes synthesis 377–379
– product selection criteria 48, 49
reactors
– batch reactors 128, 129
– biphasic 99
– continuous reactors 129, 321
– dual bed reactors 291, 292
– Fischer–Tropsch process configurations 285–288, 290, 291
– fixed-bed reactors 161, 162, 174–176, 179, 180, 291, 292
– fluidized-bed gasifiers 291–294, 297, 298, 303, 305
– heat exchangers 287
– Lignol organosolv refineries 256, 257
– multitubular reactors 161
– parallel reactors 121–149
– plasma gasifiers 292, 294, 296, 298
– types 128, 129
– – Braskem's ethanol dehydration 161, 162
– – fats/oils as raw materials 174–176, 179, 180
– – Fischer–Tropsch process 285–288
redox balance 230
reduction reactions 279, 280, 290
relative activity 327
research and development (R&D) 9, 45, 46, 121–149
resin acids, crude tall oil 321–325
response surface model design 136, 137, 139
rubber 27, 61
ruthenium catalysts 146, 148, 179, 276, 277

s
Saccharomyces cerevisiae 201
Salgema Indústrias Químicas Ltda. 154, 155
scale effects 132, 133
scaling up 111, 228–230, 279, 312
screening processes 27–50, 133, 134
Segetis Inc. 21

selection criteria
– *see also* target product selection criteria
– hydrogenation catalysts   172, 173
– reactors for Fischer–Tropsch process   285
separable deactivation model   327
sesquiterpenes   58–60
Shell International   17, 283, 284
shikimic acid   209–213, 218, 219
side product selectivity   137–139
side reactions   160, 161
silicate materials   381
silicoaluminophosphates (SAPO)   103
simultaneous saccharification and fermentation (SSF)   97
sitostanol   321, 322
sitosterol   321–325
slag   381, 382
slurry operations
– catalysts   283
– fats/oils as raw material   174–177
– fatty alcohols case study   179, 180
– FT slurry bubble column   285–288
Sobrerol   336, 341
sodium metal-catalyzed polymerization   62
solid acid catalysts   103–107, 112, 341
Solvay   39
solvents   109–112, 250, 251
– organosolv biorefining   245–267
Sorona®   44
soybeans   53, 152, 153, 172
space-time yields   137–139
special chemicals   321–325
spinning disc reactors   333
Spirulina   383
standards   55, 355, 356
star diagrams   35, 36, 96
starch   43, 44, 84, 86, 87
statistical design   132–138
Statoil technology   302–305, 307
steam cracking   197, 198, 351, 352
steam reforming   191, 298
stereoregularity   61
sterols   322
stigmasterol   322
stilbenes   222–226
storage carbohydrate sources   84, 86, 87
storage of electrical energy   380, 381
strain engineering   54–57
strong Lewis acids   63
structural carbohydrates   86–92
structure preservation   256
styrenes   219–221
succinic acid   19, 20, 39

sugar-based biomass
– aromatics from pyrolysis   192–195
– benzene–toluene–xylene bio-refinery concept   184, 185
– chemicals derived from   122
– Lignol biorefinery process   255, 256
– – pyrolysis   192–195
– plantations, labor conditions   158
– polyester of ethylene glycol route   208
– polyols, ADM   44
– sucrose sources   84, 87
– sugar cane, Brazilian ethanol production   152, 153
– sugar-beet, ethanol production   153
sulfated tin oxide catalyst   107
sulfated zirconia (SZ)   105, 106
sulfite liquor   225
sulfonated graphene oxide   107
sulfonic acid-functionalized materials   104, 105
sulfur   175, 176
sunflower oil   172
supports for cobalt FT catalysts   282
sustainability case studies   356–359
syngas   191, 248, 299, 300, 302
– *see also* Fischer–Tropsch process
system boundaries   355

t
Taiwan Chlorella Manufacturing & Co.   382
tall oil   321–325
tall oil rosin (TOR)   317
target product selection criteria   27–50
– byproduct uses   46
– carbohydrate-derived products   37, 38
– catalysis aspects   48, 49
– chemicals, outcomes comparison   40, 41
– evaluation processes   41–48
– existing biorefinery infrastructures   43, 44
– market drivers   44, 45
– outcomes comparison   40, 41
– previous activities   31–38
– research drivers   45, 46
– structures/technologies   30, 31
– terpenes   54
– validation   38–40
– value-added processing   46
technical lignins   257, 258
technology development scenarios   30, 31, 33
temperature   102, 271–312, 322, 323
tensile strength   67, 68
terephthalic acid   21, 206, 213–216

terpenes
- building blocks of commercial interest 57, 58
- catalytic processes development 53–78
- epoxidation 326
- esterification 339
- etherification 339
- hydration 336
- isomerization 318
- lubricants 68–78
- microbial strain engineering 54–57
- oxidation 319, 320, 337
- polymers 60–68
- sesquiterpenes as building blocks 58–60
α-terpineol 336
terpinolene oxide 340
tetrahydroxybenzene 221–223
thermal gravimetric analysis (TGA) 66, 67
thermal processes 6, 7, 248, 249, 338
thermodynamics 366–369
tin oxide catalyst 107
toluene, see benzene–toluene–xylene (BTX) process
Top Ten Value Added Chemicals from Biomass (US DOE) 57, 58
- 2004 Report 33, 35–39, 41, 42, 45, 46, 122
- 2007 Report 122
- 2010 Report 37, 38, 42
top-50 commodity chemicals (US) 352
torrefaction 249, 288, 289
total acid numbers (TAN) 17
Total Petrochemicals 197, 198
transesterification 372
transportation fuels 18
tri-block copolymers 386
triglycerides 169, 170
Triunfo Braskem plant 155, 156, 164
tubular fixed-bed reactors 285, 286
turpentine 319, 325–343
tyrosine ammonia lyases (TAL) 220

u

Uhde technology 295, 297
United States (US)
- 1993 Report 31
- Billion-Ton study 185
- Biological and Chemical Catalysts Technologies Program 32
- *Biomass as Feedstock for a Bioenergy and Bioproducts Industry* 228
- ethanol production 153
- *Top Ten Value Added Chemicals from Biomass* 57, 58, 205

- – 2004 Top Ten Report 33, 35–39, 41, 42, 45, 46, 122
- – 2007 Top Ten Report 122
- – 2010 Top Ten Report 37, 38, 42
- top-50 commodity chemicals 352
- transportation fuels 18
upgrading of FT hydrocarbons 306, 307
urea 369, 371, 372

v

vacuum gas oil (VGO) 131
valeric acid-based fuels 16–18
validation, target products 38–40
valine pathway 200, 201
value chain selection criteria 27–50
value-added processing
- Lignol biorefinery process 253–266
- organosolv biorefining 245–267
- product selection criterion 46
- Top listed chemicals 2004/2007 33, 35–39, 41, 42, 45, 46, 122
Van Krevelen diagrams 173, 174, 178, 180
vanadium phosphates 106
vanillic acid 217, 218
vanillin 217, 218, 222–225
vegetative biomass 10, 11, 14, 15, 245–267
Vennestrøm report 45
verbenol 319
verbenone 319, 335
vinyl carbamates 376
Virent Energy Systems 192–194, 198, 199
viscosity 74, 77

w

waste streams 10, 11
water
- see also aqueous...
- electrolysis 381
- removal, Braskem's ethanol dehydration 163
- water gas reaction 379
- water traps 369, 371
- water-gas-shift reaction 291, 379
- water-splitting 379
- water-tolerant heterogeneous catalysts 103, 112
water-soluble bio-oil (WSBO) 196, 197
waxes 276
wood extractives 255, 317–344
wood processing 84, 85
woody biomass 10, 11, 14, 15, 245–267
workflow for process design 121–149

x

XTL energy efficiency   310–312
p-xylene (PX)
– *see also* benzene–toluene–xylene (BTX) process
– biological oxidation to purified terephthalic acid   203, 204
– from isobutanol/isobutylene   199, 200
– purified terephthalic acid from   186, 199
– xylene isomerization   185, 186

xyloglucans   87, 90
xylose   98–102, 205
D-xylose dehydration   104, 105

y

yields   138, 175–177, 179, 186, 190, 192, 193, 195, 196, 199, 200, 202, 203, 210, 212

z

zeolite catalysts   160, 192, 196, 197
Ziegler catalysts   61